Understanding Topology

Understanding Topology

A Practical Introduction

SHAUN V. AULT

Valdosta State University

Johns Hopkins University Press
Baltimore

Johns Hopkins University Press
2715 North Charles Street
Baltimore, Maryland 21218-4363
www.press.jhu.edu

Library of Congress Control Number: 2017937102

ISBN 978-1-4214-2407-1 (hardback : alk)
ISBN 1-4214-2407-X (hardback : alk)
ISBN 978-1-4214-2408-8 (ebook)
ISBN 1-4214-2408-8 (ebook)

Composed using the LaTeX Document Preparation System. TeX is a trademark of the
American Mathematical Society.

Contents

Preface

This textbook grew out of the lecture notes that were prepared for a topology course developed at Valdosta State University, a primarily undergraduate teaching institution, and so the text is intended for undergraduates in math and the sciences. Topics are selected to appeal to a wide range of interests, with emphasis placed on breadth rather than depth. When possible, the mathematics is first motivated by intuition and analogy; however, the full rigor of mathematical proof then follows afterward (except where indicated in the text). You may find that some topics are not covered in as much detail as in a more traditional topology text, while other themes are present in this book that may not be found in the standard texts.

The basic flow of the text is inspired by the wonderfully readable textbook by Sue Goodman, *Beginning Topology* [Goo05], especially in its coverage of vector fields, graphs and maps, and knot theory, topics that usually do not find themselves in a topology text; however, this treatment differs in its focus and coverage. Among other topics, *Understanding Topology* also includes a short introduction to algebraic topology, modeled loosely after Hatcher [Hat02]. Some chapters contain additional sections marked with an asterisk (*) that may be of interest but are not required for understanding subsequent material. These additional sections can be used for student projects, covered in lecture as time permits, or omitted altogether. Each section has associated *Exercises*, and each chapter ends with *Supplementary Reading*. Appendix A provides a brief review of set theory and functions. Appendix B gives a terse review of essential topics in group theory and linear algebra. Following these are selected solutions to exercises and an extensive list of notations used.

Part I – *Euclidean Topology* – begins with Chapter 1, *Introduction to Topology*, with an informal taste of topological thinking via a hands-on (albeit not very rigorous) treatment of continuous deformations. Chapter 2, *Metric Topology in Euclidean Space*, is the main thrust of Part I, in which the terminology and tools of metric topology are introduced in a familiar setting. The last section of this chapter, §2.5 - *Metric Spaces in General*, serves to motivate the level of abstraction that will be explored later in the textbook. Some applications of Euclidean topology to vector fields in $\mathbb{R}^2$ are given in Chapter 3, culminating in an application of topology to the analysis of autocatalytic chemical reactions in §3.3.

The material in Part II – *Abstract Topology with Applications* – especially Chapter 4, *Abstract Point-Set Topology*, forms the heart of the textbook, introducing concepts and methods of elementary point-set topology, complete with precise definitions in the language of set theory. Then in Chapter 5, *Surfaces*, we encounter the most basic manifolds (beyond trivial zero and one-dimensional examples), which help to prepare the way toward an understanding of manifolds in general. Surfaces are analyzed using plane models and combinatorics, giving a concrete method for studying unfamiliar spaces, and making use of invariants such as Euler characteristic and orientability. Applications of topology are found in Chapter 6, *Applications in Graphs and Knots*, which also serves to reinforce combinatorial ideas in the study of topology.

Part III – *Basic Algebraic Topology* – introduces the the tools of algebraic topology by building on the notion of topological invariant. The material in this part moves quite a bit quicker than in the previous parts, and it is helpful to have prior experience with abstract algebra and linear algebra, though all necessary algebraic structures will be defined as needed. Chapter 7, *The Fundamental Group*, gives the basic idea of the first homotopy group along with a careful computation of the first "interesting" case, the circle. Then a less formal discussion shows how to use fundamental groups in the classification of compact surfaces and to study knots via their complements in $\mathbb{R}^3$. The chapter ends with a short discussion of higher homotopy groups, giving the necessary definitions and methods to begin the study of homotopy theory; however, the treatment here only scratches the surface. Chapter 8, *Introduction to Homology*, leads the reader through some of the algebraic and topological machinery needed to understand homology, with emphasis placed on combinatorial descriptions and matrix methods readily available at the undergraduate level. In defining integral homology, we take a different approach than most texts at this level. Instead of defining homology in terms of abelian groups, we use the concept of $\mathbb{Z}$-modules (informally, at least). The reason is twofold: first, since module theory plays such a large role in algebraic topology, it is advantageous to see the term early on; second, the conceptual idea of a module is closely related to that of a vector space, and indeed many methods from linear algebra carry over, including matrix reduction algorithms. Again, the treatment of these topics is woefully incomplete. Nevertheless, it is important to get a taste of what might be next.

The appendices include a *Review of Set Theory and Functions* (Appendix A), which provides a readable refresher on the essential notation and definitions used throughout the text. While an appendix on set theory within a topology textbook is usually nothing more than a list of definitions, notations, and properties, this text hopes to attract a wider audience by explaining the concepts of sets and functions at a lower, more accessible level. On the other hand, the next appendix, *Group Theory and Linear Algebra* (Appendix B), is intended as a quick review alongside the more advanced material in Part III – there is no need to reference Appendix B if only Parts I and II are covered. Both appendices include comprehensive exercise sets.

Definitions, Theorems, and Examples

The tools of topology include definitions and theorems, and examples are provided to illustrate the concepts. Each definition is highlighted by placing it in a *shadowbox*, with the defined word in bold:

> **Definition 0.0.1.** This is a **definition**.

Each theorem is placed in a *doublebox*:

> **Theorem 0.0.2.** *This is a theorem.*

Examples are set off from the main text as shown below:

> **Example 1.** This is an example.

A Note about the Style

What is mathematics? Is it a body of knowledge consisting of definitions, formulas, lemmas, and theorems? Is it a method for solving problems? Sure, these things are essential to mathematics. However, mathematics is also a *conversation*. Mathematicians rarely work in isolation. Even when research is done individually, the paper or book that comes as a result is typically shared with peers and students, forming a literature and contributing to the culture of mathematics. I have endeavored to let this text speak more freely, less formally, and in more varied tones than perhaps other textbooks might do. You'll see contractions (like that one). Maybe a sentence fragment for effect. On the other hand, the language used in definitions, theorems, and proofs generally follows formal conventions and is mathematically precise; there is always an indication in the text whenever formal precision and rigor have not been achieved.

Acknowledgments

I would like to thank my students from the Spring 2014 Topology course, Geoffrey Buie-Collard, Aaron Calvin, Daniel Drummond, and Nathaniel Jones, for their helpful feedback about the material as well as advice and suggestions for what to include in this textbook. Others whose careful reviewing helped to catch my errors and shape my writing include Scott Campbell, Iwan Elstak, Robert Everett, Christopher McClain, John McSweeney, and Arsalan Wares. A special thanks goes out to Charles Kicey for contributing exercises and examples in set theory and metric spaces, and to Robert Kane for contributing solutions to some exercises and careful proofreading. Last but not least, this book would not be possible without the love and support of my wife, Megan, who had to put up with my telling her "I can't do that right now because I'm writing" whenever she wanted me to do a chore around the house, and the inspiration of our children, Joshua, Holley, Samuel, and our newest addition, Felix.

The text of this book was prepared using LaTeX2e.[1] The figures and other illustrations in this book were obtained through a number of sources. Most were created by the author using one or more of the following tools:

- Latex packages such as pstricks, pst-plot, pst-math, pst-knot, tikz-cd, and skak.

- Image drawing and editing sofware, including GIMP[2] and inkscape.[3]

- Mathematical packages with image capabilities, such as Sagemath[4] and pplane.[5]

Other images were used with permission of the artist or under Creative Commons licenses.

---

[1] http://www.latex-project.org/.
[2] https://www.gimp.org/.
[3] https://inkscape.org/en/.
[4] Sage Group (William Stein et al.) [Ste].
[5] Java program created by Professor John C. Polking of Rice University [Pol].

Part I

Euclidean Topology

Chapter 1

Introduction to Topology

What is topology, anyway? When I talk to people about what I do, some say, "Oh, so you study maps?" To which I respond, "You may be thinking of *topography*." In retrospect, *studying maps* is not far from the mark; we do study maps of a sort. Of course, the maps that topologists study are a lot different than the one shown in Figure 1.1, though such *topographical maps* can be studied from a *topological* point of view as well. To a topologist, a **map** is a **continuous function** from one **space** to another.

Figure 1.1: A topographical map of two small hills.

From calculus, the term *continuous function* should remind us of an unbroken graph, usually representing an equation in x and y, such as the rule $y = x^2$, which transforms real numbers x into their squares, x^2. The square of $x = 2$ is 4, but notice also that if x is *close* to 2, say, $x = 1.9$ or $x = 2.003152$, then x^2 is also *close* to 4 ($1.9^2 = 3.61$ and $2.003152^2 = 4.012617935104$). Thus all points ($x$-values) in a small *neighborhood* remain relatively close together after the function has been applied; hence we say that a continuous function *preserves closeness*.

However, not every continuous map preserves the essential nature of a space. Just think of the constant function $f(x) = 0$, which takes as input any point of the number line, and gives the value 0 as output, so $f(100) = 0$, $f(-11325.42) = 0$, $f(\pi + 1) = 0$, etc. The entire output set of f consists of a single point, which is very much unlike the input set (the real number line). However, some continuous maps do less damage. We say a function is **invertible** if there is another continuous map (an **inverse** map) that *reverses* the effect of the map. For example, the function $f(x) = 3x + 1$ is invertible. It has an inverse function $g(x) = \frac{x-1}{3}$. If the two maps are composed in either order, they "cancel" each other out: $g(f(x)) = \frac{(3x+1)-1}{3} = x$, and $f(g(x)) = 3\left(\frac{x-1}{3}\right) + 1 = x$. The constant map $f(x) = 0$ is not invertible.

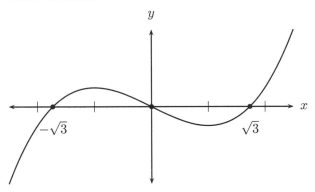

Figure 1.2: Graph of $f(x) = x^3 - 3x$, where f is a continuous function from the real number line (x-axis) to the real number line (y-axis), but f is not invertible.

Now consider $f(x) = x^3 - 3x$. This function takes as input any point of the number line, and yields any real number as output – that is, the domain and range are both $\mathbb{R}$ (see Figure 1.2). However, f is not invertible. The graph doubles back and hits the same y-values more than once; for example, $f(-\sqrt{3}) = f(0) = f(\sqrt{3}) = 0$. Suppose an inverse $g(x)$ exists; what would $g(0)$ be? If we say $g(0) = 0$, then $g(f(\sqrt{3})) = g(0) = 0 \neq \sqrt{3}$. If we say $g(0) = \sqrt{3}$, then $g(f(0)) = g(0) = \sqrt{3} \neq 0$. In the language of set theory, we would say f is not *injective*, and thus not invertible.[1]

---

[1]See Appendix A for a more thorough discussion of functions.

Topology

Ok, so what is topology? Before getting into the specific details, we can say that **topology** is the study of the qualities of a space that are preserved under invertible maps.[2] However, let's back up and look at one word in particular. The word *space* has a precise mathematical meaning that allows for myriad beautiful, complex, and useful interpretations. The spaces that we consider are not confined just to the familiar three-dimensional world we can perceive around us. Some spaces, such as the *torus* and *Klein bottle* (see Figure 1.3), bend and connect in strange ways. Others, such as the *Cantor set*, which we will encounter in §1.2, display seemingly paradoxical qualities that challenge our intuitive understanding of connectedness and size. Still others do not seem to fit in this category because we are not used to thinking of them in this way; for example, we may define the space of all integer sequences, or the space of all polynomials, or the space of all differentiable functions f such that $f(0) = 10$.

Figure 1.3: *Left*, the surface of a doughnut is a topologist's torus. *Right*, an "impossible" surface, the Klein bottle cannot exist in three-dimensional space; however, it can be modeled in higher-dimensional spaces with no self-intersection. Images courtesy of Wikimedia Commons.

I sometimes have to field the question, *What is topology good for?* While topology is mainly used in service of other branches of mathematics such as analysis, there are also a number of immediate applications. For example, *topological data analysis* and *topological structures in computational biology* have recently emerged as important fields of study. And there is one aspect of topology that most of us in the information age use and rely on every day – *network topology*. In a network of computers, routers, hubs, and servers, the physical locations of the devices are not as important as knowing which device is connected to which others. Our standard notion of distance in terms of miles from one computer to another is meaningless; two computers may be considered "close" in a network if there are relatively few intermediate networking devices connecting them, even if the two computers are separated by thousands of physical miles. By this definition of the word *close*, we may visualize the Internet as in Figure 1.4.

[2]To be more precise, invertible continuous functions whose inverses are also continuous.

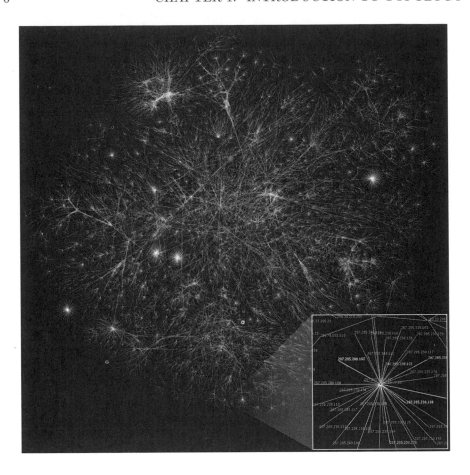

Figure 1.4: A topological map of the Internet. Each node represents a particular server. Nodes are connected to one another if they can communicate directly, regardless of the physical distance between the servers. Image courtesy of Wikimedia Commons.

Abstracting is an extremely important tool to a topologist. Complicated relationships may become much clearer when all the "unnecessary" detail is conveniently ignored. For example, Leonhard Euler used spatial abstraction in a 1735 paper [Eul36] in what many consider to be the first explicit use of topology: solving the *Seven Bridges of Königsberg Problem*. At that time in the eighteenth century, Königsberg, Prussia (now Kaliningrad, Russia) had seven bridges connecting two islands to the mainlands on both sides of the Pregel River. It was a popular pastime to stroll through the city traversing all of the bridges, and it was not known whether one could take a stroll that crossed each bridge only once. Euler observed that the details of each land mass were unimportant; only the data about which bridges connect which land masses would be necessary to analyze the problem. Euler drew what we now call a *graph* (see Chapter 6 for details about *graph theory*), in which the vertices represent the four land masses and the edges represent the bridges. Then the question could be recast in terms of finding a path in the graph that traverses each edge exactly once, as suggested by Figure 1.5. Such a path is now called *Eulerian* in honor of this great mathematician. It is fairly easy to prove that no such path exists in the Seven Bridges of Königsberg Problem (see §6.1, Exercise 5).

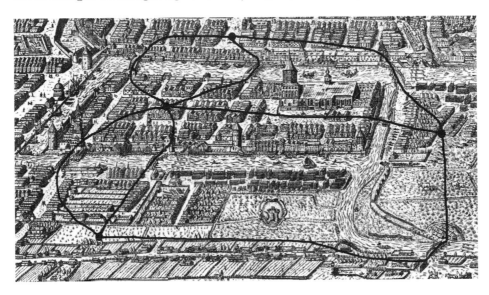

Figure 1.5: The seven bridges of Königsberg, together with a graph representing the landmasses and bridges. Image: public domain (modified).

Topology as a mathematical discipline developed in fits and starts as a series of ad hoc results in support of other branches of mathematics, such as *real* and *complex analysis*. In these applications, mathematicians needed more flexibility than standard Euclidean geometry could afford. Although topologists often deal with familiar geometric shapes such as triangles, disks, and spheres, we will see that topology is not concerned with such geometric measures as length, angle, area, volume, etc. In fact, to a topologist, the surface of a sphere is the same as (topologically equivalent to) that of a football, a brick, and even a wine glass (see Figure 1.6).

The torus is topologically equivalent to the surface of a coffee mug, as illustrated in Figure 1.7. What could possibly be useful about a study in which a coffee mug is indistinguishable from the doughnut being dunked into it? As it turns out, quite a lot. Consider a myopic[3] ant crawling along the surface of the doughnut, only able to see its immediate surroundings but not aware of the overall structure of the doughnut in three-dimensional space. All that it knows is that if it sets out in one direction, it will come back to the same spot after

---

[3]Nearsighted, only able to see a very small distance around itself.

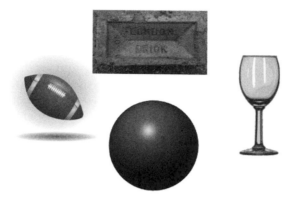

Figure 1.6: A sphere is a football is a brick is a wine glass (in the sense that their surfaces are topologically equivalent). Images courtesy of Wikimedia Commons.

Figure 1.7: A doughnut is the "same" as a coffee mug. Image courtesy of Cyrus Rua.

traveling some distance (as it travels around the center "hole," for example). Now imagine that this doughnut is made of soft clay and it gets deformed, bit by bit, until it looks like a coffee mug, as shown in Figure 1.7. So long as the transformation is *continuous*, then the little patch of the surface the ant is currently residing on does not get distorted too much. After the transformation, the ant may still observe that when it sets out in a certain direction, it will arrive back at the same spot after some time (perhaps by traveling along the "handle" of the mug). The most basic nature of the ant's world has not changed, and there may be no way for it (as an extremely myopic ant) to realize that any tranformation has occurred at all.

Now imagine the world we live in. Compared to the vast size of the universe, we are myopic ants indeed. If we want to study certain general properties of our universe, then we need a way to ignore any nonessential, overly specific qualities, focusing instead on the properties of space that would remain unchanged under small, continuous modifications. Topology provides a way to strip down to the basics in order to answer questions such as *Does the universe ever fold back in on itself?* or *Are there paths through space that cause a mirror reversal in the traveler?* Indeed, topology (along with a fair amount of geometry) can begin to answer the question *What is the shape of our universe?* We begin our study with an informal introduction to topological equivalence of spaces through malleability and deformations.

Onward, brave myopic ants!

1.1 Deformations

In geometry, two objects, A and B, are considered to be **equivalent** if they are **congruent** (notated $A \cong B$), that is, if all the lengths, angles, etc. of one object match those of the second object. Congruence may be thought of as the quality of an object that remains unchanged as it undergoes *rigid motion*, such as translation, rotation, or reflection. Thus geometry is the study of properties of objects that remain unchanged by rigid motions (or, to use the technical term, **isometry**). In topology, much more freedom is allowed.

Topological Equivalence

Two objects, X and Y, are considered to be **topologically equivalent** or **homeomorphic**, notated in this text[4] by $X \approx Y$, if *there is a continuous invertible function from X to Y, whose inverse is also continuous*. It takes some time to define the terms so that homeomorphism can be properly understood. In this introductory section, let's informally explore a particular type of homeomorphism: invertible *deformation*.

Imagine an object made out of soft clay. The clay can be molded in various ways: bending it, smoothing it out, pulling it into a long filament. The clay is **malleable**. However, we want to avoid separating or tearing parts of the clay. We also do not allow joining parts of the clay to itself or to other lumps of clay. So if the lump of clay originally looks like the doughnut in Figure 1.7, then it can be molded to any shape that also possesses a "hole," such as the coffee mug. This type of transformation is an example of a **deformation**.[5] Going forward, we will now use the more traditional term **(topological) space** instead of *object*.

In this section, we only consider **invertible** deformations, those that can be reversed. A line segment may be deformed into a semicircle, because there is a deformation that reverses it (unbending the semicircle into a straight segment). The segment cannot be deformed into a circle because that involves bringing together two points that were originally separated (the endpoints of the segment). The deformation that shrinks the segment to a single point seems to fit the definition, but it is not *invertible*, for much the same reason that the constant function $f(x) = 0$ is not invertible.

---

[4]The notation is not standard. Many authors use $\cong$, while others use $\sim$ or $\equiv$ for topological equivalence.

[5]A more precise term would be *ambient isotopy*.

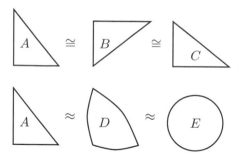

Figure 1.8: Congruence vs. homeomorphism. Triangles A, B, and C are congruent to each other but not to shapes D and E. However, all of the shapes are homeomorphic to one another.

When one space can be transformed into the other by an invertible deformation, then we may consider those objects homeomorphic, that is, topologically the same.[6] The line segment is homeomorphic to a semicircle but not homeomorphic to a point or a circle. Figure 1.8 illustrates the increased freedom we have in topology as compared to geometry. The triangle and circle are topologically equivalent (homeomorphic) because the triangle can be deformed into the shape of a circle, and vice versa. In fact, any polygon or simple closed curve *is* a circle (topologically, that is), and we may label any one of these by $\mathbb{S}^1$, which is the notation used in this text for the circle.[7] The superscript 1 on $\mathbb{S}^1$ refers to the *dimension* of the circle. It's essentially one dimensional, even though we typically draw it in two-dimensional space – a myopic ant living on the circle may only see a small arc that looks like a part of a line – a geometrical object that has only one dimension, *length*.

In a similar way, when we look outside, we see only a small portion of the surface of the Earth. On a perfectly clear day, with an unobstructed view, a person 2 meters tall can see the surface extending up to about 5 kilometers away.[8] The small disk that can be experienced from your vantage point seems to be flat and two dimensional, possessing no curvature at all. Based on this myopic view alone, it would not be unreasonable to guess that the whole surface of the Earth is flat – except that we know better.[9] Not only do we have the *local* information based on what our eyes can see, but we also have access to

---

[6]The idea of deformation implies that there is a step-by-step process to transform one space to another. This is actually too restrictive for topological equivalence, but it will take some time to arrive at the precise definitions needed for topology.

[7]A number of recent topology texts, such as Goodman [Goo05], use this "blackboard bold" notation for common families of spaces. The more traditional notation for the circle is S^1.

[8]Of course, some tall features such as buildings and mountain peaks can be seen from even greater distances.

[9]Well, *most* of us know better. There are some people living in the twenty-first century who *still* insist that the Earth is flat. The convoluted logic required to entertain such a position can be found at http://www.theflatearthsociety.org/home/.

Figure 1.9: A clay block in the shape of A may be deformed into a D, but not into a B or C.

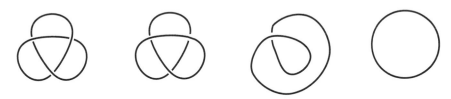

Figure 1.10: The trefoil knot deforms to $\mathbb{S}^1$. The upper right crossing changes from over to under by allowing one strand to move into a fourth spatial dimension, then back again. Afterward, the knot can easily be unraveled.

global information. Photos taken from space and careful measurements made from various points on the Earth show that the surface of Earth is essentially a model for $\mathbb{S}^2$, the two-dimensional surface of a sphere. We will return to the concept of dimension in later chapters.

Let us return to deformation and topological equivalence. Figure 1.9 demonstrates equivalence through deformation. However, one major drawback to our intuitive understanding of deformation is that it can lead us to the wrong conclusion. For example, the *trefoil knot*, pictured in Figure 1.10, is homeomorphic to $\mathbb{S}^1$, even though there appears to be no way to untangle it. An ant on the trefoil experiences the same kind of one-dimensional "universe" that an ant on a circle would. From the point of view of the two ants, the spaces are equivalent, regardless of the way the spaces happen to be positioned within a larger three-dimensional space (which the ants cannot physically interact with anyway). The difference in viewpoint is that an ant can only experience *intrinsic* properties of the space (one dimensionality, no self-intersections, ability to return to the same point after setting out in one direction), while the *extrinsic* properties (how the space is situated within a larger space) are irrelevant in topology. Certainly we cannot untangle the trefoil by deforming it within our familiar three-dimensional world; however, if we are willing to entertain the notion of four spatial dimensions, then the trefoil readily deforms into $\mathbb{S}^1$. (This type of deformation is analogous to moving a point from within a circle in $\mathbb{R}^2$ by "lifting" the point up into a third dimension, carrying it some distance, then dropping it back onto the plane on the outside of the circle.)

Another serious problem with the intuitive definition of deformation is that it is unclear how to apply it to spaces with infinite extent. For example, the real line $\mathbb{R}^1$ – or, in interval notation, $(-\infty, \infty)$ – extends infinitely in both the positive and negative directions, and yet is topologically equivalent to the open ray $(0, \infty)$ as well as to the finite-length open interval $(0, 1)$. We shall need the more precise definition of homeomorphism given in §2.2 to prove this.

Topological Invariants

Now that we have spent some time discussing topological equivalence, we should say something about topological *in*equivalence. If any polygon is just the same as a circle (topologically), then the natural question arises: What spaces are *not* the same? It seems intuitively clear that $\mathbb{S}^1$ cannot be continuously deformed into the open interval $(0, 1)$, because we would have to "break" the circle in order to straighten it out. On the other hand, our intuition often fails us, so how can we be sure? In the coming chapters, we will develop certain properties of spaces called *topological invariants* that help to distinguish one space from another.

> **Definition 1.1.1.** Suppose for every topological space there is a measurable quantity or property $\mathscr{Q}$. If the value of $\mathscr{Q}$ on any space X is the same as for all topological spaces Y homeomorphic to X, then $\mathscr{Q}$ is called a **topological invariant**.

Once you have proven that $\mathscr{Q}$ is a topological invariant, then you can use $\mathscr{Q}$ to *distinguish* nonhomeomorphic spaces. The following theorem is perhaps the most important tool in topology.

> **Theorem 1.1.2.** *Suppose $\mathscr{Q}$ is a topological invariant and X, Y are two spaces. If the value of $\mathscr{Q}$ on X differs from the value of $\mathscr{Q}$ on Y, then X and Y are not homeomorphic.*

On the other hand, invariants *cannot generally* be used to show that two spaces are homeomorphic. In other words, if X and Y both have the same value for some invariant $\mathscr{Q}$, then it is not automatically true that $X \approx Y$. As a small example, let's assume that the quality of *number of pieces* is a topological invariant (see §4.4 for a more formal treatment in terms of *connectedness*). Notice that both $\mathbb{R}^1$ and $\mathbb{S}^1$ have the same number of pieces, namely, 1. However, $\mathbb{R}^1$ is not homeomorphic to $\mathbb{S}^1$. Indeed, suppose any point x is removed from $\mathbb{R}^1$; the result is a space having two pieces, $(-\infty, x) \cup (x, \infty)$. On the other hand, if any point is removed from $\mathbb{S}^1$, the result still has only one piece. This argument would be sufficient to prove $\mathbb{R}^1 \not\approx \mathbb{S}^1$.

One very important topological invariant that has been around since before topology was ever a full-fledged subject is the **Euler characteristic**.[10] For

---

[10]Named after the same Leonhard Euler that pondered the Seven Bridges of Königsberg Problem.

now, we define the Euler characteristic only for surfaces of *polyhedra.*[11] Recall that a **polyhedron** is a three-dimensional solid consisting of flat *faces* joined at straight *edges*, which meet at *vertices*. We assume that the polyhedron surface is closed, in the sense that if the solid were filled with water, the water could not escape. You may be aware of the five **Platonic solids** (shown in Figure 1.11). These are the only completely *regular* polyhedra (in the sense that all faces are congruent, and every vertex has the same number of edges, all meeting at exactly the same angle).

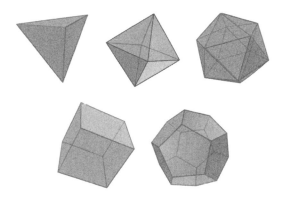

Figure 1.11: The five Platonic solids. *Across the top*, the tetrahedron, octahedron, and icosahedron; *across the bottom*, the cube and dodecahedron. Image courtesy of Wikimedia Commons.

Definition 1.1.3. Suppose P is a polyhedron with f faces, e edges, and v vertices. The **Euler characteristic** of P, denoted $\chi(P)$, is defined by

$$\chi(P) = v - e + f.$$

Example 2. Find $\chi(P)$ for all the Platonic solids.

Solution:

P	v	e	f	$\chi(P)$
tetrahedron	4	6	4	2
octahedron	6	12	8	2
icosahedron	12	30	20	2
cube	8	12	6	2
dodecahedron	20	30	12	2

---

[11]There are generalizations of the Euler characteristic to a wider class of spaces called *cell complexes* (see §5.4), which in turn lead to the concept of *homology groups* (see Chapter 8).

If you find it difficult to count the vertices, edges, and faces of any of these polyhedra, try "flattening out" the picture. Imagine the edges and faces are infinitely flexible, and deform the solid so that all edges spread out on the page. Be careful, though, as one face will be hidden as a result. Figure 1.12 shows a "flattened" diagram of the dodecahedron.

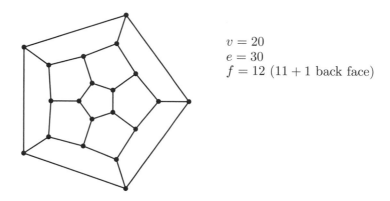

$$v = 20$$
$$e = 30$$
$$f = 12 \ (11 + 1 \ \text{back face})$$

Figure 1.12: A dodecahedron that has been "flattened."

The fact that all five Platonic solids can be deformed into the sphere $\mathbb{S}^2$ and the fact that $\chi(P)$ is a topological invariant (which we haven't proven yet) show why the Euler characteristic is always 2. In fact, $\chi(P) = 2$ for *any* polyhedron P, regular or not, just so long as the surface of P is topologically equivalent to the sphere. On the other hand, consider the polyhedron shown in Figure 1.13, which is a sort of "triangular doughnut," or three-dimensional block form of the capital Greek letter Δ (*Delta*). Carefully counting all faces, edges, and vertices, we find

$$\chi(T) = v - e + f = 12 - 24 + 12 = 0.$$

Thus the surface of T is *not* topologically equivalent to the surface of any *convex*[12] polyhedron, or to $\mathbb{S}^2$. By the way, T *is* homeomorphic to the torus (surface of a doughnut), so although a topologist can't tell the difference between a doughnut and a coffee mug, there is no difficulty in distinguishing a doughnut (torus) and a wine glass (sphere). *Cheers!*

Homotopy

Suppose now we lift part of the restrictions for reversible deformation. A continuous deformation still must preserve *closeness*. However, distinct points may gradually be brought closer to one another and may merge together so long as the process can be made continuous; that is, at some time t_1, the space must be

---

[12]A solid is **convex** if for any two points P and Q in the solid, the line segment $\overline{PQ}$ is entirely contained in the solid. We shall discuss convexity again in §2.4.

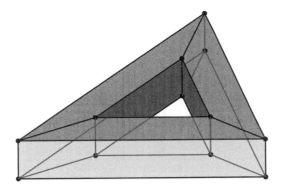

Figure 1.13: A polyhedron with a "hole."

essentially the same[13] as at a time $t_1 + \epsilon$ for small ϵ. For now, let's see an example or two. We shall call two spaces **homotopic** if there is such a deformation from one to the other. For example, see Figure 1.14; the line segment may be continuously shrunk down to a point. This is not to say that the segment and the point are topologically equivalent (they're not), but they *are* homotopic. On the other hand, the segment is not homotopic to $\mathbb{S}^1$. Any deformation will eventually involve bringing together the two endpoints. If that happens at time $t = t_1$, then the space at time $t_1 - \epsilon$ is *essentially* different than the space at time t_1, the former having no "loops," while the latter does.

Figure 1.14: The segment and point are homotopic spaces. The segment and circle are not.

We shall find that if two spaces are homeomorphic, then they are also homotopic to one another, but as the example above shows, the reverse is not true in general. So why bother with homotopy at all? It may not be clear at this point, but the concept of homotopy is incredibly powerful and much easier to work with in practice than homeomorphism. If our job is to classify all the spaces we encounter in mathematics, then classification by homotopy seems to be more tractable than by homeomorphism.

---

[13]We must be vague at this point, having not seen any formal definitions. However, one intuitive criterion that helps to understand this new type of equivalence is that if a new loop is created, then the spaces are no longer *essentially the same*. We defer the details and precision until Chapter 7.

To round out this section, I offer a practical application of homotopy: solving mazes. Mazes have fascinated humans for ages,[14] and there are already a number of efficient, effective maze-solving algorithms. What follows is not a particularly efficient technique, but it is simple enough to implement by drawing on the maze itself and is guaranteed to work so long as the maze has a solution. The method relies on homotopy equivalence to progressively deform the walls of the maze until a path emerges. Assume first that the maze walls are segments of a square grid, as shown in Figure 1.15.

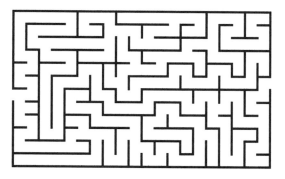

Figure 1.15: A maze, created using `http://www.mazegenerator.net/`.

1. Anywhere in the maze, if there is a square that has three walls (a *dead-end*), then fill the square. This reflects the topological fact that the three wall configuration (⊔) is homotopic to the filled-in square (■) via a homotopy that progressively thickens the segments and expands to fill the area of the whole square.

2. Repeat step 1 until it becomes impossible to do so, at which point the unfilled squares contain a solution path. If there is only one solution, then that path goes through all of the unfilled squares.

That's it – that's the entirety of the algorithm. More general mazes, even those with curving walls, can be tackled in the same way so long as the *dead-ends* can be identified.

Example 3. Let's see how the method works on a small example maze.

---

[14]There is a short but very interesting article on symmetries in ancient mazes by Tony Phillips, "Hidden Symmetries of Labyrinths from Antiquity and the Middle Ages," currently available online at `http://www.ams.org/samplings/feature-column/fc-2015-10`.

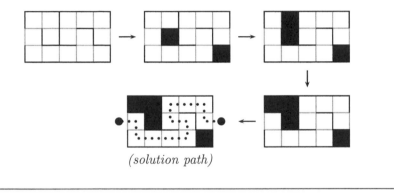

(solution path)

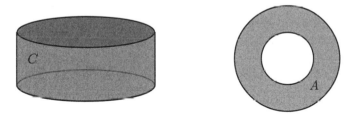

Exercises

1. Consider the capital letters of the alphabet written in block form and thickened to three-dimensional objects (as in Figure 1.9). Group the letters into sets based on topological equivalence.

2. Repeat Exercise 1 with the capital letters of the Greek alphabet[15] (in block form, as three-dimensional objects).

3. Let C be a (hollow) *cylinder* with no top or bottom. Let A be an *annulus*, which is defined as a region in the plane between two concentric circles (see below). Is A homeomorphic to C via deformation? If so, show an intermediate step. If not, explain why not.

4. Figure 1.16 shows two surfaces that are topologically equivalent. In fact, one can be deformed to the other through a sequence of moves entirely within $\mathbb{R}^3$. Draw a sequence of pictures demonstrating how one space deforms into the other.

5. Find the Euler characteristic of a solid cube with three holes drilled through the middle, as shown in Figure 1.17. *Be careful*; you will have to draw additional edges so that each face contains no hole.

6. Use the homotopy algorithm to solve the maze in Figure 1.15.

---

[15]See https://en.wikipedia.org/wiki/Greek_alphabet#Letters.

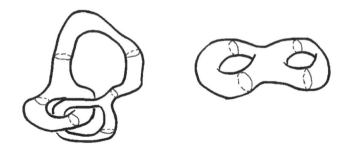

Figure 1.16: Figure for Exercise 4.

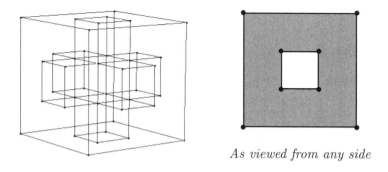

As viewed from any side

Figure 1.17: Figure for Exerise 5.

1.2   Topological Spaces

Now let's explore a few topological spaces. This section is intended only to introduce the spaces in a hands-on way without delving too deeply into their topological structure.

Intervals and the Cantor Set

The **real line** is a visualization of the set $\mathbb{R}$ of real numbers. What could be simpler than an unbroken line? It turns out that there is a lot of deep mathematical structure hidden within this line and its subsets. Beginning in Chapter 2, we find that one of the key concepts in topology is understanding what it means for a set to be *open*. An interval such as $A = (a, b)$ is called *open* because every point x in A can be surrounded by a small interval of the form $(x - \epsilon, x + \epsilon)$ contained entirely within A. For example, if $A = (0, 1)$, and $x = 0.9$, then one could choose $\epsilon = 0.1$ so that the interval $(0.9 - 0.1, 0.9 + 0.1) = (0.8, 1)$ is contained within A.

Open intervals exhibit a curious property though: while the interval (a, b) does not contain either a or b, the interval does contain every real number x, $a < x < b$, that is *arbitrarily close* to a or b, respectively. For example, 1 is not an element of $A = (0, 1)$. But 0.9 is, and so is 0.99, and so is 0.999, etc. In fact, the number

$$0.999\,999\,999\,999\,999\,999\,999\,999\,999\,999\,999\,999\,999\,999\,999\,999\,999\,999\,999\,999,$$

which is within 10^{-60} of the value 1, is a member of the interval $(0, 1)$. However, note that the number $z = 0.\overline{9} = 0.9999\ldots$ (digit 9 repeating forever) is *not* in $(0, 1)$, because z is actually equal to 1. Here's a quick argument, but this strange result can be proven more rigorously using *geometric series* (see Exercise 1).

$$
\begin{aligned}
z &= 0.\overline{9} \\
10z &= 9.\overline{9} \\
10z - z &= 9.\overline{9} - 0.9 \\
9z &= 9.\overline{0} = 9 \\
z &= 9/9 = 1
\end{aligned}
$$

By contrast, an interval such as $B = [a, b]$ is called *closed* because any real number that is arbitrarily close to A is actually contained in A. Intervals such as $[a, b)$ or $(a, b]$ are considered neither open nor closed.[16] The concepts of open and closed will be explained more formally in later chapters, but one closed interval in particular will be important in the coming chapters: the **unit interval**, $\mathbb{I}$. By definition,

$$\mathbb{I} = [0, 1] = \{x \in \mathbb{R} \mid 0 \leq x \leq 1\}. \tag{1.1}$$

The **Cantor set** is a particular subset of the unit interval, obtained by a process of removing the "open middle third" intervals forever. To construct the Cantor set, do the following:

Step 0: Begin with the unit interval $\mathbb{I} = [0, 1]$.

Step 1: Delete the open middle third, $(1/3, 2/3)$, leaving the two disconnected closed intervals: $[0, 1/3]$ and $[2/3, 1]$.

Step 2: Delete the open middle thirds from each segment, leaving four disconnected closed intervals: $[0, 1/9]$, $[2/9, 1/3]$, $[2/3, 7/9]$, and $[8/9, 1]$.

$\ldots$

Step n: Delete the open middle thirds from each segment, leaving 2^n disconnected closed intervals.

$\ldots$

---

[16]Some textbooks call intervals of this type *half-open*, which is something of a misnomer that we avoid in this text.

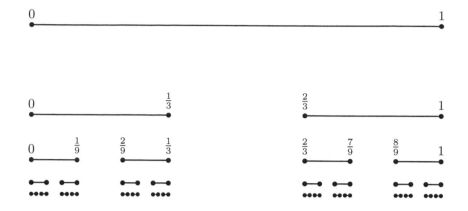

Figure 1.18: Starting with $\mathbb{I}$, remove the middle third. Then remove the middle third of the remaining segments iteratively forever. The end result is called the Cantor set.

The Cantor set C is what remains of $\mathbb{I}$ after infinitely many steps. Figure 1.18 illustrates the first few steps of the process. Leaving the technical details aside for now, you can imagine that there is some ideal set C, which we can *approximate* to any desired accuracy by following n steps of the construction procedure; the higher the n, the more accurate the picture. How large is the Cantor set? In one sense, there's barely anything there at all. The total length of the C is 0. You can see this by adding up the lengths of the deleted middle thirds. In Step 1, exactly $1/3$ has been removed. In Step 2, exactly $2 \cdot (1/9) = 2/9$ has been removed. In Step 3, $4 \cdot (1/27) = 4/27$ has been removed. The total length L removed is equal to

$$L = \frac{1}{3} + \frac{2}{3^2} + \frac{2^2}{3^3} + \cdots + \frac{2^{n-1}}{3^n} + \cdots = \frac{1}{3} \sum_{n=0}^{\infty} \left(\frac{2}{3} \right)^n . \qquad (1.2)$$

You may recognize (1.2) as a convergent *geometric series*. Recall that if $|r| < 1$, then

$$\sum_{n=0}^{\infty} r^n = \frac{1}{1-r}. \qquad (1.3)$$

Therefore the sum (1.2) works out to:

$$L = \frac{1}{3} \sum_{n=0}^{\infty} \left(\frac{2}{3} \right)^n = \frac{1}{3} \cdot \frac{1}{1 - \frac{2}{3}} = 1.$$

But the length of the unit interval is also 1, so the length of C must be $1 - L = 1 - 1 = 0$ units. Ok, so maybe that's not so surprising, given the amount of material that was taken out at each step. What is quite remarkable is the amount of "stuff" that remains in C. There are more points in C than in the entire set $\mathbb{Q}$ of rational numbers, in the sense that C is *uncountable*, while $\mathbb{Q}$ is *countably infinite*.[17]

Exploring Surfaces

We live on the surface of the Earth. The page you're reading right now is a model of a surface. Surfaces are fairly easy to understand because we live in a three-dimensional world that gives us the vantage point to look at a given surface from "above" (though some topological surfaces, such as the *Klein bottle*, do not actually admit a well-defined notion of what is "above" or "below" the surface). As mentioned before, the surface of a doughnut is called a torus. To better understand the torus, and surfaces in general, we may rely on what's called a *plane model* of the surface, which is a "cut and flattened" representation of the space. The plane model of a torus, as shown in Figure 5.12 (Chapter 5), is a square with *identifications* along its boundary, indicating that if an object traveled to the right edge of the square, then it would reappear on the left edge, and if it traveled to the top edge, then it would reappear on the bottom. An equivalent model of the torus is as an infinite repeating grid of squares, as shown in Figure 1.19. The *Klein bottle* can be defined analogously.

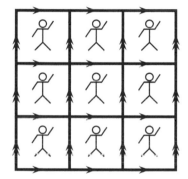 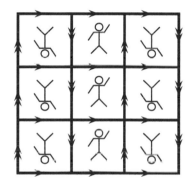

Figure 1.19: Representing the torus (*left*) and the Klein bottle (*right*) as an infinite grid of identical squares. Sammy the Stick Figure is waving to show the how the squares must fit together to make each surface.

Let's have some fun. Imagine a game of tic-tac-toe, checkers, or chess in which the board takes up the whole surface of the torus or Klein bottle. A "win" in tic-tac-toe may involve connecting three in a row across an edge. Chess pieces may exit one side of the board and reappear on the other.[18] Let's explore a few examples together.

---

[17] We refer the reader to any standard text on mathematical analysis or set theory for more information on countability. These terms will not be used significantly in this text.

[18] Check out www.geometrygames.org, by Jeffrey Weeks, for computer versions of these and other games on a variety of surfaces.

Example 4. Determine whether the configuration below is a win for either X or O if the game is being played on a (a) torus or (b) Klein bottle.

Solution: It's easy to verify using the infinite grid representations (torus shown on left, and Klein bottle on right).

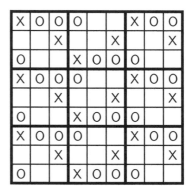

(a) On the torus, neither X nor O has the win. (But the top center space would be X's next play for the win.)

(b) On the Klein bottle, O is the winner.

Example 5. On a traditional chessboard, a bishop must remain on its own color throughout the game because it can only move diagonally. The same is true on the torus, but not on the Klein bottle, as illustrated in Figure 1.20.

Connections to Calculus: Path Integrals

What does topology have to do with calculus? It turns out that much of the motivation for the early development of topology stems from certain observations and abstractions from calculus, and especially from the analysis of vector fields. The domain set of a vector field may be thought of as a topological space, and the vector field itself is something like a continuous map on the space. A path through the vector field can also be thought of as a continuous map, as we shall see in Chapter 2. If you haven't seen vector calculus yet, feel free to skip this section.

In many cases, the domain space of a vector field is simply the familiar Euclidean plane $\mathbb{R}^2$, but in other instances, the domain may be a subset of the

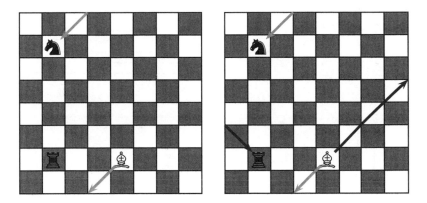

Figure 1.20: *Left*, the bishop on a torus remains on its color, and can attack the knight but not the rook. *Right*, the bishop on a Klein bottle may find itself on the opposite color; this bishop can attack either the knight or the rook.

plane, a more general surface, or even more exotic higher-dimensional spaces called *manifolds*. The topic of vector fields will return in Chapter 3. Recall that a **vector field V** in $\mathbb{R}^2$ is a vector-valued function of two variables,

$$\mathbf{V}(x,y) = M(x,y)\mathbf{i} + N(x,y)\mathbf{j}. \tag{1.4}$$

Here, **i** and **j** are the unit vectors in the positive x- and y-directions, respectively. The field **V** represents something like the direction and magnitude of force around a bar magnet, or the specific direction and strength data of wind at each point of a region. At any rate, **V** may be visualized as a collection of arrows (vectors), one arrow attached to each point of the plane.

Now imagine that a particle takes a journey through this field. If it follows the general direction of the arrows, then the journey will not take much effort (the field will be doing *positive work* on the particle), but we want to consider arbitrary paths that may not follow the arrows at all. We track the progress of our particle mathematically by way of a parametrized curve. Recall that a **parametrized curve**, C, in $\mathbb{R}^2$ is specified by a vector-valued function **r** of one variable, t (for *time*).

$$\mathbf{r}(t) = u(t)\mathbf{i} + v(t)\mathbf{j}, \quad a \leq t \leq b \tag{1.5}$$

See Figure 1.21 for an example of a path within a vector field. (Of course these definitions may be extended to $\mathbb{R}^n$ for any natural number n.)

Now we may define a quantity called *work* that measures how the field interacts with a particle moving along a given path.

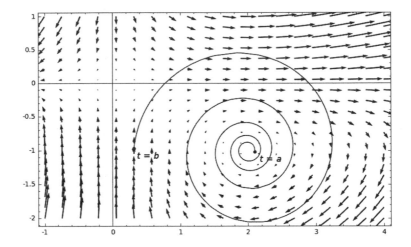

Figure 1.21: A parametrized curve within a vector field. The curve does not necessarily follow the flow lines. Created using Sagemath.

Definition 1.2.1. If $\mathbf{V}$ is a vector field and $\mathbf{r}$ parametrizes a smooth curve, C, then the **work**, W, done by the field in moving an object from the beginning of C to the end of C is

$$W = \int_C \mathbf{V} \cdot d\mathbf{r}.$$

In many important real-world applications, only the beginning point, $A = \mathbf{r}(a)$, and ending point, $B = \mathbf{r}(b)$, matter to the value of W, while the shape of the path C has no effect at all. A vector field having this quality of path independence on an open domain set D is called **conservative** on D (see Figure 1.22). In fact, if $\mathbf{V}$ is conservative and $A = B$, so that C is a closed path, then $\int_C \mathbf{V} \cdot dr = 0$. To use the language of topology, if C is homeomorphic to the circle $\mathbb{S}^1$, then the work done along C is 0 in a conservative field.

It becomes important to be able to identify whether a field is conservative in order to apply path independence. Again, topology will come into play. If the domain set of the vector field satisfies a certain topological property called *simply connectedness*, then there is an easy criterion to check.

Theorem 1.2.2. *Let* $\mathbf{V} = u(x,y)\mathbf{i} + v(x,y)\mathbf{j}$ *be a vector field on a simply connected open domain set* U, *and suppose* u *and* v *have continuous first partial derivatives. Then* $\mathbf{V}$ *is conservative if and only if* $\dfrac{\partial v}{\partial x} - \dfrac{\partial u}{\partial y} = 0$.

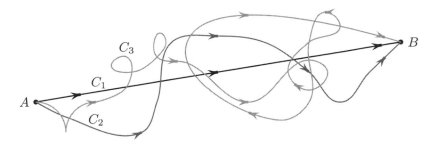

Figure 1.22: If $\mathbf{V}$ is conservative, then $\int_{C_1} \mathbf{V} \cdot d\mathbf{r} = \int_{C_2} \mathbf{V} \cdot d\mathbf{r} = \int_{C_3} \mathbf{V} \cdot d\mathbf{r}$.

In $\mathbb{R}^2$, an open domain set U is called *simply connected* if U is homeomorphic to the interior of a disk.[19] Intuitively, a simply connected space does not have any "holes." Thus the entire plane $\mathbb{R}^2$ is simply connected, as is any region in the plane bounded by a simple closed curve; however, the annulus, $\{(x, y) \mid 1 < x^2 + y^2 < 2\}$, is *not* simply connected, nor is the punctured disk, $\{(x, y) \mid 0 < x^2 + y^2 < 1\}$. It turns out that *simply connectedness* is another example of a quality that remains the same for homeomorphic spaces; in other words, it is a topological invariant.

The proof of Theorem 1.2.2 may be found in many standard calculus textbooks. Observe that the result may be false if U is not simply connected.

Example 6. *(This example requires multivariable calculus.)* Consider the field

$$\mathbf{V}(x, y) = \frac{-y}{x^2 + y^2}\mathbf{i} + \frac{x}{x^2 + y^2}\mathbf{j}, \tag{1.6}$$

defined on the punctured plane $U = \{(x, y) \in \mathbb{R}^2 \mid (x, y) \neq (0, 0)\}$. Note that $\mathbf{V}$ cannot be extended continuously to include the origin $(0, 0)$. We have $\frac{\partial v}{\partial x} - \frac{\partial u}{\partial y} = 0$, as verified below.

$$\frac{\partial v}{\partial x} - \frac{\partial u}{\partial y} = \frac{-x^2 + y^2}{(x^2 + y^2)^2} - \frac{-x^2 + y^2}{(x^2 + y^2)^2}$$
$$= 0$$

Let C be the unit circle, traversed once counterclockwise from $(1, 0)$, which may be parametrized by $\mathbf{r}(t) = (\cos 2\pi t)\mathbf{i} + (\sin 2\pi t)\mathbf{j}$, where $0 \leq t \leq 1$. Then $\mathbf{V}(\mathbf{r}(t)) = (-\sin 2\pi t)\mathbf{i} + (\cos 2\pi t)\mathbf{j}$ and $d\mathbf{r} = [(-2\pi \sin 2\pi t)\mathbf{i} + (2\pi \cos 2\pi t)\mathbf{j}]\, dt$; hence

$$\int_C \mathbf{V} \cdot d\mathbf{r} = \int_a^b \mathbf{V}(\mathbf{r}(t)) \cdot \frac{d\mathbf{r}}{dt}\, dt = \int_0^1 2\pi(\sin^2 2\pi t + \cos^2 2\pi t)\, dt = 2\pi.$$

The integral is nonzero, so $\mathbf{V}$ fails to be conservative. On the other hand, it can be shown that for an arbitrary piecewise smooth, closed curve C avoiding the

---

[19]This definition only works in $\mathbb{R}^2$. The general definition of simply connected involves the *fundamental group*, $\pi_1(U)$ (see Chapter 7).

origin, the value of $\frac{1}{2\pi}\int_C \mathbf{V}\cdot d\mathbf{r}$ computes the net number of times C makes a complete counterclockwise rotation about the origin.

Exercises

1. Use (1.3) for the sum of a geometric series to prove that $z = 0.\overline{9} = \frac{9}{10} + \frac{9}{100} + \frac{9}{1000} + \cdots$ is indeed equal to 1.

2. There is an analog of the Cantor set in higher dimensions, called *Cantor dust*. Read up on Cantor dust at `http://www.2dcurves.com/fractal/fractald.html`. Draw an approximate picture of Cantor dust in the plane, and determine the "area" contained within Cantor dust.

3. Play a few games of tic-tac-toe on both the torus and the Klein bottle with a classmate. Then try checkers, being careful to define precisely what you mean by "forward" moves.

4. The usual starting position of chess will not work on a torus or Klein bottle since the two opposing kings would begin the game in mutual check. Consider the starting position shown in Figure 1.23 (from Jeffrey Weeks, *The Shape of Space* [Wee02]).

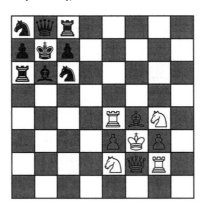

Figure 1.23: One possible initial position for torus chess.

   (a) Play a few games of torus chess using this starting position. You'll have to decide appropriate rules for the movement of the pawns.

   (b) Try the variant of *besiege chess* played on a torus (`http://www.chessvariants.com/shape.dir/toruschess.html`). Then play besiege chess on the Klein bottle.

5. On a traditional chessboard, the knight always alternates between black and white squares with each move.

   (a) Show, by drawing a few pictures, that a knight must alternate colors on a torus chessboard.

   (b) Provide a picture showing that the knight may remain on the same-color square when moving on a Klein bottle chessboard.

6. Consider the black knight, as shown in Figure 1.20. We say that a piece *controls* a square if it could attack that square on its next move. Highlight all of the squares that the black knight controls if the board is:

   (a) Traditional   (b) A torus   (c) A Klein bottle

7. Let $\mathbf{V}$ be the vector field defined by (1.6) in Example 6. Fix an arbitrary integer n, and consider the curve C_n parametrized by $\mathbf{r}(t) = (\cos 2\pi nt)\mathbf{i} + (\sin 2\pi nt)\mathbf{j}$, where $0 \leq t \leq 1$.

   (a) Describe the curve C_n for $n = 1, 2, 3$ and $n = -1, -2, -3$. In words, what does C_n do?

   (b) Compute $\frac{1}{2\pi} \int_{C_n} \mathbf{V} \cdot d\mathbf{r}$. How does this result relate to your answer to part (a)?

Supplemental Reading

- Barr [Bar64], Chapter 1.

- Bartle and Sherbert [BS11], §11.1. Cantor set.

- Milnor [Mil15] provides a concise survey of important topological milestones.

- Prasolov [Pra95], Chapter 1. Discussion of malleability and deformations.

- Thomas et al. [TWH10], Chapter 16. Review of path integrals and vector fields in $\mathbb{R}^2$.

- Weeks [Wee02], Part I. Understanding the properties of a topological surface through "games" and other visualizations.

Chapter 2

Metric Topology in Euclidean Space

Distance is a concept central to our life. It's important to know the number of miles to the next gas station, the shortest flight distance from New York to Tokyo, or whether there might be a "shortcut" path through space and time that would allow humans to travel to distant galaxies within a single lifetime. The trouble is that our intuitive notion of distance (which is known as *Euclidean* distance) may be fine for getting us to the nearest gas station, but fails us in problems on a grander scale. For instance, the shortest distance from one point to another on the surface of a sphere is along a *great circle*, which may be defined as a circle that separates the sphere into two equal parts. When we view the globe as a flat map, many great circles look like highly curved paths (see Figure 2.1), challenging our understanding of Euclidean geometry, in which the shortest path between two points is a straight line. What's more, if we want to answer questions about distance between locations in the universe, we are further at a loss. It is not even clear what the *shape* of space is – for example, does it curve around on itself so that two points that look far apart are really next door to one another? In other words, are there *wormholes* (as in Figure 2.2)? Being myopic ants, we can really only experience distances that are quite small, and on this level, Euclidean geometry is an appropriate and useful tool. Thus we begin in this chapter to understand the fundamentals of Euclidean distance.

2.1 Distance

In this chapter, let $n \in \mathbb{N}$. We assume the reader has some familiarity with the *Euclidean* spaces $\mathbb{R}^n$, and certain properties of the real numbers $\mathbb{R}$, namely, that $\mathbb{R}$ is the set of all points on the number line, including all rational and irrational

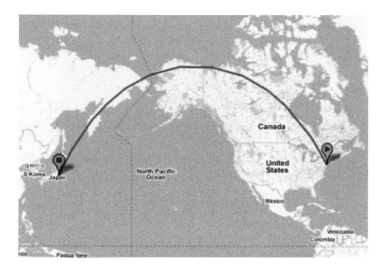

Figure 2.1: The shortest distance between New York and Tokyo lies along a *great circle*. Image generated using Google Maps (note that even though the picture shows a series of line segments, this is simply an artifact of how Google Maps generates the path – the actual shortest path should be smooth.)

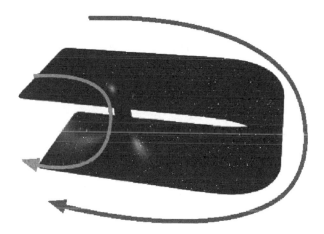

Figure 2.2: There may be "shortcuts" through space-time, called *wormholes*. Our current understanding of the shape of the cosmos does not rule out these structures. Image courtesy of Wikimedia Commons.

numbers.[1] Elements of $\mathbb{R}^n$ may be thought of as either points or vectors. Points of $\mathbb{R}^n$ will typically be written in *italic* type (e.g., x, y, z, p, q, etc.) as opposed

---

[1] For a deeper understanding of $\mathbb{R}$, see any standard text on real analysis, e.g., [BS11].

to the typical **bold** type (e.g., $\mathbf{x}$, $\mathbf{y}$, $\mathbf{z}$, $\mathbf{p}$, $\mathbf{q}$, etc.), although the origin will always be denoted $\mathbf{0} \in \mathbb{R}^n$ to distinguish it from the scalar value $0 \in \mathbb{R}$. Points (vectors) of $\mathbb{R}^n$ may also be identified by their *coordinates*; hence

$$\mathbb{R}^n = \{x = (x_1, x_2, \ldots, x_n) \mid x_i \in \mathbb{R} \text{ for each } i = 1, 2, \ldots, n\} \qquad (2.1)$$

$$= \left\{ x = \begin{pmatrix} x_1 \\ x_2 \\ \vdots \\ x_n \end{pmatrix} \ \middle| \ x_i \in \mathbb{R} \text{ for each } i = 1, 2, \ldots, n \right\}. \qquad (2.2)$$

While a point x is (by definition) a dimensionless element, we may also regard x as the *vector* from the origin $\mathbf{0} = (0, 0, \ldots, 0)$ to the point $x = (x_1, x_2, \ldots, x_n)$. As such, x has length. Recall that the standard *dot product* of two vectors is defined by

$$x \cdot y = (x_1, x_2, \ldots, x_n) \cdot (y_1, y_2, \ldots, y_n) = x_1 y_1 + x_2 y_2 + \ldots + x_n y_n,$$

and the **length** (or **magnitude**, or **norm**) of x is defined by

$$\|x\| = \sqrt{x \cdot x} = \sqrt{x_1^2 + x_2^2 + \cdots + x_n^2}.$$

Note that for $n = 1$, a vector or point in $\mathbb{R}^1 = \mathbb{R}$ is just a real number x. The length of x simply boils down to its absolute value: $\|x\| = \sqrt{x^2} = |x|$, and so distance from x to y in $\mathbb{R}$ is defined by $|x - y|$. Distance between points in Euclidean space $\mathbb{R}^n$ is defined by the following formula (and illustrated by Figure 2.3).

Definition 2.1.1. The **distance** between points $x = (x_1, x_2, \ldots, x_n)$ and $y = (y_1, y_2, \ldots, y_n)$ in $\mathbb{R}^n$ is denoted $d(x, y)$ and defined by the formula

$$d(x, y) = \sqrt{(x_1 - y_1)^2 + (x_2 - y_2)^2 + \cdots + (x_n - y_n)^2},$$

or, in terms of vectors,

$$d(x, y) = \|x - y\|.$$

Note that $\|x\| = d(x, \mathbf{0})$. If we need to know the angle θ between two vectors x and y, then we could use the relation $x \cdot y = \|x\| \|y\| \cos \theta$.

Once we know how to compute distance, we may define spheres. For any $n \geq 0$, define the **unit sphere** $\mathbb{S}^n \subseteq \mathbb{R}^{n+1}$ by

$$\mathbb{S}^n = \{x \in \mathbb{R}^{n+1} \mid d(x, \mathbf{0}) = 1\}.$$

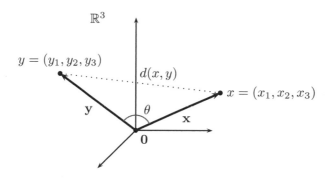

Figure 2.3: The distance between points x and y (or vectors $\mathbf{x}$ and $\mathbf{y}$) and the angle θ between the vectors in a Euclidean space.

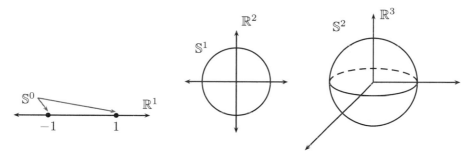

Figure 2.4: Low-dimensional unit spheres in Euclidean spaces.

Example 7. Figure 2.4 illustrates the unit sphere $\mathbb{S}^n$ for $n = 0, 1, 2$.

- $\mathbb{S}^0 \subseteq \mathbb{R}$ consists of just two points, -1 and 1, because $|-1-0| = |1-0| = 1$, and no other $x \in \mathbb{R}$ satisfies $|x - 0| = 1$.

- $\mathbb{S}^1$ is another name for the **unit circle** in $\mathbb{R}^2$; that is, $\mathbb{S}^1 = \{(x,y) \in \mathbb{R}^2 \mid d((x,y),(0,0)) = 1\} = \{(x,y) \in \mathbb{R}^2 \mid x^2 + y^2 = 1\}$. Only the points on the circumference of the circle are part of $\mathbb{S}^1$.

- $\mathbb{S}^2$ is the unit sphere in $\mathbb{R}^3$, $\{(x,y,z) \in \mathbb{R}^2 \mid x^2 + y^2 + z^2 = 1\}$, which is the set of points on the surface of a "ball."

- For $n \geq 3$, it is hard to visualize what $\mathbb{S}^n$ could look like because our only experience is with three-dimensional space (even $\mathbb{S}^3$ lives in $\mathbb{R}^4$, for example). So we cannot quite "see" the so-called **hyperspheres**.[2] We

---

[2]See Edwin Abbott's *Flatland* [Abb84] for an account of how one might "perceive" higher-dimensional objects by their intersections with the lower-dimensional spaces. See also Dante's

only experience these mathematical constructs by their defining property (distance from the origin is equal to 1), or all solutions to the equation $x_1^2 + x_2^2 + \cdots + x_n^2 = 1$.

The distance function d satisfies four important properties:

Proposition 2.1.2. *Let $n \geq 1$. The Euclidean distance function $d :$ $\mathbb{R}^n \times \mathbb{R}^n \to \mathbb{R}$ satisfies:*

1. *[Nonnegativity] $d(x,y) \geq 0$, $\forall x, y \in \mathbb{R}^n$.*

2. *[Zero-Distance Rule] $d(x,y) = 0$ if and only if $x = y$.*

3. *[Symmetry] $d(x,y) = d(y,x)$, $\forall x, y \in \mathbb{R}^n$.*

4. *[Triangle Inequality] $d(x,z) \leq d(x,y) + d(y,z)$, $\forall x, y, z \in \mathbb{R}^n$.*

The proofs of 1–3 are quite easy, relying on properties of the real-valued functions $f(x) = x^2$ and $g(x) = \sqrt{x}$. The last part, the *Triangle Inequality*, states that any side of a Euclidean triangle is no longer than the sum of the other two sides. While this seems obvious to anyone with some knowledge of geometry, it must be proven. The proof is taken up in greater generality in §2.5.

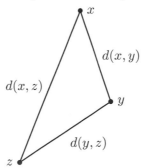

Open Balls

In $\mathbb{R}^n$, if we want to talk about the immediate vicinity of a point, we usually consider a small open set surrounding it. It is natural to consider points up to a specified distance away from the point in question.

Definition 2.1.3. If $x \in \mathbb{R}^n$ and $\epsilon \in \mathbb{R}^+$ (i.e., $\epsilon > 0$), the ϵ-**ball around** x is the set
$$B_\epsilon(x) = \{y \in \mathbb{R}^n \mid d(x,y) < \epsilon\}.$$

Divine Comedy, in which the various spheres that make up the Earth, heaven, hell, and purgatory can arguably be modeled within $\mathbb{S}^3$.

An ϵ-ball around x is a type of *neighborhood* of x. More generally, we define a neighborhood as follows:

> **Definition 2.1.4.** A set N is called a **neighborhood** of a point $x \in \mathbb{R}^n$ if there is a value of $\epsilon \in \mathbb{R}^+$ such that $B_\epsilon(x) \subseteq N$.

In this section we focus on the ϵ-balls; however, the more general idea of neighborhoods will become useful later on.

Example 8. In $\mathbb{R}^1$, an ϵ-ball is simply an open interval around a point on the number line:

$$B_\epsilon(x) = \{y \in \mathbb{R} \mid |x - y| < \epsilon\} = \{y \in \mathbb{R} \mid x - \epsilon < y < x + \epsilon\},$$

or in interval notation, $B_\epsilon(x) = (x - \epsilon, x + \epsilon)$.

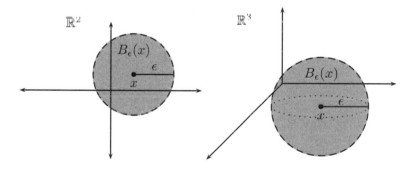

In $\mathbb{R}^2$, an ϵ-ball is a disk of radius ϵ with no boundary (we will call this an **open disk** even though we haven't formally defined *open* yet).

$$B_\epsilon(x) = B_\epsilon((x_1, x_2)) = \left\{(y_1, y_2) \in \mathbb{R}^2 \mid \sqrt{(x_1 - y_1)^2 + (x_2 - y_2)^2} < \epsilon\right\}$$

In $\mathbb{R}^n$ in general, an ϵ-ball is the set of points interior to a (hyper-)sphere of radius ϵ, and not including the boundary (see Figure 2.5).

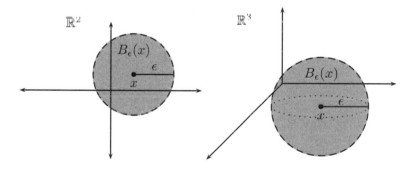

Figure 2.5: *Left*, an open ϵ-ball (disk) in $\mathbb{R}^2$. *Right*, an open ϵ-ball in $\mathbb{R}^3$.

Notice that each example ϵ-ball above "takes up space," in the following sense: The open interval in $\mathbb{R}$ has nonzero length; the open disk in $\mathbb{R}^2$ has nonzero area; and the open ball in $\mathbb{R}^3$ has nonzero volume. Intuitively, we should consider an *open set* of $\mathbb{R}^n$ as one that takes up some nontrivial amount of space in $\mathbb{R}^n$ and does not include its own boundary points. Every point of such a space should have a (perhaps tiny) distance around it containing only its fellow points in the set. To be most precise, we use the following definition.

Definition 2.1.5. A subset $U \subseteq \mathbb{R}^n$ is **open** if for every $x \in U$, there is an $\epsilon \in \mathbb{R}^+$ such that $B_\epsilon(x) \subseteq U$. Equivalently, $U \subseteq \mathbb{R}^n$ is open if for every $x \in U$ there is a neighborhood N of x such that $N \subseteq U$.

Example 9. Fix $n \in \mathbb{N}$. Two special subsets of $\mathbb{R}^n$ are easily shown to be open, the empty set and the entire set.

- By definition, the empty set, $\emptyset$, is considered an open subset of $\mathbb{R}^n$. Because there are no points $x \in \emptyset$, there is nothing to check, and so Definition 2.1.5 is *vacuously* true for $U = \emptyset$.

- The whole set $\mathbb{R}^n$ is open, since any point $x \in \mathbb{R}^n$ can be surrounded by the ϵ-ball $B_\epsilon(x)$ for any choice of $\epsilon \in \mathbb{R}^+$.

Example 10. Let $U = \{(x, y) \in \mathbb{R}^2 \mid |x| < 1 \text{ and } |y| < 1\}$, and let $V = \{(x, y) \in \mathbb{R}^2 \mid |x| \leq 1 \text{ and } |y| < 1\}$. Show that U is open and V is not open.

Solution: Refer to Figure 2.6. First we prove that U is open.

Let $(x, y) \in U$ be arbitrary. We have to find an ϵ-ball around (x, y) contained entirely within U. Without loss of generality, assume that $x, y \geq 0$ (the other cases may be proved separately, or we may rely on the symmetry of the square to argue that the other cases are equivalent). Since $|x| < 1$ and $|y| < 1$, we have $0 \leq x < 1$ and $0 \leq y < 1$. Let $\epsilon_1 = 1 - x$ and $\epsilon_2 = 1 - y$. Observe that both $0 < \epsilon_1 \leq 1$ and $0 < \epsilon_2 \leq 1$. Let $\epsilon = \min\{\epsilon_1, \epsilon_2\}$.

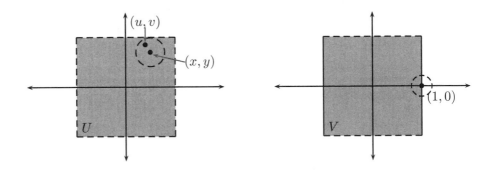

Figure 2.6: *Left,* U is open in $\mathbb{R}^2$. *Right,* V is not open.

Claim: $B_\epsilon((x,y)) \subseteq U$. *Proof of claim:* Let $(u,v) \in B_\epsilon((x,y))$ be arbitrary. Then by definition, $d((x,y),(u,v)) = \sqrt{(x-u)^2 + (y-v)^2} < \epsilon$. Now using the fact that $(y-v)^2 \geq 0$ and the fact that $\epsilon \leq \epsilon_1$,

$$|x - u| = \sqrt{(x-u)^2} \leq \sqrt{(x-u)^2 + (y-v)^2} < \epsilon \leq \epsilon_1.$$

This shows that $x - \epsilon_1 < u < x + \epsilon_1$. But since $x \geq 0$ and $\epsilon_1 \leq 1$, we find that $x - \epsilon_1 \geq 0 - \epsilon_1 \geq -1$. Moreover, since $\epsilon_1 = 1 - x$, we have $x + \epsilon_1 = 1$. Substituting into the inequality produces $-1 < u < 1$, or $|u| < 1$.

Similarly, $|v - y| < \epsilon_2$ implies that $v \in (-1, 1)$. This proves that $(u, v) \in U$ (since $|u| < 1$ and $|v| < 1$), and hence $B_\epsilon((x,y)) \subseteq U$. Thus U is open.

Now, to show that V is not open, we just have to find a point in V that cannot be surrounded by any ϵ-ball within V. Consider $(1, 0) \in V$. For any $\epsilon \in \mathbb{R}^+$, if $B_\epsilon((1, 0)) \subseteq V$, then $(1 + \epsilon/2, 0) \in V$, but this is false since $1 + \epsilon/2 > 1$. This contradiction shows that $B_\epsilon((1,0)) \not\subseteq V$. Thus V is not open.

What about the ϵ-balls themselves? Is $B_\epsilon(x)$ open? Notice that Definition 2.1.5 does not immediately imply that $B_\epsilon(x)$ is open – we need to establish that every single point y in the ball has its *own ball centered at y*, $B_\delta(y)$, contained within $B_\epsilon(x)$. If y is near the boundary, then δ may have to be very small indeed, but the important thing is that an appropriate $\delta > 0$ can always be found. The proof hinges on the triangle inequality.

Proposition 2.1.6. *For every $\epsilon \in \mathbb{R}^1$ and $x \in \mathbb{R}^n$, $B_\epsilon(x)$, is an open subset of $\mathbb{R}^n$.*

Proof. Let $y \in B_\epsilon(x)$; hence $d(x, y) < \epsilon$. Let $\delta = \epsilon - d(x, y)$, which is a positive number.

Claim: $B_\delta(y) \subseteq B_\epsilon(x)$ (as illustrated by Figure 2.7). *Proof of claim:* In order to prove the set inclusion, choose an arbitrary element $z \in B_\delta(y)$, so $d(y, z) < \delta$ by definition. Then by the triangle inequality,

$$d(x, z) \leq d(x, y) + d(y, z) < d(x, y) + \delta = d(x, y) + \epsilon - d(x, y) = \epsilon.$$

This shows that $z \in B_\epsilon(x)$; hence $B_\delta(y) \subseteq B_\epsilon(x)$ as required. By definition then, $B_\epsilon(x)$ is open. $\qquad\square$

Now suppose U and V are both open sets of $\mathbb{R}^n$. Then for every point in their union, $x \in U \cup V$, either $x \in U$ or $x \in V$ (or both). So there is a value $\epsilon > 0$ such that $B_\epsilon(x) \subseteq U$ or $B_\epsilon(x) \subseteq V$. In either case, $x \in B_\epsilon(x) \subseteq U \cup V$, proving that the union $U \cup V$ of open sets is also open. It is also straightforward to show that the intersection of open sets is open. By induction, then, the union and intersection of any finite number of open sets must also be open. In fact, arbitrary unions of open sets are open, but arbitrary intersections may not be. These essential properties, together with the fact that the whole space and the

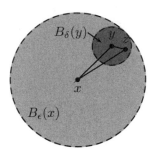

Figure 2.7: An ϵ-ball is open because every point of the ball can be surrounded by its own smaller ball contained within the set.

empty set are both open, form the basis of the abstraction of *topological space* that we will encounter in Chapter 4.

Proposition 2.1.7.

(a) Suppose U and V are open subsets of $\mathbb{R}^n$. Then $U \cap V$ is open.

(b) Suppose that for each $k \in \mathcal{I}$, $U_k \subseteq \mathbb{R}^n$ is open. Then $\bigcup_{k \in \mathcal{I}} U_k$ is open.

Proof. See Exercise 3.   □

Closed Sets and Limit Points

Definition 2.1.8. A subset $C \subseteq \mathbb{R}^n$ is **closed** if its complement, $\mathbb{R}^n \setminus C$, is open.

Example 11. Since the set U from Example 10 is open, its complement $\mathbb{R}^2 \setminus U = \{(x, y) \in \mathbb{R}^2 \mid |x| \geq 1 \text{ and } |y| \geq 1\}$ is closed. On the other hand, $\mathbb{R}^2 \setminus V$ is not closed.

Caution: the terms *open* and *closed* are not opposites of one another. A set that is not open may not necessarily be closed. For example, V from Example 10 is neither open nor closed.

Example 12. In §1.2, we used the terms *open* and *closed* in reference to the intervals (a, b) and $[a, b]$, respectively, where $a < b$ are real numbers. Indeed, since $(a, b) = B_\epsilon(x)$ for $\epsilon = \frac{b-a}{2}$ and $x = \frac{a+b}{2}$, the interval $(a, b) \subseteq \mathbb{R}$ is open by Proposition 2.1.6. On the other hand, $\mathbb{R} \setminus (a, b) = (\infty, a] \cup [b, \infty)$ is not open, because neither of the points $x = a$ nor $x = b$ can be surrounded by an ϵ-ball within the set. Thus (a, b) is open but not closed. The interval $[a, b]$ is not open (no ϵ-ball can surround $x = a$ or $x = b$), but $\mathbb{R} \setminus [a, b] = (-\infty, a) \cup (b, \infty)$ is open. (*Why?*) Thus by definition $[a, b]$ is closed but not open in $\mathbb{R}$.

Subsets that are both open and closed are sometimes called **clopen** subsets, though we rarely use the term in this text. For any n, $\mathbb{R}^n$ possesses only two clopen sets, $\emptyset$ and $\mathbb{R}^n$ itself. However, we will soon explore more exotic topological spaces that could contain many clopen sets.

Example 13. For each $\epsilon \in \mathbb{R}^+$, define the ϵ-**disk** (or **closed ball** of radius ϵ) centered at $x \in \mathbb{R}^n$ by:

$$D_\epsilon(x) = \{y \in \mathbb{R}^n \mid d(x, y) \leq \epsilon\}.$$

Show that the $D_\epsilon(x)$ is closed.

Solution: We must show that $U = \mathbb{R}^n \setminus D_\epsilon(x)$ is open. Let $y \in U$; so $y \notin D_\epsilon(x)$, which implies $d(x, y) > \epsilon$. Let $\delta = d(x, y) - \epsilon > 0$, which is equivalent to saying $d(x, y) = \delta + \epsilon$.

Claim: $B_\delta(y) \subseteq U$ (see Figure 2.8). *Proof of claim:* Choose an arbitrary element $z \in B_\delta(y)$ (so $d(y, z) < \delta$). Then by the triangle inequality,

$$d(x, y) \leq d(x, z) + d(z, y) \quad \Longrightarrow \quad d(x, z) \geq d(x, y) - d(z, y). \quad (2.3)$$

Using what we know about the distances,

$$d(x, y) - d(z, y) = \delta + \epsilon - d(y, z) > \delta + \epsilon - \delta = \epsilon. \quad (2.4)$$

Taken together, (2.3) and (2.4) imply that $d(x, z) > \epsilon$, and so $z \notin B_\epsilon(x)$. Thus $B_\delta(y) \subseteq U$, as required. Since y was arbitrary, we have U open, and thus by definition, $D_\epsilon(x)$ is closed.

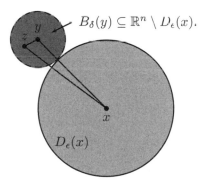

$B_\delta(y) \subseteq \mathbb{R}^n \setminus D_\epsilon(x)$.

Figure 2.8: An ϵ-disk is closed since its complement is open.

Limit Points

You may have noticed that closed sets seem to include "boundary" points, while open sets do not. Indeed, the term *closed interval* traditionally refers to an interval of the number line that includes its boundary points, such as $[a, b]$, while an *open interval* is one that does not, such as (a, b). Consider the open interval $U = (0, \infty) = \{x \in \mathbb{R} \mid x > 0\}$ and the closed interval $V = [0, \infty) = \{x \in \mathbb{R} \mid x \geq 0\}$. Every ϵ-ball around the point $z = 0$, $B_\epsilon(0) = (-\epsilon, \epsilon)$, intersects both U and V *nontrivially*, that is, intersects at a point other than z itself. Specifically, for any choice of $\epsilon \in \mathbb{R}^+$, we have $\epsilon/2 \in B_\epsilon(0) \cap U$ and $\epsilon/2 \in B_\epsilon(0) \cap V$. For this reason, we can say that $z = 0$ is a *limit point* of either set U or V. Now $0 \in V$ but $0 \notin U$. We shall see that a set is closed if and only if it contains all of its limit points.

> **Definition 2.1.9.** Let $A \subseteq \mathbb{R}^n$. A point $x \in \mathbb{R}^n$ is called a **limit point** of A if for every $\epsilon \in \mathbb{R}^+$, $B_\epsilon(x)$ meets A in at least one point other than x. That is, $\forall \epsilon \in \mathbb{R}^+, B_\epsilon(x) \cap (A \setminus \{x\}) \neq \emptyset$.

As the following example demonstrates, the limit points of a set may include part or all of the set, as well as points not in the set. The set of limit points may even be disjoint from the original set.

Example 14.   Determine the set of all limit points of the following.

(a) The unit disk, $A = \{x \in \mathbb{R}^2 \mid d(x, \mathbf{0}) \leq 1\} = D_1(\mathbf{0})$.

(b) The unit disk without boundary, $B = \{x \in \mathbb{R}^2 \mid d(x, \mathbf{0}) < 1\} = B_1(\mathbf{0})$.

(c) $C = B_1(\mathbf{0}) \cup \{(2, 0)\}$.

(d) $D = \left\{ \frac{1}{k} \mid k \in \mathbb{N} \right\} = \left\{ 1, \frac{1}{2}, \frac{1}{3}, \frac{1}{4}, \frac{1}{5}, \dots \right\} \subseteq \mathbb{R}$.

(a) Let $x \in A$ be arbitrary. For any $\epsilon \in \mathbb{R}^+$, $B_\epsilon(x) \cap (A \setminus \{x\}) \neq \emptyset$. Therefore every point of A is a limit point of A. Next, consider any point $y \notin A$. Then $d(y, \mathbf{0}) > 1$. Let $\delta = d(y, \mathbf{0}) - 1 > 0$. The open ball $B_\delta(y)$ does not intersect A at all. So no points outside of A are limit points of A. Thus the set of all limit points of A is A itself.

(b) As above, if $x \in B$, then x is a limit point of B. Now if $y \notin B$, then there are two possibilities. First, if $d(y, \mathbf{0}) > 1$, then the same argument as above shows that y cannot be a limit point of B. However, if $d(y, \mathbf{0}) = 1$, then any open ball $B_\epsilon(y)$ will intersect B. Thus the set of limit points of B is the (closed) unit disk, $D_1(\mathbf{0})$.

(c) C is the set B together with the point $P = (2, 0)$. Observe that $B_1(P) \cap (C \setminus \{P\}) = \emptyset$. We call P an **isolated** point because it can be surrounded by an open ball containing no other points of the subset. By definition, isolated points can never be limit points, so the set of all limit points of C is again the (closed) unit disk, $D_1(\mathbf{0})$.

(d) Consider any point $x = \frac{1}{k} \in D$. Here x is separated from the other points of D by at least $\left| \frac{1}{k} - \frac{1}{k+1} \right| = \frac{1}{k(k+1)}$ units, so *every* point $x \in D$ is isolated. This shows that no point of D is a limit point of D. However, we can see that every open ball $B_\epsilon(0)$ does intersect D: so long as $k \geq \frac{1}{\epsilon}$, then $\frac{1}{k} \leq \epsilon$, implying that $\frac{1}{k} \in B_\epsilon(0)$. Thus the set of limit points of D is the singleton $\{0\}$.

The following theorem provides an important link connecting limit points with the concept of closed set.

> **Theorem 2.1.10.** *$C \subseteq \mathbb{R}^n$ is closed if and only if C contains all of its limit points.*

Note that since the theorem statement is an equivalence ($\iff$), both directions ($\implies$ and $\impliedby$) must be proved.

Proof. ($\implies$): Suppose C is closed. Let $x \in \mathbb{R}^n$ be any limit point of C. We want to show that $x \in C$. Suppose to the contrary that $x \notin C$. By definition of closed, $\mathbb{R}^n \setminus C$ is open. Since $x \in \mathbb{R}^n \setminus C$, there is an ϵ-ball $B_\epsilon(x) \subseteq \mathbb{R}^n \setminus C$. Therefore $B_\epsilon(x) \cap C = \emptyset$. But this shows that x cannot be a limit point of C, because if it were, then $B_\epsilon(x) \cap (C \setminus \{x\}) \neq \emptyset$. This contradiction implies that $x \in C$.

($\impliedby$): Suppose that C contains all of its limit points. We want to show that C is closed, or equivalently, $\mathbb{R}^n \setminus C$ is open. Suppose to the contrary that $\mathbb{R}^n \setminus C$ is not open. Then there exists a point $x \in \mathbb{R}^n \setminus C$ such that for every $\epsilon \in \mathbb{R}^+$, the open ball $B_\epsilon(x)$ is not entirely contained in $\mathbb{R}^n \setminus C$. So for every $\epsilon > 0$, $B_\epsilon(x) \cap C \neq \emptyset$. Now since $x \notin C$, we could write $B_\epsilon(x) \cap (C \setminus \{x\}) \neq \emptyset$. Thus by definition, x is a limit point of C. However, it was assumed that C contains all of its limit points, contradicting the fact that $x \notin C$. The contradiction implies $\mathbb{R}^n \setminus C$ must be open, or C is closed. $\square$

Interior, Exterior, and Boundary Points

Given a subset A of $\mathbb{R}^n$, the question as to what is in A or not in A is a question of set theory: x is *in* A if $x \in A$ (by definition). Consider the half-open interval $A = (0, 1] \subseteq \mathbb{R}$. Clearly, points like $1/2$, 0.814253, and $\ln(2)$ are in A, while points like -2, 1.1, and π are not in A. The point 1 is in A, while 0 is not. However, there is something distinctive about the points 1 and 0. We might say that 1 is *just barely* in the set A: any point $x > 1$ would not be in A no matter how close x is to 1, so 1 is not *interior* to A but serves as a *boundary* point between interior and exterior. Similarly, we might say that 0 is *just barely* not in

A, so 0 is not really *exterior* to A but serves as a *boundary* point. The following definitions make these concepts precise.

Definition 2.1.11. Let $A \subseteq \mathbb{R}^n$.

- A point $x \in \mathbb{R}^n$ is an **interior point** of A if there is an $\epsilon \in \mathbb{R}^+$ such that $B_\epsilon(x) \subseteq A$.

- A point $x \in \mathbb{R}^n$ is an **exterior point** of A if there is an $\epsilon \in \mathbb{R}^+$ such that $B_\epsilon(x) \subseteq \mathbb{R}^n \setminus A$.

- A point $x \in \mathbb{R}^n$ is a **boundary point** of A if x is neither an interior nor exterior point of A.

If x is neither interior nor exterior, then no open ball surrounding x can consist of only points in A or only points not in A. Equivalently,

a point $x \in \mathbb{R}^n$ is a **boundary point** of A if every neighborhood N of x contains at least one point in A and at least one point not in A.

We note here that boundary points and limit points are distinct concepts. Not every limit point is a boundary point, and not every boundary point is a limit point in a given set (see Exercise 14).

Definition 2.1.12. Let $A \subseteq \mathbb{R}^n$.

- The **interior** of A is the set $\text{int}(A)$ of all interior points of A.

- The **exterior** of A if the set $\text{ext}(A)$ of all exterior points of A.

- The **boundary** of A is the set ∂A of all boundary points of A.

Definition 2.1.11 implies that $\text{ext}(A) = \text{int}(\mathbb{R}^n \setminus A)$. In other words, the exterior of a set is the interior of its complement. Because of this symmetry, it follows that $\partial A = \partial(\mathbb{R}^n \setminus A)$.

Example 15. Let $a < b$ be real numbers, and let $W = (a, b)$, $X = (a, b]$, $Y = [a, b)$, $Z = [a, b]$. Despite the fact that no pair of these sets is equal, they share the same interior, exterior, and boundary.

$$
\begin{aligned}
\text{int}(W) = \text{int}(X) = \text{int}(Y) = \text{int}(Z) &= (a, b) \\
\text{ext}(W) = \text{ext}(X) = \text{ext}(Y) = \text{ext}(Z) &= (-\infty, a) \cup (b, \infty) \\
\partial W = \partial X = \partial Y = \partial Z &= \{a, b\}
\end{aligned}
$$

Example 16. Suppose $X \subseteq \mathbb{R}^n$ has an isolated point x. Show that $x \in \partial X$.

Solution: By definition, x is isolated if there is an $\epsilon \in \mathbb{R}^+$ such that $B_\epsilon(x) \cap X = \{x\}$. Clearly, then, every neighborhood of x contains at least one point in X (namely, x itself), and at least one point not in X (any point that is some positive distance away from x in the neighborhood). Thus $x \in \partial X$.

Caution: the definitions of interior and exterior do not always coincide with our notions of *inside* and *outside*.

Example 17. Find the interior, exterior, and boundary of $A = \{(x, y) \in \mathbb{R}^2 \mid y < x^2\}$.

Solution: A is the set of points below the graph of $y = x^2$. Here $\text{int}(A) = A$, $\text{ext}(A) = \{(x, y) \in \mathbb{R}^2 \mid y > x^2\}$, and ∂A is the graph itself, that is, $\partial A = \{(x, y) \in \mathbb{R}^2 \mid y = x^2\}$. The three sets are illustrated in Figure 2.9.

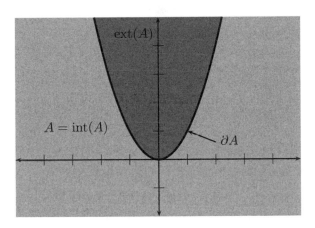

Figure 2.9: The sets $\text{int}(A)$, $\text{ext}(A)$, and ∂A, for the set A from Example 17.

Definition 2.1.13. Let $A \subseteq \mathbb{R}^n$. The **closure** of A is the set,

$$\overline{A} = A \cup \partial A.$$

It is obvious (by definition) that $\text{int}(A) \subseteq A \subseteq \overline{A}$. It may not be obvious (though it is true) that $\text{int}(A)$ is the *largest* open set contained in A, and $\overline{A}$ is the *smallest* closed set containing A. The following proposition displays identities that will become useful later in the text.

> **Proposition 2.1.14.** *Fix $n \in \mathbb{N}$. Let $\epsilon \in \mathbb{R}^+$ and $x \in \mathbb{R}^n$.*
>
> *(a)* $\text{int}(D_\epsilon(x)) = B_\epsilon(x)$ *(c)* $\partial B_1(\mathbf{0}) = \partial D_1(\mathbf{0}) = \mathbb{S}^{n-1}$
>
> *(b)* $\overline{B_\epsilon(x)} = D_\epsilon(x)$ *(d)* $D_1(\mathbf{0}) \setminus B_1(\mathbf{0}) = \mathbb{S}^{n-1}$

Exercises

1. Prove Proposition 2.1.2, parts 1–3.

2. Let $x, y \in \mathbb{R}^n$ be arbitrary distinct points. Determine the largest value of $\epsilon \in \mathbb{R}^+$ such that $B_\epsilon(x) \cap B_\epsilon(y) = \emptyset$.

3. Prove both parts of Proposition 2.1.7.

4. For each $k \in \mathbb{N}$, let $U_k = \left(-1, \frac{1}{k}\right)$. Show that the infinite intersection, $\bigcap_{k \in \mathbb{N}} U_k$, is neither open nor closed.

5. Use De Morgan's Law (see Theorem A.1.15) and Exercise 3 to prove that the union or intersection of two closed sets is closed. Then prove that the arbitrary intersection of closed sets is closed (*Hint:* see Exercise 12 of §A.1). What can be said of the arbitrary union of closed sets?

6. Let $a < b$ be real numbers. Consider the following intervals in $\mathbb{R}$: $(a, b]$, $[a, b)$, (a, ∞), $[a, \infty)$, $(-\infty, b)$, $(-\infty, b]$, and $(-\infty, \infty)$.

   (a) Which intervals are open?

   (b) Which intervals are closed?

   (c) Determine the set of limit points for each interval.

7. What are the limit points of $\mathbb{Z}$ in $\mathbb{R}$? What is the boundary of $\mathbb{Z}$ in $\mathbb{R}$?

8. (a) Prove that every rational number x is a limit point of $\mathbb{Q}$. (*Hint:* consider points $x \pm \frac{1}{n} \in \mathbb{Q}$.)

   (b) Prove that π is a limit point of $\mathbb{Q}$ in $\mathbb{R}$ (*Hint:* the hint for part (a) no longer works. Use the decimal expansion $\pi = 3.141592654\ldots$ to find rational numbers that approach π.)

   (c) Make a conjecture about the set of limit points of $\mathbb{Q}$ in $\mathbb{R}$.

9. Find the interior, exterior, boundary, and closure of each interval mentioned in Exercise 6.

10. Let $A \subseteq \mathbb{R}^n$ be arbitrary. Prove the following.

    (a) $\text{int}(A)$, $\text{ext}(A)$, and ∂A partition $\mathbb{R}^n$

    (b) $\text{int}(A)$ and $\text{ext}(A)$ are open, while ∂A is closed in $\mathbb{R}^n$

11. Prove that int(A) is the largest open subset of A.

12. Prove that $\overline{A}$ is the smallest closed superset of A.

13. Prove all statements in Proposition 2.1.14.

14. (a) Prove that every interior point of a subset $X \subseteq \mathbb{R}^n$ is a limit point of X (thus not every limit point is a boundary point).

    (b) Prove that every boundary point of a subset $X \subseteq \mathbb{R}^n$ is either a limit point of X or an isolated point in X (thus not every boundary point is a limit point).

2.2   Continuity and Homeomorphism

Consider the curve $y = x^2$ graphed in $\mathbb{R}^2$, as in Figure 2.10. More precisely, let $C = \{(x, x^2) \mid x \in \mathbb{R}\}$ be the set of points of this graph. A myopic ant may find that walking along C is not much different than taking a stroll along the line $\mathbb{R}^1$; there are essentially two directions, backward and forward, and the end can never be reached in either direction. Intuitively, $\mathbb{R}^1 \approx C$, but to be sure, we should construct an explicit mapping that takes the points of $\mathbb{R}^1$ to those of C in a way that preserves all of the local structure. Consider the function $f : \mathbb{R}^1 \to C$ defined by $f(x) = (x, x^2) \in C$.

(i) All points sufficiently near a given point p on the the line are mapped to points that are near its image $f(p) = q$ on the parabola.

(ii) The function f is a bijective correspondence between the points on the x-axis and the points on the parabola. Because f is bijective, there is an inverse mapping, namely, the mapping $g : C \to \mathbb{R}^1$ sending $(x, x^2) \mapsto x$.

(iii) The inverse mapping g also preserves *nearness*.

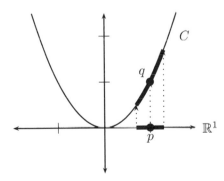

Figure 2.10: The graph of $y = x^2$ is homeomorphic to $\mathbb{R}^1$.

Such a mapping f will be called a **homeomorphism**. Note that if f is a homeomorphism, then by definition so is its inverse function $g = f^{-1}$.

The parabola defined by $y = x^2$ may also be regarded as a function in and of itself. This time, the mapping is from $\mathbb{R}$ to $\mathbb{R}$ (see Figure 2.11). Using the terminology of Chapter 2, we find that $f(x) = x^2$ is neither injective nor surjective, so it does not satisfy condition (ii) and thus cannot satisfy (iii) either. However, f still satisfies condition (i), a property we call *continuity*.

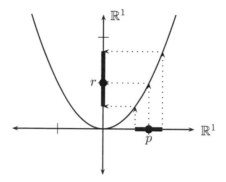

Figure 2.11: Here $f(x) = x^2$ is a continuous function from $\mathbb{R}$ to $\mathbb{R}$.

Continuous Functions

There is a standard definition for continuity of functions that appears in most calculus textbooks, tied to the concept of *limits*. A function $y = f(x)$ is **continuous** at a point $x = c$ if $\lim_{x \to c} f(x) = f(c)$, and f is continuous on an interval if it is continuous on all points c in that interval. Now, together with the standard definition of a limit, we obtain:

> **Definition 2.2.1.** Let $f : D \to \mathbb{R}$, where $D \subseteq \mathbb{R}$ is a subset of the domain of f; f is **continuous** on D if:
>
> For every $c \in D$, and $\epsilon > 0$, there exists $\delta > 0$ (depending on c and ϵ), such that if $x \in D$ with $|x - c| < \delta$, then $|f(x) - f(c)| < \epsilon$.

Let's unpack this definition using some notation we have developed in §2.1. To say that $x \in D$ with $|x - c| < \delta$ means that $x \in B_\delta(c) \cap D$. Similarly, to say that $|f(x) - f(c)| < \epsilon$ means that $f(x) \in B_\epsilon(f(c))$. Then, using the concepts of *forward* and *inverse image* (see Definition A.2.8), we can dispense with the mention of "x" altogether, by observing that the statement "If $x \in B_\delta(c) \cap D$, then $f(x) \in B_\epsilon(f(c))$" is equivalent to "$f[B_\delta(c) \cap D] \subseteq B_\epsilon(f(c))$," which in turn is equivalent to "$B_\delta(c) \cap D \subseteq f^{-1}[B_\epsilon(f(c))]$" (see Figure 2.12). Moreover, this move to open balls allows us to generalize Definition 2.2.1 to functions from $\mathbb{R}^m$ to $\mathbb{R}^n$ for arbitrary $m, n > 0$.

> **Definition 2.2.2.** Let $f : D \to \mathbb{R}^n$, where $D \subseteq \mathbb{R}^m$ is a subset of the domain of f; f is **continuous** on D if:
>
> For every $c \in D$, and $\epsilon > 0$, there exists $\delta > 0$ (depending on c and ϵ), such that $B_\delta(c) \cap D \subseteq f^{-1}[B_\epsilon(f(c))]$.

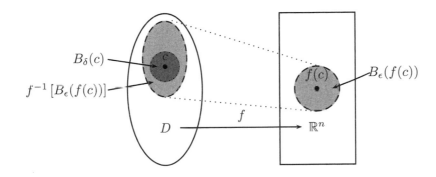

Figure 2.12: A function $f : D \to \mathbb{R}^n$ is continuous at $x = c$ if and only if the inverse image of every open ball $B_\epsilon(f(c))$ contains an open ball $B_\delta(c)$.

Example 18. Let $n \in \mathbb{N}$. Show that the norm function, $x \mapsto \|x\|$, is a continuous function $\mathbb{R}^n \to \mathbb{R}$.

Let $f(x) = \|x\|$, and suppose $c \in \mathbb{R}^n$ and $\epsilon \in \mathbb{R}^+$ are arbitrary. Then $B_\epsilon(f(c)) = B_\epsilon(\|c\|)$ is an open interval $(\|c\| - \epsilon, \|c\| + \epsilon) \subseteq \mathbb{R}^1$. The inverse image of this interval under the norm map is the set

$$A = f^{-1}[B_\epsilon(f(c))] = \{y \in \mathbb{R}^n \mid \|c\| - \epsilon < \|y\| < \|c\| + \epsilon\}. \tag{2.5}$$

Though it doesn't matter to this proof, we note that A is an open annulus if $\|c\| - \epsilon > 0$, an open punctured disk if $\|c\| - \epsilon = 0$, or open disk if $\|c\| - \epsilon < 0$ (observe that $\|c\| + \epsilon$ must always be positive). The case in which A is an annulus is shown in Figure 2.13. Consider $\delta = \epsilon$. Let $y \in B_\delta(c) = B_\epsilon(c)$ be arbitrary. By the triangle inequality, we have

$$\|y\| = d(y, \mathbf{0}) \leq d(y, c) + d(c, \mathbf{0}) < \epsilon + \|c\|$$

and

$$\|c\| = d(c, \mathbf{0}) \leq d(c, y) + d(y, \mathbf{0}) < \epsilon + \|y\|$$
$$\implies \|y\| > \|c\| - \epsilon.$$

Taken together, we have $\|c\| - \epsilon < \|y\| < \|c\| + \epsilon$, that is, $y \in A$. Since $y \in B_\delta(c)$ was arbitrary, we have found a $\delta > 0$ such that $B_\delta(c)$ is contained in the inverse image of $B_\epsilon(\|c\|)$, as required. Finally, since $c \in \mathbb{R}^n$ was arbitrary, we find that $f(x) = \|x\|$ is continuous on all of $\mathbb{R}^n$.

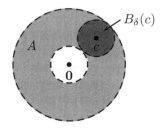

Figure 2.13: The inverse image of $B_\epsilon(\|c\|)$ under the norm map, with the set $B_\delta(c)$ shown ($\delta = \epsilon$). The annulus A has inner radius $\|c\| - \epsilon$ and outer radius $\|c\| + \epsilon$.

Compositions of continuous functions are continuous (see Exercise 1); however, we will find in Chapter 4 that the proof is more natural in the abstract language of topology.

Standard Transformations

The following functions are examples of continuous functions. These *standard transformations* often serve as building blocks for putting together more complicated functions.

- **Translation.** Let $r \in \mathbb{R}^n$. Translation by r is a function $T_r : \mathbb{R}^n \to \mathbb{R}^n$ defined by $T_r(x) = x + r$ (by vector addition when $n \geq 2$).

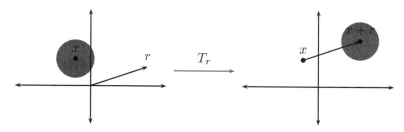

We will prove that T_r is continuous. For any $c \in \mathbb{R}^n$, and $\epsilon \in \mathbb{R}^+$, $B_\epsilon(T_r(c)) = B_\epsilon(c + r)$. Since translation preserves the size and shape of geometric figures, the ϵ-ball is preserved under forward and inverse images. Thus $T_r^{-1}[B_\epsilon(c + r)] = B_\epsilon(c)$, so Definition 2.2.2 is satisfied with $\delta = \epsilon$.

- **Scaling.** Let $k \in \mathbb{R}^+$. Scaling by a factor of k is a function $S_k : \mathbb{R}^n \to \mathbb{R}^n$ defined by $S_k(x) = kx$ (scalar multiplication). When $k > 1$, objects are expanded by a factor of k; when $0 < k < 1$, objects are compressed. Since $k \neq 0$, the inverse transformation can be defined by $S_k^{-1}(x) = \frac{1}{k}x$; hence the inverse image of an ϵ-ball under the map S_k will be an $\frac{\epsilon}{k}$-ball. This observation implies that S_k is continuous.

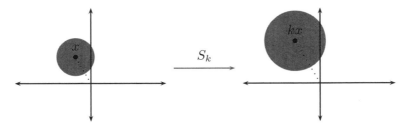

- **Reflection**. Suppose $\ell \subseteq \mathbb{R}^2$ is a line. Reflection in ℓ is a function $\rho_\ell : \mathbb{R}^2 \to \mathbb{R}^2$ that maps any point a on one side of ℓ to its *mirror image* $a' = \rho_\ell(a)$ on the other side of ℓ. The figure below shows reflection across the y-axis, which sends $(x, y) \mapsto (-x, y)$.

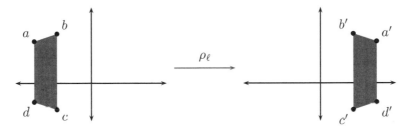

Reflections in $\mathbb{R}^3$ are defined by choosing a *plane* with respect to which points are reflected. All reflections are continuous because the size and shape of geometric figures are preserved.

- **Rotation**. In $\mathbb{R}^2$, points may be rotated about any center point. Rotation about the origin by an angle θ, $R_\theta : \mathbb{R}^2 \to \mathbb{R}^2$, is easily expressed in matrix notation:

$$R_\theta \left(\begin{pmatrix} x_1 \\ x_2 \end{pmatrix} \right) = \begin{pmatrix} \cos\theta & -\sin\theta \\ \sin\theta & \cos\theta \end{pmatrix} \begin{pmatrix} x_1 \\ x_2 \end{pmatrix} = \begin{pmatrix} x_1\cos\theta - x_2\sin\theta \\ x_1\sin\theta + x_2\cos\theta \end{pmatrix}.$$

Rotations in $\mathbb{R}^3$ are defined by choosing a *line* around which to rotate. Again, since size and shape are preserved, all rotations are continuous.

- **Projection.** Projection is typically defined from a space of higher dimension onto a space of lower dimension, for example, when a solid object casts a flat shadow. The figure below illustrates projection onto the xy-plane, which can be interpreted as a function $P_{xy} : \mathbb{R}^3 \to \mathbb{R}^2$ defined by $P_{xy}(x, y, z) = (x, y)$. Projections are generally not injective; however, they are continuous. (Can you see why?)

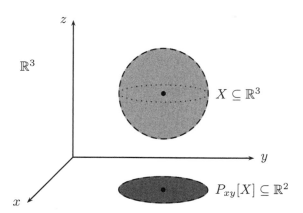

- **Inversion.** Consider the function $I : \mathbb{R}^n \setminus \{\mathbf{0}\} \to \mathbb{R}^n \setminus \{\mathbf{0}\}$ defined by the rule

$$I(x) = \frac{x}{\|x\|^2}.$$

Since $I(x)$ is a positive scalar multiple of x, the map I preserves rays from the origin. Now if $\|x\| = 1$, then $I(x) = x$. In other words, inversion leaves every point on the unit sphere, $\mathbb{S}^{n-1}$, fixed. The map I takes points from $\operatorname{ext}(B_1(\mathbf{0}))$ to points in $\operatorname{int}(B_1(\mathbf{0}))$, and vice versa. It can be shown that inversion maps open balls into unbounded open sets – either exteriors of open balls or open half-spaces – and I is self-inverse, $I^{-1} = I$. These facts together imply that I is continuous.

Example 19. Let ℓ be the line parametrized by $x = 2t, y = t + 1, t \in \mathbb{R}$. Describe the image $I[\ell]$, where $I : \mathbb{R}^2 \setminus \{\mathbf{0}\} \to \mathbb{R}^2 \setminus \{\mathbf{0}\}$ is inversion in the circle.

$$I(x, y) = I(2t, t + 1) = \left(\frac{2t}{(2t)^2 + (t + 1)^2}, \frac{t + 1}{(2t)^2 + (t + 1)^2} \right)$$

Graphing the result shows that $I[\ell]$ is a circle missing one point (the origin). Though it is not obvious, it can be shown that I always maps circles to circles, and lines not going through the origin to circles missing a point at the origin.

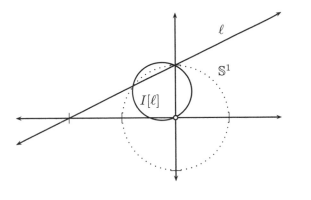

- **Stereographic Projection.** Have you ever noticed that on some world
maps familiar continents seem distorted? Certain regions look to be much
larger or smaller than they should. This is a side effect of *projection.*[3] One
simple type of projection from the sphere to the plane is called *stereographic.*
To visualize stereographic projection, imagine a sphere ($\mathbb{S}^2$) placed on the
xy-plane so that its south pole S is at the origin. For each point $P \in \mathbb{S}^2$
except $P = N$, extend the line $\overline{NP}$ until it intersects the plane at a point
Q. The rule $F(P) = Q$ defines a map $F : \mathbb{S}^2 \setminus \{N\} \to \mathbb{R}^2$. For example,
$F(S) = (0, 0)$. See Figure 2.14.

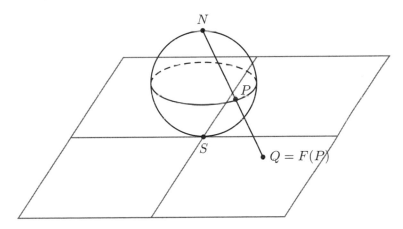

Figure 2.14: Stereographic projection.

---

[3]There are many types of projections from the sphere to the plane, but each has *distortions*
of some type because it is impossible to preserve the curved geometry of the sphere when
mapping it to a flat plane. See, for example, `http://en.wikipedia.org/wiki/World_map`.

Homeomorphism

In this section we will extend our definition of continuity, Definition 2.2.2, to include more general codomains. In particular, we want to consider functions $f : X \to Y$ for which $X \subset \mathbb{R}^m$ and $Y \subset \mathbb{R}^n$ are arbitrary. A consequence of this, for example, is that we would be able to prove that a circle in the plane is topologically equivalent to a knot in $\mathbb{R}^3$ (recall Figure 1.10 from Chapter 1). The key point in the following definition is to work with intersections of X and Y with open balls in the ambient Euclidean space.

Definition 2.2.3. A function $f : X \to Y$, where $X \subseteq \mathbb{R}^m$ and $Y \subseteq \mathbb{R}^n$, is **continuous** if:

For every $x \in X$, and $\epsilon \in \mathbb{R}^+$, there exists $\delta > 0$ (depending on x and ϵ), such that $B_\delta(x) \cap X \subseteq f^{-1}\left[B_\epsilon(f(x)) \cap Y\right]$.

Definition 2.2.4. A function $f : X \to Y$ is called a **homeomorphism** if all three of the following are true:

- f is continuous,

- f is bijective, and

- $f^{-1} : Y \to X$ is continuous.

Note that bijectivity of f is required for the existence of the inverse function f^{-1}. A homeomorphism is a bijective, **bicontinuous** function, meaning that both f and f^{-1} must be continuous. This term *homeomorphism* is related to the term *homeomorphic* used back in §1.1, but now we have a precise enough definition to finally address the question, *When are two spaces topologically the same?* (at least in the case that the spaces are part of our familiar Euclidean space).

Definition 2.2.5. Two sets X and Y are called **homeomorphic** or **topologically equivalent**, and we write $X \approx Y$ if there is a homeomorphism $f : X \to Y$.

Example 20. The squaring function $f : \mathbb{R} \to \mathbb{R}$, defined by $f(x) = x^2$, is continuous but not bijective – and therefore is *not* a homeomorphism. However, if the domain and codomain are both *restricted* to $[0, \infty)$, then f becomes a homeomorphism, with inverse $f^{-1}(x) = \sqrt{x}$.

On the other hand, the cubing function $f : \mathbb{R} \to \mathbb{R}$, defined by $f(x) = x^3$, is a homeomorphism on its natural domain.

Example 21. Consider the tangent function $\tan(x)$, restricted to the interval $I = (-\pi/2, \pi/2)$. On this interval, $\tan : I \to \mathbb{R}^1$ is continuous and has continuous inverse $\tan^{-1} : \mathbb{R}^1 \to I$. Therefore $\tan$ is a homeomorphism, which shows that $(-\pi/2, \pi/2) \approx \mathbb{R}^1$.

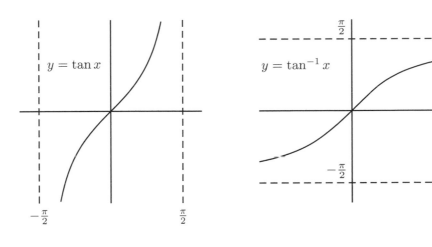

Example 22. Stereographic projection has a well-defined inverse function. Referring to Figure 2.14, for any point Q on the xy-plane, draw the line $\overline{NQ}$. Then the point of intersection of this line with the sphere defines $F^{-1}(Q)$. It can be shown F^{-1} is continuous, and so there is a homeomorphism $\mathbb{S}^2 \setminus \{N\} \approx \mathbb{R}^2$. Thus, while the sphere and plane have much different *geometric* properties, the punctured sphere is *topologically equivalent* to the plane.

If we append one additional point to $\mathbb{R}^2$, called the **point at infinity**, denoted by ∞, then F could be extended to a map $F : \mathbb{S}^2 \to \mathbb{R}^2 \cup \{\infty\}$. In order to make the map continuous, we define an open ϵ-ball around ∞ by $B_\epsilon(\infty) = \text{ext}(B_{1/\epsilon}(\mathbf{0}))$. In fact, F becomes a homeomorphism $\mathbb{S}^2 \approx \mathbb{R}^2 \cup \{\infty\}$. This idea works in higher dimensions as well. For each $n \in \mathbb{N}$, there are homeomorphisms $\mathbb{S}^n \approx \mathbb{R}^n \cup \{\infty\}$ and $\mathbb{S}^n \setminus \{z_0\} \approx \mathbb{R}^n$, where $z_0 \in \mathbb{S}^n$ is any point.

Typically, proving $X \approx Y$ is quite difficult because it requires finding an explicit bicontinuous function from X to Y. On the other hand, proving $X \not\approx Y$ generally amounts to finding a *topological invariant* that has differing values on X and Y (recall Definition 1.1.1 and Theorem 1.1.2). So how do we find topological invariants? By definition, any proposed invariant $\mathscr{Q}$ must take the same value on X and Y if $X \approx Y$. So if $f : X \to Y$ is *any* homeomorphism (which is guaranteed to exist if $X \approx Y$), and if f *preserves* the value of $\mathscr{Q}$, then we know that $\mathscr{Q}$ is an invariant. We will spend the next few sections discussing a number of topological invariants.

Exercises

1. Use Definition 2.2.2 to prove the following important **composition theorem**: Suppose $f : D \to \mathbb{R}^n$ and $g : E \to \mathbb{R}^p$, where D is an open subset of $\mathbb{R}^m$, E is an open subset of $\mathbb{R}^n$, and $f[D] \subseteq E$. Then, if both f and g are continuous, so is the composition $g \circ f : D \to \mathbb{R}^p$.

2. Prove that Definition 2.2.2 is equivalent to the following:

   > Let $f : D \to \mathbb{R}^n$, where $D \subseteq \mathbb{R}^m$ is an open subset of the domain of f; f is **continuous** on D if and only if the inverse image of any open set (in $\mathbb{R}^n$) is again open (in D).

3. Prove that $f : D \to \mathbb{R}^n$ is continuous if and only if the inverse image of any closed set (in $\mathbb{R}^n$) is again closed (in D).

4. Let $n > 0$ be arbitrary. How should reflections be defined in $\mathbb{R}^n$? How should rotations be defined in $\mathbb{R}^n$?

5. Show that the inversion map satisfies $(I \circ I)(x) = x, \forall x \in \mathbb{R}^n \setminus \{\mathbf{0}\}$.

6. Which of the standard transformations are homeomorphisms? Justify your responses by finding a continuous inverse in each case.

7. (a) Show that Definition 2.2.5 is *symmetric*. In other words, show that $X \approx Y$ if and only if $Y \approx X$.

   (b) Show that Definition 2.2.5 is *transitive*. In other words, show that if $X \approx Y$ and $Y \approx Z$, then $X \approx Z$ (*Hint:* Use Exercise 1).

8. Prove that $(0, \infty) \approx \mathbb{R}^1$. *Hint:* Find a specific homeomorphism.

9. Using Example 21, Exercise 8, and standard transformations, prove that $\mathbb{R}^1 \approx (a, b) \approx (a, \infty) \approx (-\infty, b)$ for any $a < b$.

10. Let $x_0 \in \mathbb{R}^n$ be fixed, and define a function $g : \mathbb{R}^n \to \mathbb{R}$ by $g(x) = d(x_0, x)$. Prove that g is continuous.

2.3   Compactness and Limits

You may have already encountered the term *compact* in an advanced calculus or analysis course. If so, it may have been defined as *closed and bounded*. This is sufficient for subsets of $\mathbb{R}^n$, but the concept of *bounded* requires measuring distances. We eventually want to generalize to any topological space, even those in which there is no well-defined concept of distance (see §4.4). Toward this goal, we define compactness in terms of open sets.

Open Covers and Compactness

What does the word *compact* bring to mind? Small and close together, perhaps? As huge as Manhattan is, it still seems very *compact* to me, because there is not a square inch of space wasted and yet everything is confined to a single island. Every point of a compact set must be fairly "close" to all the others – if not in the same neighborhood, then perhaps just a few neighborhoods away. Moreover, the number of neighborhoods in a compact set should be limited in some way. If infinitely many neighborhoods are required to cover a set, then some points in that set could be arbitrarily "far" from one another. The precise definition involves the concept of an *open cover*.

Definition 2.3.1. An **open cover** of a set $X \subseteq \mathbb{R}^n$ is a collection $\mathscr{U}$ of open sets such that

$$X \subseteq \bigcup_{U \in \mathscr{U}} U.$$

The collection $\mathscr{U}$ may include infinitely many sets in general. However, a set X is called *compact* if you never *need* infinitely many open sets to cover X. Let's make this more precise. Suppose $\mathscr{U}$ is an open cover of X. Any subset $\mathscr{U}' \subseteq \mathscr{U}$ is called a **subcover** if $\mathscr{U}'$ also covers X. As a concrete example, suppose you have a set $\mathscr{U}$ of ten blankets on your bed arranged in various ways so that your body is entirely covered. Then suppose your roommate comes in and takes away three of the blankets without disturbing the positions of the rest. If the remaining set $\mathscr{U}'$ of seven blankets still cover every part of your body, then $\mathscr{U}'$ is a *subcover* of $\mathscr{U}$. Figure 2.15 illustrates a simple cover and subcover of a region of the plane.

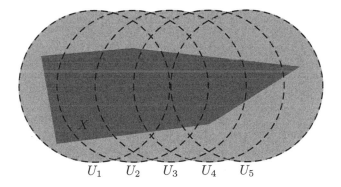

$$U_1 \quad U_2 \quad U_3 \quad U_4 \quad U_5$$

Figure 2.15: In this diagram, each U_k is an open disk in the plane. $\mathscr{U} = \{U_1, U_2, U_3, U_4, U_5\}$ is a cover of X (dark gray polygon). $\mathscr{U}' = \{U_1, U_3, U_5\} \subseteq \mathscr{U}$ is a subcover since $X \subseteq U_1 \cup U_3 \cup U_5$. What other subcovers can you find?

Definition 2.3.2. A set $X \subseteq \mathbb{R}^n$ is **compact** if *every* open cover $\mathscr{U}$ has a *finite* subcover.

That is, X is compact if for any choice of open cover $\mathscr{U}$ one can find finitely many open sets $U_1, U_2, \ldots, U_k \in \mathscr{U}$ such that $X \subseteq U_1 \cup U_2 \cup \cdots \cup U_k$.

Example 23. Prove that the set $X = \{0, 1, \frac{1}{2}, \frac{1}{3}, \frac{1}{4}, \ldots, \frac{1}{n}, \ldots\}$ is a compact subset of $\mathbb{R}^1$.

Suppose $\mathscr{U}$ is any open cover of X. Then there is some open set $U_0 \in \mathscr{U}$ such that $0 \in U_0$. Now since U_0 is open, there is some open ball $B_\epsilon(0) \subseteq U_0$. So long as $n > \frac{1}{\epsilon}$, we have $\frac{1}{n} \in B_\epsilon(0)$, so there are only finitely many points possibly not covered by U_0, namely, $1, \frac{1}{2}, \frac{1}{3}, \ldots, \frac{1}{m}$ where $m < \frac{1}{\epsilon}$. For each of these points $\frac{1}{k}$, there is at least one open set $U_k \in \mathscr{U}$ such that $\frac{1}{k} \in U_k$. Thus there exists a finite subcover, $U_0 \cup U_1 \cup U_2 \cup \ldots \cup U_m$. By definition, X is compact.

Example 24. Unbounded and nonclosed sets in $\mathbb{R}^n$ fail to be compact.

- $\mathbb{R}^1$ is not compact. To show this, we only need to find an infinite open cover that cannot be reduced to a finite subcover. Consider $\mathscr{U} = \{(n - \frac{3}{4}, n + \frac{3}{4}) \mid n \in \mathbb{Z}\}$. Clearly $\mathscr{U}$ covers $\mathbb{R}^1$, and if any set is removed from $\mathscr{U}$, the remaining sets no longer cover $\mathbb{R}^1$.

- The open interval $(0, 1)$ is not compact. Consider $\mathscr{U} = \{(\frac{1}{n}, 1) \mid n \in \mathbb{N}\}$. For any $x \in (0, 1)$, we have $x > \frac{1}{n}$ for some sufficiently large $n \in \mathbb{N}$, so $x \in (\frac{1}{n}, 1)$. This shows that $\mathscr{U}$ covers $(0, 1)$. Now suppose there is a finite subcover, $\{U_1, U_2, U_3, \ldots, U_m\}$, where each $U_k = (\frac{1}{x_k}, 1)$, for some $x_k \in \mathbb{N}$. Let $a = \min\{\frac{1}{x_1}, \frac{1}{x_2}, \ldots, \frac{1}{x_m}\}$. The minimum of finitely many positive numbers is again positive, so $a \in (0, 1)$. But $a \notin U_k$ for any $k = 1, 2, \ldots, m$. Thus no finite subcover of $\mathscr{U}$ exists, proving $(0, 1)$ to be noncompact.

Compactness is a powerful property, ensuring the existence of limit points in Euclidean spaces.

Theorem 2.3.3. *Any infinite subset of a compact subset $X \subseteq \mathbb{R}^n$ must have a limit point in X.*

Proof. Let $X \subseteq \mathbb{R}^n$ be compact and $A \subseteq X$ be an infinite subset. Suppose to the contrary that A has no limit points in X. For any $x \in X$, since x is not a limit point of A, there is an $\epsilon(x) \in \mathbb{R}^+$ such that the open ball $B_{\epsilon(x)}(x)$ has no points in common with A except possibly x itself. Set $U_x = B_{\epsilon(x)}(x)$. The collection $\mathscr{U} = \{U_x \mid x \in X\}$ is an open cover of X, and since X is compact, there is a finite subcover $\mathscr{U}' = \{U_{x_1}, U_{x_2}, \dots, U_{x_n}\}$. However, by construction, we know that at most n points of A could be contained in $U_{x_1} \cup U_{x_2} \cup \cdots \cup U_{x_n}$. But then the infinite set $A \subseteq X$ *cannot* be covered by $\mathscr{U}'$, a contradiction. $\square$

The Heine-Borel Theorem

The fact that *compact* is equivalent to *closed and bounded* in Euclidean spaces is extremely important and useful, a result known as the **Heine-Borel Theorem**. First let's define precisely what we mean by *bounded*.

> **Definition 2.3.4.** A subset $X \subseteq \mathbb{R}^n$ is **bounded** if there is a radius $r \in \mathbb{R}^+$ such that $X \subseteq B_r(\mathbf{0})$.

Example 25. For $n \in \mathbb{N}$, the Euclidean space $\mathbb{R}^n$ is not bounded since no open ball $B_r(\mathbf{0})$ of any finite radius contains all of $\mathbb{R}^n$. Similarly, no line or plane is bounded (as a subset of $\mathbb{R}^n$), since no finite radius ball can contain a line or a plane.

Example 26. Let $n \in \mathbb{N}$, $\epsilon \in \mathbb{R}^+$, and $x \in \mathbb{R}^n$. Prove that the open ball $B_\epsilon(x)$ is bounded.

Solution: Let $y \in B_\epsilon(x)$. So we have $d(x, y) < \epsilon$. By the triangle inequality, $d(\mathbf{0}, y) \leq d(\mathbf{0}, x) + d(x, y) < \|x\| + \epsilon$. This shows that $y \in B_{\|x\|+\epsilon}(\mathbf{0})$. Since y was chosen arbitrarily, $B_\epsilon(x) \subseteq B_{\|x\|+\epsilon}(\mathbf{0})$, proving boundedness.

> **Theorem 2.3.5** (Heine-Borel). *A subset $X \subseteq \mathbb{R}^n$ is compact if and only if X is closed and bounded.*

We must prove both "directions" of the statement. The proof of the Heine-Borel theorem is quite technical, and we will not actually prove the reverse direction ($\Longleftarrow$) in full generality until Chapter 4.

Proof. ($\Longrightarrow$): Suppose X is compact. Consider the open cover of X by unit-radius balls centered at each $x \in X$. That is, $\mathscr{U} = \{B_1(x) \mid x \in X\}$. Since X is compact, there exists a finite subcover,

$$X \subseteq B_1(x_1) \cup B_1(x_2) \cup \cdots \cup B_1(x_n).$$

Let $x \in X$ be arbitrary. Then $x \in B_1(x_i)$ for some $i \in \{1, 2, \dots, n\}$. Now by the triangle inequality,

$$d(x, \mathbf{0}) \leq d(x, x_i) + d(x_i, \mathbf{0}) < 1 + \|x_i\|.$$

Thus $X \subseteq B_r(\mathbf{0})$, where $r = \max\{\|x_i\| + 1 \mid i = 1, 2, \ldots, n\}$, proving that X is bounded.

To show that X is closed, it is equivalent to show that $\mathbb{R}^n \setminus X$ is open. Let $y \in \mathbb{R}^n \setminus X$ be arbitrary. For each $x \in X$, let $\epsilon(x) = \frac{1}{2}d(x, y)$, and define open balls $U_x = B_{\epsilon(x)}(x)$ and $V_x = B_{\epsilon(x)}(y)$, so that $U_x \cap V_x = \emptyset$ (see §2.1, Exercise 2). Now the collection of sets $\mathscr{U} = \{U_x \mid x \in X\}$ is an open cover of X. Since X is compact, there is a finite subcover, $\mathscr{U}' = \{U_{x_1}, U_{x_2}, \ldots, U_{x_m}\}$. Let $\mathscr{V}' = \{V_{x_1}, V_{x_2}, \ldots, V_{x_m}\}$ be the corresponding collection of open balls around y. Since there are only finitely many of these, their intersection, $V = \bigcap_{k=1}^{m} V_{x_k}$, is also an open ball around y, whose radius r is the minimum of the radii of the balls V_{x_k}. By construction, $V \cap X = \emptyset$, and so $V = B_r(y) \subseteq \mathbb{R}^n \setminus X$. Finally, since y was arbitrary, we have shown that $\mathbb{R}^n \setminus X$ is open; hence X is closed.

($\Longleftarrow$): We want to show that if $X \subseteq \mathbb{R}^n$ is closed and bounded, then X is compact. We will only prove the result for a closed and bounded interval in $\mathbb{R}^1$, referring the reader to §4.5, Example 83, for the general case. Let $a, b \in \mathbb{R}$ with $a < b$, and consider the interval $[a, b] \subseteq \mathbb{R}^1$. Suppose $\mathscr{U}$ is a cover of $[a, b]$ by open sets. Define a subset C by

$$C = \{x \in (a, b] \mid [a, x] \text{ has a finite subcover } \mathscr{U}' \subseteq \mathscr{U}\}.$$

First, C is nonempty. To verify this, let $U \in \mathscr{U}$ be any set for which $a \in U$. (there's guaranteed to be at least one such U since $\mathscr{U}$ covers $[a, b]$). By definition of *open*, there exists an $\epsilon \in \mathbb{R}^+$ such that $B_\epsilon(a) \subseteq U$. Let $x \in (a, a + \epsilon)$. Then $[a, x] \subseteq (a - \epsilon, a + \epsilon) \subseteq U$; therefore $\mathscr{U}' = \{U\} \subseteq \mathscr{U}$ is a subcover for $[a, x]$ having only *one* open set. This implies $x \in C$.

Note that for any $y \in C$, it follows that $x \in C$ for every $a < x < y$. This is because the finite open subcover of $[a, y]$ would also cover $[a, x]$. Our goal now is to show that b is a member of C (so that a finite subcover would exist for the entire interval $[a, b]$). We will need the following useful **completeness property** of $\mathbb{R}$, which is typically discussed in a mathematical analysis course:

> If a subset $X \subseteq \mathbb{R}$ is nonempty and bounded, then there is both a **supremum** (or **least upper bound**) and an **infimum** (or **greatest lower bound**) for X in $\mathbb{R}$.

- We say b is a supremum for X if $b \geq x$, $\forall x \in X$, and if there is any other $y \geq x$, $\forall x \in X$, then $b \leq y$.
- We say a is an infimum for X if $a \leq x$, $\forall x \in X$, and if there is any other $w \leq x$, $\forall x \in X$, then $a \geq w$.

The bounded nonempty set C has a supremum $c \in \mathbb{R}$. Consider the subinterval $[a, c]$, and choose $U \in \mathscr{U}$ containing c. As before, we can find $\epsilon > 0$ such that $(c - \epsilon, c + \epsilon) \subseteq U$. Let $y \in (c - \epsilon, c)$, and since $y < c$, it follows that $y \in C$ (otherwise, c could not be the supremum of C). This implies that $[a, y]$ has a finite open subcover $\mathscr{U}' \subseteq \mathscr{U}$. But since $[y, c] \subseteq U$, the finite collection $\mathscr{U}' \cup U$ covers $[a, y] \cup [y, c] = [a, c]$. This proves that $c \in C$.

Now with this same c, if there is any $z \in (c, c + \epsilon) \cap [a, b]$ (implying $z > c$), then there is a finite open subcover for $[a, z]$, which puts $z \in C$, contradicting the fact that $c \geq z$ (since c is the supremum of C). Thus the only option is that $(c, c + \epsilon) \cap [a, b] = \emptyset$, but since $c \in [a, b]$, this means that $c = b$, which is what we needed to show.[4]   □

Example 27.

- Consider $\mathbb{I} = [0, 1] \subseteq \mathbb{R}^1$. Since $\mathbb{I}$ is closed and bounded, $\mathbb{I}$ is compact. Similarly, the unit square $\mathbb{I}^2 \subseteq \mathbb{R}^2$ and unit cube $\mathbb{I}^3 \subseteq \mathbb{R}^3$ are both closed and bounded, and hence compact. In fact, for any $n \in \mathbb{N}$, the closed unit **hypercube** $\mathbb{I}^n \subseteq \mathbb{R}^n$ is compact.

- Let $n \in \mathbb{N}$. The unit sphere $\mathbb{S}^n$ of dimension n is compact because it is closed and bounded.

- Consider the Cantor set C, as defined in §1.2. Since $C \subseteq \mathbb{I}$, C is bounded. We will show C is also closed. The easiest way to do this is to provide an equivalent definition of C in terms of the sets that must be removed from $\mathbb{I}$ to make the Cantor set:

$$C = \mathbb{I} \setminus \bigcup_{k \geq 1} U_k, \quad \text{where} \quad U_k = \bigcup_{\ell=0}^{3^{k-1}-1} \left(\frac{3\ell + 1}{3^k}, \frac{3\ell + 2}{3^k} \right). \tag{2.6}$$

It is an instructive exercise to write out the first few sets U_k to compare this construction with the "deleting middle thirds" construction.[5] Since each U_k is a union of open sets, U_k itself is open (see Proposition 2.1.7). For the same reason, $U = \bigcup_{k \geq 1} U_k$ is an open set. Thus $C = \mathbb{I} \setminus U$ is closed.

Since the Cantor set is both closed and bounded, it is compact by Heine-Borel.

The Extreme Value Theorem

As mentioned above, every bounded subset $X \subseteq \mathbb{R}$ has both a supremum and an infimum. According to the definitions, a supremum or infimum x may or may not be part of the set X; however, if $x \notin X$, then x must be a limit point of X (if you have seen some mathematical analysis, try to prove this statement). Thus, if X is also closed, it contains its limit points, so the supremum and infimum must be in the set X. This implies that compact subsets of $\mathbb{R}$ must contain their **extreme values**. For example, the extreme values of the interval $[a, b]$ are simply the endpoints a and b.

---

[4]Proof adapted from Munkres [Mun00].

[5]The U_k's are not disjoint, so some intervals seem to get removed "more than once." Can you see why this poses no problem?

Recall the **Extreme Value Theorem** (EVT):

> If a real-valued function f is continuous on a closed and bounded interval $[a, b]$, then f attains both an absolute maximum M and absolute minimum value m for some values $x_M, x_m \in [a, b]$.

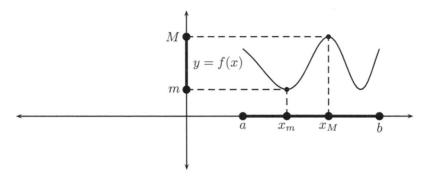

The EVT is a direct result of the fact that the image of a compact space under a continuous function is compact (see §4.4, Theorem 4.4.5). The idea is that $W = f\big[[a, b]\big]$ is compact in $\mathbb{R}$, and so the supremum M and infimum m are in W, which implies that there are values $x_M, x_m \in [a, b]$ such that $f(x_M) = M$ and $f(x_m) = m$ are the extreme values of f on the interval.

Sequences and Limits

A (real) sequence is typically defined as an infinite ordered list of real numbers,

$$(x_1, x_2, x_3, \ldots, x_k, \ldots).$$

In other words, a sequence is an element of $\mathbb{R}^\infty$. Another way to define a sequence is to say that it is a function $x : \mathbb{N} \to \mathbb{R}$, defined by the rule $x(k) = x_k$ for each natural number k (see §4.5, Example 84). This definition may be extended to sequences in higher-dimensional Euclidean spaces as well.

Definition 2.3.6. A **sequence** in $\mathbb{R}^n$ is a function $x : \mathbb{N} \to \mathbb{R}^n$.

A common notation used for a sequence $x : \mathbb{N} \to \mathbb{R}^n$ is $(x_k)_{k \in \mathbb{N}}$, or just (x_k) when context allows. For example, the **harmonic sequence** is defined by

$$(x_k) = \left(\frac{1}{k}\right) = \left(1, \frac{1}{2}, \frac{1}{3}, \frac{1}{4}, \frac{1}{5}, \ldots\right).$$

The variable k is called the **index variable**, which could be given any letter label without changing the sequence.[6] Thus the harmonic sequence could just as well have been defined by $(x_n) = \left(\frac{1}{n}\right)$ or $(x_\beta) = \left(\frac{1}{\beta}\right)$.

---

[6]For this reason the index variable is often called a **dummy variable**.

Note that a *sequence* is not the same thing as a *set*. By definition, a sequence has a particular ordering: x_1 first, then x_2, then x_3, etc. A set has no such ordering. Moreover, the terms of a sequence could repeat, while a given element is either in or not in a given set (repetition is irrelevent). When we want to refer to the set of elements of a sequence $(x_k)_{k \in \mathbb{N}}$, we use the usual set-builder notation, $\{x_k \mid k \in \mathbb{N}\}$, or when context is clear, we may abbreviate to $\{x_k\}$.

Example 28. For each $k \in \mathbb{N}$, let $x_k = (-1)^k$. The *sequence* (x_k) is the infinite list, $(-1, 1, -1, 1, -1, \ldots)$, which forever oscillates between the values -1 and 1. The *set* $\{x_k\}$ has only two elements, $\{-1, 1\}$.

A sequence (x_k) in $\mathbb{R}^n$ may or may not have a *limit*, that is, a point $x \in \mathbb{R}^n$ to which x_k gets arbitrarily close. As we have seen before, "close" ultimately means "within an open ϵ-ball."

Definition 2.3.7. A point $x \in \mathbb{R}^n$ is a **limit** of a sequence (x_k) if for every $\epsilon \in \mathbb{R}^+$ there is a number $N \in \mathbb{N}$ such that $x_k \in B_\epsilon(x)$ for all $k \geq N$. If such a point exists, then we call the sequence **convergent**.

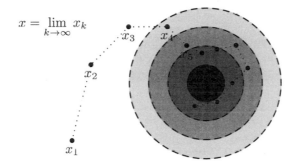

Figure 2.16: In this example, all terms x_k for $k \geq 4$ are contained within the lightest gray disk. All terms x_k for $k \geq 5$ are contained within the next smaller disk. Eventually, all terms beyond a certain threshhold index are contained in the smallest disk. No matter how small a radius given, there is a number N so that every x_k for $k \geq N$ is contained in the disk of that given radius.

Figure 2.16 illustrates how Definition 2.3.7 might be visualized. Note that the definition of limit does not in itself imply that a convergent sequence must have only *one* limit. In §4.2 we shall see that there are topological spaces in which a sequence may converge to two or more distinct points. Fortunately, in

$\mathbb{R}^n$, limits are indeed unique (which requires proof; see Exercise 6). If such a point x exists (and is unique), then we write $x = \lim_{k \to \infty} x_k$.

Subsequences and the Bolzano-Weierstrass Theorem

While an arbitrary sequence $(x_k)_{k \in \mathbb{N}}$ may fail to have a limit, there are situations in which a portion of the sequence must converge. As a trivial example, consider the sequence defined by $x_k = (-1)^k$ (recall Example 28). This sequence has no limit because when $\epsilon \leq 2$, the numbers 1 and -1 cannot be in the same ball $B_\epsilon(x)$ for any $x \in \mathbb{R}$. However, the *subsequence* of even indexed terms, $((-1)^{2m})_{m \in \mathbb{N}} = (1, 1, 1, \ldots)$, is a constant sequence converging to 1. Similarly, the subsequence of odd indexed terms, $((-1)^{2m-1})_{m \in \mathbb{N}} = (-1, -1, -1, \ldots)$, converges to -1.

Definition 2.3.8. Suppose $(x_k)_{k \in \mathbb{N}}$ is a sequence. A **subsequence** of (x_k) is any sequence of the form

$$(x_{k_m})_{m \in \mathbb{N}} = (x_{k_1}, x_{k_2}, x_{k_3}, \ldots),$$

where $k_1 < k_2 < k_3 < \ldots$ in $\mathbb{N}$.

It turns out that compactness plays a role in the existence of convergent subsequences. In Euclidean spaces, the relationship is very straightforward.

Theorem 2.3.9 (Bolzano-Weierstrass). *Every sequence in a compact subset of $\mathbb{R}^n$ has a convergent subsequence.*

Proof. Suppose that $X \subseteq \mathbb{R}^n$ is compact, and let $(x_k)_{k \in \mathbb{N}}$ be a sequence in X. If there is a constant subsequence, $x_{k_1} = x_{k_2} = x_{k_3} = \cdots = L$, then (x_{k_m}) is a convergent subsequence in X (with limit L). Now suppose no subsequence is constant. Then the set $A = \{x_k \mid k \in \mathbb{N}\}$ is an infinite subset of X. By Theorem 2.3.3, A has a limit point $x \in X$. We construct a convergent subsequence as follows. Let k_1 be the least integer such that $x_{k_1} \in B_1(x)$. Then let k_2 be the least integer greater than k_1 such that $x_{k_2} \in B_{1/2}(x)$. In general, once $x_{k_1}, \ldots, x_{k_{m-1}}$ have been chosen, let k_m be the least integer greater than k_{m-1} such that $x_{k_m} \in B_{1/m}(x)$. By construction, $(x_{k_m})_{m \in \mathbb{N}}$ is a subsequence of (x_k) that converges to $x \in X$. $\square$

The Bolzano-Weierstrass Theorem also implies that if a sequence $(x_k)_{k \in \mathbb{N}}$ in $\mathbb{R}^n$ is bounded, then it has a convergent subsequence. Boundedness implies $(x_k) \subseteq B_r(\mathbf{0})$ for some $r \in \mathbb{R}^+$, and so (x_k) is a sequence in the closed and bounded – hence compact – set $\overline{B_r(\mathbf{0})}$.

Example 29. Consider the sequence $(\cos k)_{k \in \mathbb{N}}$ in $\mathbb{R}$. The first 10 terms of the sequence are shown below.

$$
\begin{aligned}
\cos 1 &\approx 0.540302305868140 \\
\cos 2 &\approx -0.416146836547142 \\
\cos 3 &\approx -0.989992496600445 \\
\cos 4 &\approx -0.653643620863612 \\
\cos 5 &\approx 0.283662185463226 \\
\cos 6 &\approx 0.960170286650366 \\
\cos 7 &\approx 0.753902254343305 \\
\cos 8 &\approx -0.145500033808614 \\
\cos 9 &\approx -0.911130261884677 \\
\cos 10 &\approx -0.839071529076452
\end{aligned}
$$

There is no apparent pattern to the values, and indeed the sequence has no limit, in the sense that $\lim_{k \to \infty} \cos k$ does not exist. However, because $-1 \le \cos k \le 1$ for all $k \in \mathbb{N}$, Bolzano-Weierstrass *guarantees* that a convergent *sub*sequence can be found within this bounded sequence. In fact, it can be shown that every point of $[-1, 1]$ is a limit point of the set $\{\cos k \mid k \in \mathbb{N}\}$, so there is a subsequence converging to any given point $-1 \le L \le 1$. For example, the first few terms of a subsequence[7] converging to 0 have $k_1 = 1, k_2 = 11, k_3 = 344, k_4 = 40459, k_5 = 51109, \ldots$, as shown below.

$$
\begin{aligned}
\cos k_1 = \cos 1 &\approx 0.540302305868140 \\
\cos k_2 = \cos 11 &\approx 0.00442569798805079 \\
\cos k_3 = \cos 344 &\approx -0.00439555392789772 \\
\cos k_4 = \cos 40459 &\approx 0.000989255990869298 \\
\cos k_5 = \cos 51109 &\approx 0.0000849255511978441
\end{aligned}
$$

Exercises

1. Show that Definition 2.3.4 is equivalent to the following:

   A subset $X \subseteq \mathbb{R}^n$ is **bounded** if there is a radius $r \in \mathbb{R}^+$ and a point $x \in \mathbb{R}^n$ such that $X \subseteq B_r(x)$.

2. Classify each of the various types of intervals of $\mathbb{R}^1$, listed here as compact or not, and bounded or not: (a, b), $(a, b]$, $[a, b)$, $[a, b]$, (a, ∞), $[a, \infty)$, $(-\infty, b)$, $(-\infty, b]$, $(-\infty, \infty)$.

---

[7]These particular indices are such that the jth term is within 10^{-j+1} of 0. They were found using a script written in Sage [Ste].

3. Determine whether each subset of Euclidean space is compact or not. If not compact, state whether the subset is closed, bounded, or neither.

(a) $B_\epsilon(x) \subseteq \mathbb{R}^n$

(b) $D_\epsilon(x) \subseteq \mathbb{R}^n$

(c) $\{(x,y) \in \mathbb{R}^2 \mid |x| \leq 1, |y| \leq 1\}$

(d) $\{(x,y) \in \mathbb{R}^2 \mid |x| \leq 1, |y| < 1\}$

(e) $\{(x,y) \in \mathbb{R}^2 \mid |x| \leq 1, |y| \geq 1\}$

(f) $\{1, \frac{1}{2}, \frac{1}{3}, \frac{1}{4}, \ldots\}$

4. Consider the alternate definition of the Cantor set shown in (2.6). Determine U_1, U_2, and U_3 explicitly.

5. Prove: If $f : D \to \mathbb{R}^n$ is continuous, and if $\lim_{n\to\infty} x_n = x$, then $\lim_{n\to\infty} f(x_n) = f(x)$. In other words, show that continuous functions *commute* with taking limits.

$$\lim_{n\to\infty} f(x_n) = f\left(\lim_{n\to\infty} x_n\right) = f(x)$$

6. Let (x_k) be a sequence in $\mathbb{R}^n$. Suppose there are two points $x, y \in \mathbb{R}^n$ that are both limits for (x_k) (i.e., both x and y satisfy Definition 2.3.7). Prove that $x = y$.

7. Let (x_k) be a sequence in $\mathbb{R}^n$.

(a) Show that if $x = \lim_{k\to\infty} x_k$, then either $x = x_k$ for all k large enough, or x is the only limit point of the set $\{x_k \mid k \in \mathbb{N}\}$.

(b) Construct a sequence $(x_k)_{k\in\mathbb{N}}$ such that the set $\{x_k \mid k \in \mathbb{N}\}$ has more than one limit point (thus showing that *limit points* of a set are not necessarily *limits* of the corresponding sequence).

2.4 Connectedness

You can't get there from here. Have you ever stopped for directions in an unfamiliar place and heard these discouraging words? We know what is meant: *It's not easy to get there from here*, rather than *It's physically impossible to get there from here*. Every point on the surface of the Earth is connected to every other point by some route – in some cases perhaps an extremely difficult or dangerous one. Now if you had been asking directions to Mare Tranquillitatis (the Sea of Tranquility on the moon), then you really *can't* get there from here since the surface of the Earth and surface of the moon are not connected.[8]

---

[8]Of course, you *could* get to Mare Tranquillitatis by traveling through space, but let's stick to Earth-bound transportation for this illustration.

Arc-Connectedness

Intuitively, a space is *arc-connected*[9] if an ant could walk from any point to any other point in *finite time*. The plane $\mathbb{R}^2$ is certainly arc-connected because any two points can be joined by a straight line of finite length. The sphere and torus are both arc-connected. The plane excluding a point is arc-connected, because the missing point can be avoided, but the plane excluding a line is not arc-connected (see Figure 2.17).

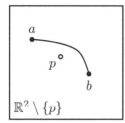 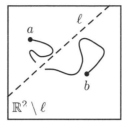

Figure 2.17: *Left*, $\mathbb{R}^2 \setminus \{p\}$ is arc-connected. *Right*, $\mathbb{R}^2 \setminus \ell$ is not arc-connected, as no arc exists from a to b, since any such line would cross the line of "missing" points.

As always, a good mathematical concept requires a precise definition. We define an *arc* between two points in much the same way as a parametrized curve (Equation (1.5) of Chapter 1 may be regarded as a continuous function $\mathbf{r} : [a,b] \to \mathbb{R}^2$). In the following definition, recall that $\mathbb{I} = [0,1]$ is the unit closed interval.

Definition 2.4.1. An **arc** γ in a set X is a continuous function,

$$\gamma : \mathbb{I} \to X.$$

X is called **arc-connected** if for every pair of points $x, y \in X$ there exists an arc γ in X such that $\gamma(0) = x$ and $\gamma(1) = y$.

According to Definition 2.4.1, an arc is just a function γ, but we often visualize γ by its image in X, that is, $\gamma[\mathbb{I}] = \{\gamma(t) \mid t \in \mathbb{I}\} \subseteq X$. However, it is important to realize that an arc is not just a static set of points in X, but more like a *journey* from a starting point $\gamma(0)$ to an ending point $\gamma(1)$ (think of the journey of an ant along the arc parametrized by the function γ). Some arcs intersect themselves (i.e., $\gamma(t_1) = \gamma(t_2)$ for some pair of values $t_1 \neq t_2$, as

---

[9]We use the terms *arc* and *arc-connected* where many texts use *path* and *path-connected*. There are two reasons for this. First, it frees up the term *path* for a more specific kind of arc in graph theory (see §6.1). Second, the term (*simple*) *arc* is traditionally found in the literature in connection with the Jordan Curve Theorem, and so using the term here seems appropriate.

in Figure 2.18); some retrace the same set of points multiple times (see Example 30); and some arcs, called **trivial** arcs, simply stay at the same point for all t (i.e., γ may be a *constant* function $\gamma(t) = x_0$ so the image of γ is the single point $\{x_0\}$).

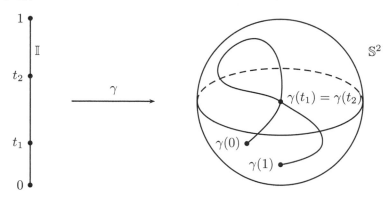

Figure 2.18: An arc $\gamma : \mathbb{I} \to \mathbb{S}^2$ whose image crosses itself.

Example 30. Let $\gamma : \mathbb{I} \to \mathbb{R}^2$ by $\gamma(t) = (\cos 2\pi t, \sin 2\pi t)$, and $\eta : \mathbb{I} \to \mathbb{R}^2$ by $\eta(t) = (\cos 4\pi t, \sin 4\pi t)$. As t ranges from 0 to 1, γ traces out the unit circle in a counterclockwise direction, while η traces out the unit circle *twice* counterclockwise (see Figure 2.19).

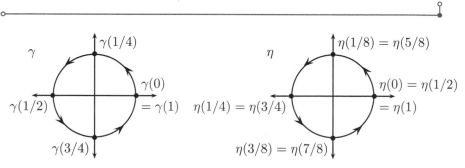

Figure 2.19: Here γ wraps once around the circle, while η maps twice.

Example 31. Let's introduce a third dimension. Define $\theta : \mathbb{I} \to \mathbb{R}^3$ by

$$\theta(t) = (\cos 4\pi t, \sin 4\pi t, t).$$

The arc θ is called a helix (with two complete turns), as shown in Figure 2.20.

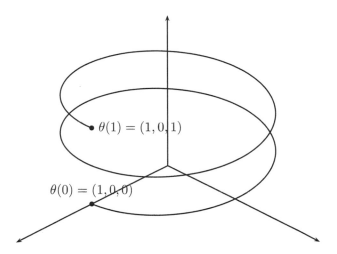

Figure 2.20: An arc in $\mathbb{R}^3$ whose image is part of a helix.

Arc-connectedness is preserved under continuous maps. It is assumed in this chapter that all spaces are subsets of Euclidean space, but the proof of Theorem 2.4.2 does not use anything other than definitions and properties of continuous functions.

Theorem 2.4.2. *If X is arc-connected, and $f : X \to Y$ is continuous, then the image $f[X] \subseteq Y$ is arc-connected.*

Proof. Consider any two points $y_1, y_2 \in f[X]$. Let $x_1 \in f^{-1}[\{y_1\}]$ and $x_2 \in f^{-1}[\{y_2\}]$ in X (these preimage sets are nonempty — *why?*). Now by arc-connectedness, there is an arc $\gamma : \mathbb{I} \to X$ such that $\gamma(0) = x_1$ and $\gamma(1) = x_2$. Consider the composition $f \circ \gamma : \mathbb{I} \to Y$. Since both f and γ are continuous, so is their composition. Thus $f \circ \gamma$ is an arc in Y, with $(f \circ \gamma)(0) = f(\gamma(0)) = f(x_1) = y_1$ and $(f \circ \gamma)(1) = y_2$, which proves that Y is arc-connected. $\qquad\square$

Theorem 2.4.3. *Arc-connectedness is a topological invariant.*

Proof. To prove that something is a topological invariant, we just have to show that it is preserved by homeomorphism. Suppose $X \approx Y$, and let $f : X \to Y$ be a homeomorphism. If X is arc-connected, then Theorem 2.4.2 implies that $f[X]$ is arc-connected, but since f is surjective, we have $f[X] = Y$ is arc-connected. $\quad\square$

It is actually quite challenging to prove rigorously that a given set is or is not arc-connected. Even a set that *seems* to be all "one piece" may nevertheless fail to be arc-connected (see Exercise 1). On the other hand, if a set consists of pieces that are "clearly" separated from one another, then it is not arc-connected. Of course, we require a precise definition for *separated* before we can use the term in mathematical proofs. To that end we turn our attention to the related but weaker topological invariant, *connectedness*.

Connectedness

Our intuitive notion of *connected* may be something like *not separated*. But what does it mean for two sets to be *separated*? Certainly we mean more than simply disjoint. Consider four intervals in $\mathbb{R}^1$: $I = (-\infty, 0)$, $J = [0, \infty)$, $K = [1, \infty)$, and $L = (0, \infty)$. Even though I and J are disjoint, we would not consider them to be separated because I and J *touch* at 0. Indeed, $I \cup J = \mathbb{R}^1$, which is clearly one piece. On the other hand, I and K are clearly separated by a nonzero distance. What about I and L? This time, there is no positive distance separating the two intervals; however, $I \cup L \neq \mathbb{R}^1$, and in fact, $I \cup L = \mathbb{R}^1 \setminus \{0\}$. When we remove a point from $\mathbb{R}^1$, we expect that the hole separates the number line into two pieces. These distinctions make us realize the profound subtlety in our concepts of *connected* vs. *separated*. The topological definition of connectedness begins with a precise definition of separation.

> **Definition 2.4.4.** A **separation** of a set $X \subseteq \mathbb{R}^n$ is a disjoint pair of open sets $U, V \subseteq \mathbb{R}^n$, with $U \cap X$ and $V \cap X$ nonempty, such that $X \subseteq U \cup V$.

> **Definition 2.4.5.** If X has a separation, then X is called **disconnected**; otherwise, X is **connected**.

Example 32.

- $\mathbb{R}^1 \setminus \{a\}$ has a separation $U = (-\infty, a)$, $V = (a, \infty)$. Therefore $\mathbb{R}^1 \setminus \{a\}$ is disconnected.

- Let $U = (-\infty, a]$, and $V = (a, \infty)$. Even though $U \cap V = \emptyset$ and $U \cup V = \mathbb{R}^1$, the sets U and V are not a separation of $\mathbb{R}^1$ because U is not open. In fact, we will take for granted that $\mathbb{R}^1$ has no separation at all and so is connected. Similarly, we may assume that every interval, (a, b), $(a, b]$, $[a, b)$, $[a, b]$, (a, ∞), $[a, \infty)$, $(-\infty, b)$, $(-\infty, b]$, is connected.

- The set of integers $\mathbb{Z} \subseteq \mathbb{R}^1$ is disconnected. Indeed, every pair of distinct points, $n, m \in \mathbb{Z}$, $n \neq m$, are separated (a quality we call *totally disconnected*). What's more, the open sets $U_n = (n - 1/2, n + 1/2)$ for all $n \in \mathbb{Z}$ are mutually disjoint while $\mathbb{Z} \subseteq \bigcup_{n \in \mathbb{Z}} U_n$. This makes $\mathbb{Z}$ a *discrete* set (more about discrete sets in §2.5 and §4.1).

- The set of rational numbers $\mathbb{Q} \subseteq \mathbb{R}^1$ is also disconnected. For any fixed $x \in \mathbb{R} \setminus \mathbb{Q}$ (an irrational number), the sets $U = (-\infty, x)$ and (x, ∞) satisfy: $U \cap V = \emptyset$; $U \cap \mathbb{Q} \neq \emptyset$ and $V \cap \mathbb{Q} \neq \emptyset$; and $\mathbb{Q} \subseteq U \cup V$. Now from mathematical analysis we learn that between any two rational numbers there exists an irrational number. Thus any two points of $\mathbb{Q}$ can be separated by open sets, showing that $\mathbb{Q}$ is *totally disconnected*, but $\mathbb{Q}$ is not *discrete*.

Example 33. The graph of $y = x^2$ is connected, while the graph of $y = 1/x^2$ is disconnected. For example, the open sets $U = \{(x, y) \mid x < 0\}$ and $V = \{(x, y) \mid x > 0\}$ separate the graph of $y = 1/x^2$.

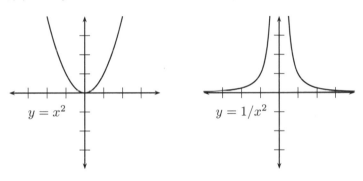

Connectedness is a *topological invariant* (see §4.4), so it can be used to distinguish topological spaces. Also, as the name suggests, there is a direct relationship between the terms connected and arc-connected.

Theorem 2.4.6. *If X is arc-connected, then X is connected.*

Proof. Suppose X is not connected. Then there is a separation U, V for X. Let $x \in U \cap X$ and $y \in V \cap X$ (which is possible since neither $U \cap X$ nor $V \cap X$ is empty), and suppose an arc $\gamma : \mathbb{I} \to X$ exists from x to y. Consider $U_0 = \gamma^{-1}[U]$ and $V_0 = \gamma^{-1}[V]$, both of which are open (*why?*). We have $0 \in U_0$ and $1 \in V_0$ (since $\gamma(0) = x \in U$ and $\gamma(1) = y \in V$). Moreover, $\mathbb{I} \subseteq U_0 \cup V_0$ (since $U \cup V$ contains all of X; hence $\gamma[\mathbb{I}] \subseteq X$). Finally, $U_0 \cap V_0 = \emptyset$, since otherwise there would be a point $w \in U_0 \cap V_0$, implying that $\gamma(w) \in U \cap V$ (contradicting $U \cap V = \emptyset$). These observations show that U_0, V_0 is a separation of the connected interval $\mathbb{I}$, a contradiction. Thus no separation exists in X, and so by definition, X is connected. $\square$

Caution: the converse of Theorem 2.4.6 does not hold in general, as we shall see in Example 34. Before delving into that example, though, we should first develop a few important properties of connectedness.

Properties and Consequences of Connectedness

The following properties help to identify connected subsets. These properties remain true in more general topological spaces.

Proposition 2.4.7. *Suppose $X \subseteq \mathbb{R}^n$.*

1. *If U, V is a separation of X, and if $A \subseteq X$ is connected, then either $A \subseteq U$ or $A \subseteq V$.*

2. *If $\{A_k\}_{k \in \mathcal{I}}$ is a collection of connected subsets of X, and if there is a point x_0 common to every A_k, then the subset $A = \bigcup_{k \in \mathcal{I}} A_k$ is connected.*

Proof.   1. See Exercise 4.

2. Suppose $\{A_k\}$ and x_0 are as indicated in the proposition. Then $x_0 \in \bigcap_{k \in \mathcal{I}} A_k$. Suppose (to the contrary) that there is a separation U, V for $A = \bigcup_{k \in \mathcal{I}} A_k$. Either $x_0 \in U$ or $x_0 \in V$. Without loss of generality, assume $x_0 \in U$. Since $\{x_0\}$ is a connected subset of each connected set A_k, part (1) shows that $A_k \subseteq U$ for every $k \in \mathcal{I}$. Thus all of A is contained within U, leaving $V \cap A$ empty. Thus no separation can exist, and A must be connected. $\square$

Example 34. Let $X = \{\mathbf{0}\} \cup \bigcup_{n=1}^{\infty} \overline{PQ_n} \subseteq \mathbb{R}^2$, where $P = (0, 1)$ and $Q_n = (1/n, 0)$ (pictured below). X is not arc-connected because there is no arc from $\mathbf{0}$ to any other point in X (recall that an arc must lie entirely within the set X).

However, X *is* connected. To see this, note first that the set $\bigcup_{n=1}^{\infty} \overline{PQ_n}$ is arc-connected. If $x \in \overline{PQ_n}$ and $y \in \overline{PQ_m}$, the arc connecting x and y is obtained by traveling through the common point P. Thus, if a separation $U \cup V$ of X exists at all, we can assume $\mathbf{0} \in U$ and $\bigcup_{n=1}^{\infty} \overline{PQ_n} \subseteq V$. U is open, so there exists $\epsilon \in \mathbb{R}^+$ such that $\mathbf{0} \in B_\epsilon(\mathbf{0}) \subseteq U$. Let $n > 1/\epsilon$. Then $Q_n = (1/n, 0) \in B_\epsilon(\mathbf{0})$, contradicting $U \cap V = \emptyset$.

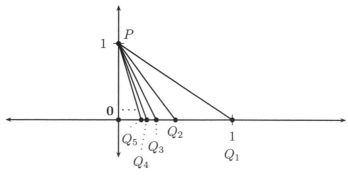

Connected subsets of the real number line $\mathbb{R}$ must satisfy a certain *completeness* condition. If $C \subseteq \mathbb{R}$ is connected, and if $a, b \in C$ such that $a < b$, then every $c \in \mathbb{R}$ such that $a < c < b$ must also be in C (this is why $\mathbb{R}$ and intervals of $\mathbb{R}$ are connected). We will have more to say about completeness in §2.5. Recall the **Intermediate Value Theorem** (IVT) from calculus:

> If a real-valued function f is continuous on a closed interval $[a, b]$, and L is any number between $f(a)$ and $f(b)$, then there is at least one value $x = c$ between a and b such that $f(c) = L$.

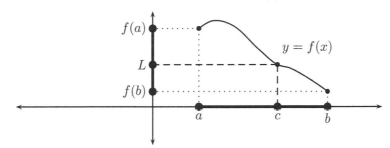

The closed interval $[a, b]$ is connected; hence the IVT can be interpreted as saying the forward image $f\big[[a, b]\big]$ is also connected. In general, the continuous image of a connected space is connected, as we shall see in §4.4, Theorem 4.4.10.

We end this section with an important theorem that states what may seem at first to be a self-evident truth.

Theorem 2.4.8 (Jordan Curve Theorem). *If $C \subseteq \mathbb{R}^2$ is a simple closed curve (i.e., a loop with no self-intersections), then $\mathbb{R}^2 \setminus C$ has precisely two arc-connected components, one that is bounded (the inside) and one that is unbounded (the outside).*

While the concept seems so simple and intuitively obvious, the proof is quite technical.[10] Of course, the result is not *really* obvious and is flat-out false in other surfaces such as the Möbius strip, torus, etc. Here we provide a proof of the Jordan Curve Theorem for the simplest of all Jordan curves, a circle.

Proof. Suppose C is a circle centered at $x_0 \in \mathbb{R}^2$, so $C = \{x \in \mathbb{R}^2 \mid d(x_0, x) = r\}$ for some fixed $r > 0$. Consider the two sets defined below.

$$Y = \{x \in \mathbb{R}^2 \mid d(x_0, x) < r\}$$
$$Z = \{x \in \mathbb{R}^2 \mid d(x_0, x) > r\}$$

Clearly, C, Y, and Z partition $\mathbb{R}^2$. That is, these three sets are mutually disjoint, and every point of $\mathbb{R}^2$ is contained in one of the three (since for every $x \in \mathbb{R}^2$, one

---

[10]See Munkres [Mun00], Chapter 10, or Armstrong [Arm10], §5.6, for proofs that employ the *fundamental group* (which we will define in Chapter 7). See also [Cai51, Tho92] for more "elementary" proofs.

and only one of the three relations can hold, either $d(x_0, x) < r$, $d(x_0, x) = r$, or $d(x_0, x) > r$). Thus we have $\mathbb{R}^2 \setminus C = Y \cup Z$ with $Y \cap Z = \emptyset$. Observe that Y is bounded and Z is unbounded.

Now both Y and Z are arc-connected. Every point $y \in Y$ can be joined to the center x_0 by a radial segment; any two points in Z can be connected by an arc of constant radius with respect to x_0, followed by a radial segment (see Figure 2.21).

Finally, we must establish that there is no arc connecting any point in Y to a point in Z lying within $Y \cup Z$. Suppose $y \in Y$ and $z \in Z$ are joined by an arc $\gamma : \mathbb{I} \to \mathbb{R}^2$. Let $g : \mathbb{R}^2 \to \mathbb{R}$ be the function defined by $g(x) = d(x_0, x)$. The composition $g \circ \gamma : \mathbb{I} \to \mathbb{R}$ is continuous since both γ and g are continuous (see Exercise 10 of §2.2).

$$
\begin{aligned}
(g \circ \gamma)(0) &= g(\gamma(0)) = g(y) = d(x_0, y) < r \\
(g \circ \gamma)(1) &= g(\gamma(1)) = g(z) = d(x_0, z) > r
\end{aligned}
$$

Hence, by the Intermediate Value Theorem, there must be a value $t \in \mathbb{I}$ such that $(g \circ \gamma)(t) = r$, but this implies $g(\gamma(t)) = d(x_0, \gamma(t)) = r$. In other words, the arc γ must cross the curve C at $\gamma(t)$ and so cannot be in $Y \cup Z$. This contradiction proves that $Y \cup Z$ is not arc-connected. $\qquad \square$

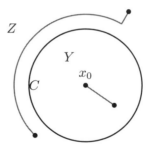

Figure 2.21: If C is a circle, then C satisfies the Jordan Curve Theorem. Both Y and Z are arc-connected, and any arc joining a point in Y to a point in Z must intersect the circle C.

The above proof can easily be extended to the case in which C encloses a *convex* region. A region X is call **convex** if for every pair of points $x, y \in X$, the line segment $\overline{xy}$ lies in X. In fact, it is not difficult to prove the Jordan Curve Theorem for curves that enclose so-called *radially convex* (or *star-shaped*) regions. A region X is called **radially convex** if there is a point $x_0 \in X$ such that for any other point $x \in X$, we have $\overline{x_0 x} \subseteq X$ (in particular, every *convex* region is radially convex). Figure 2.22 demonstrates a few such regions. The proof of Jordan Curve Theorem in this situation relies on constructing a function $f : \mathbb{R}^2 \to \mathbb{R}^2$ that deforms the region X homeomorphically to a circular disk by

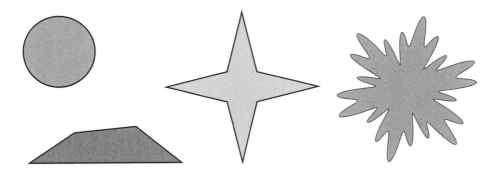

Figure 2.22: Radially convex regions.

rescaling each $x_1 \in C$ along its "radial" line segment until every point on C is the same distance from x_0.

Where the Jordan Curve Theorem becomes tricky is in situations in which the "inside" of a region is not easy to identify (remember, the proof must work for absolutely every simple curve, not just those that we can draw easily on paper). For example, try to determine which points are inside the curve shown in Figure 2.23.

Exercises

1. Define the *topologist's sine curve* $X \subseteq \mathbb{R}^2$ by:

$$X = \{(0,y) \mid y \in [-1,1]\} \cup \left\{ \left(x, \sin \frac{1}{x} \right) \mid x \in (0, \infty) \right\}.$$

   Draw the set X as accurately as you can. Be especially careful near the origin. Explain why X is not arc-connected. Is X connected?

2. Using Example 30 as a guide, define an arc $\gamma_n : \mathbb{I} \to \mathbb{R}^2$ that traces out the unit circle n times counterclockwise. Define a similar family of arcs that trace out the unit circle n time clockwise.

3. Classify each unit sphere $\mathbb{S}^n$ (for $n = 0, 1, 2, \ldots$) as connected or not connected. If not connected, find a separation.

4. Prove Proposition 2.4.7(1).

5. Let $C \subseteq \mathbb{R}$. Suppose $a, b \in C$ such that $a < b$, and let c be a real number between a and b. Show that if C is connected, then $c \in C$.

6. Determine whether each of the following subsets of $\mathbb{R}^n$ is connected or not connected. If not connected, find a separation. Be careful, as sometimes the outcome depends on n.

Figure 2.23: A Jordan curve. What is inside and outside? Image courtesy of Wikimedia Commons (modified).

(a) $\{x\}$

(b) $\{x, y\}$, if $x \neq y$

(c) $B_\epsilon(x)$, $\epsilon \in \mathbb{R}^+$

(d) $D_\epsilon(x)$, $\epsilon \in \mathbb{R}^+$

(e) $\text{ext}(B_\epsilon(x))$, $\epsilon \in \mathbb{R}^+$

(f) $B_\epsilon(x) \setminus \{x\}$, $\epsilon \in \mathbb{R}^+$

7. Determine whether each of the points A, B, C, and D in Figure 2.23 lies on the inside or outside of the curve.

2.5   Metric Spaces in General

Distance is such a fundamental concept to Euclidean space that it deserves to be studied in greater detail. We often think of distance as something concrete that can be calculated once you know exactly where two points are in relation to one another. However, there are other ways in which we use this term. We might say that the string of letters "teh" is very *close* to "the." Your computer may notice how close the two strings are and change your typo, "teh," into a real word, "the." On the other hand, you wouldn't want your computer changing "tho" into "the" (maybe "tho" was a typo for "though"). How can we measure that "the" is closer to "teh" than to "tho"? After all, strings of letters do not look much like points or vectors in $\mathbb{R}^n$, so it is unclear how to measure the "distance" between them. We need a more abstract concept of *distance*.

When abstracting away from the familiar realm of Euclidean space, it's important to know what to keep and what to ignore. It makes sense to ignore vector-like properties such as *dot product, angles, coordinates,* and even the *length* of individual elements. Only distances between two elements should matter. But what should we keep? That's a tricky question, but let's agree that our distance functions should still satisfy the fundamental properties of Euclidean distance found in Proposition 2.1.2.

Metric Functions

A *metric* provides a way to measure abstract distance between elements of a set. Metrics can be used to define how close elements should be to one another, such as when two sequences or functions are considered "close" to each other, or when a word is "close" to another.[11]

Definition 2.5.1. A **metric** (or **metric function**) on a set X is a function $d : X \times X \to \mathbb{R}$ satisfying the following axioms:

1. [Nonnegativity] $d(x,y) \geq 0$, $\forall x, y \in X$.

2. [Zero-Distance Rule] $d(x,y) = 0$ if and only if $x = y$.

3. [Symmetry] $d(x,y) = d(y,x)$, $\forall x, y \in X$.

4. [Triangle Inequality] $d(x,z) \leq d(x,y) + d(y,z)$, $\forall x, y, z \in X$.

A set X that has a metric d defined on it is called a **metric space** and is often denoted by (X, d) to indicate the metric function.

The Euclidean distance function, $d(x,y) = \|x-y\|$, defined on the set $\mathbb{R}^n$ as in Definition 2.1.1, is a metric function as a direct consequence of Proposition 2.1.2. Parts 1–3 were part of your exercises. The hardest part to show is the triangle inequality. In fact, we will show that the triangle inequality holds in spaces called *inner product spaces*.

A (real) **inner product space** is a (real) vector space X together with a function $X \times X \to \mathbb{R}$ called an **inner product**. The inner product of $x, y \in X$ is typically denoted $x \cdot y$ or $\langle x, y \rangle$ and must satisfy:

- [Symmetry] $x \cdot y = y \cdot x$, $\forall x, y \in X$.

---

[11] There is a surprisingly wide array of useful metric functions defined on various sets. For an impressive list of metrics, see the *Encyclopedia of Distances* by Michel and Elena Deza [DD09].

- [Linearity] $(\lambda x + y) \cdot z = \lambda(x \cdot z) + y \cdot z$, $\forall x, y, z \in X, \lambda \in \mathbb{R}$.

- [Positive-Definiteness] $x \cdot x \geq 0$, $\forall x \in X$, and $x \cdot x = 0 \iff x = \mathbf{0}$.

Owing to symmetry, the inner product is also linear in the second component, which implies that the inner product *distributes* over addition from the left or from the right. Once an inner product is defined, then a norm and metric function can be defined via $\|x\| = \sqrt{x \cdot x}$, and $d(x, y) = \|x - y\|$, which makes every inner product space a metric space (but not vice versa).

Now in any inner product space, the famous **Cauchy-Schwarz Inequality** holds:

$$|x \cdot y| \leq \|x\| \|y\|, \qquad \forall x, y \in X. \tag{2.7}$$

For vectors $x, y \in \mathbb{R}^n$, (2.7) follows easily from the formula $x \cdot y = \|x\| \|y\| \cos \theta$, where θ is the angle between the vectors x and y. However, we shall prove the Cauchy-Schwarz Inequality using only the properties of the inner product. This result is the bedrock providing part of the foundation for analysis, linear algebra, probability, and numerous other fields of mathematics; its importance cannot be overstated.

Observe that if either x or y is equal to $\mathbf{0}$, then (2.7) is trivially satisfied. Fix $x, y \in X$, with $y \neq \mathbf{0}$, and consider the quantity below, which is nonnegative by positive definiteness:

$$\left\| x - \left(\frac{x \cdot y}{y \cdot y} \right) y \right\| \geq 0. \tag{2.8}$$

Squaring the lefthand side of (2.8), and using the fact that $\|x\|^2 = x \cdot x$ together with linearity of the inner product, we obtain:

$$\begin{aligned} \left\| x - \left(\frac{x \cdot y}{y \cdot y} \right) y \right\|^2 &= \left(x - \left(\frac{x \cdot y}{y \cdot y} \right) y \right) \cdot \left(x - \left(\frac{x \cdot y}{y \cdot y} \right) y \right) \\ &= (x \cdot x) - 2(x \cdot y)\frac{x \cdot y}{y \cdot y} + \frac{(x \cdot y)^2}{(y \cdot y)^2}(y \cdot y) \\ &= (x \cdot x) - \frac{(x \cdot y)^2}{y \cdot y}. \end{aligned} \tag{2.9}$$

Combining (2.9) with (2.8),

$$\begin{aligned} \|x\|^2 - \frac{(x \cdot y)^2}{\|y\|^2} &\geq 0 \\ \|x\|^2 \|y\|^2 &\geq (x \cdot y)^2. \end{aligned}$$

Taking principle square roots on both sides, we arrive at (2.7).

Now we can show that the metric defined by $d(x, y) = \|x - y\|$ in an inner product space satisfies the triangle inequality. That is, we want to show that

$$\|x - z\| \leq \|x - y\| + \|y - z\|, \; \forall x, y, z \in X.$$

Letting $u = x - y$ and $v = y - z$, so that $u + v = x - z$, it is equivalent to show that

$$\|u + v\| \le \|u\| + \|v\|, \ \forall u, v \in X. \tag{2.10}$$

In what follows, note that $u \cdot v \le |u \cdot v|$ and use (2.7):

$$\begin{aligned}
\|u + v\|^2 &= (u + v) \cdot (u + v) \\
&= \|u\|^2 + 2(u \cdot v) + \|v\|^2 \\
&\le \|u\|^2 + 2\|u\|\|v\| + \|v\|^2 \\
&= (\|u\| + \|v\|)^2.
\end{aligned}$$

Inequality (2.10) follows by taking principle square roots.

Keep in mind that not all metric spaces are inner product spaces. We now assume X is a general metric space with metric function d. From Definition 2.5.1 we can reconstruct the ϵ-balls, which are so important in defining everything else that follows, though $B_\epsilon(x)$ may look very different with respect to different metric functions.

Definition 2.5.2. Suppose (X, d) is a metric space. Let $x \in X$ and $\epsilon \in \mathbb{R}^+$. The ϵ-**ball around** x is the set

$$B_{d,\epsilon}(x) = B_\epsilon(x) = \{y \in X \mid d(x, y) < \epsilon\}.$$

Once we have defined ϵ-balls, then we immediately get definitions for *neighborhoods*, *open* and *closed* sets, *limit points*, *compact*, *bounded*, and *connected* subsets, and other terms that only depend on distance or on open sets (which are defined in metric space in terms of open balls and so, ultimately, distance). Note that the definition for *bounded* must be expressed without using the idea of the *origin*, which may have no analog in an arbitrary metric space (see Exercise 1 in §2.3). We collect the general definitions here. These apply to any metric space, including the familiar Euclidean spaces $\mathbb{R}^n$. Note that if (X, d) is a metric space, then any subset $A \subseteq X$ is automatically a metric space with the restricted metric function $d|_A : A \times A \to \mathbb{R}$. So every subset of $\mathbb{R}^n$ is a metric space under the Euclidean metric function.

Suppose (X, d) is a metric space.

- $N \subseteq X$ is a *neighborhood* of $x \in X$ $\iff$ $\exists \epsilon \in \mathbb{R}^+$ such that $B_\epsilon(x) \subseteq N$.

- $U \subseteq X$ is *open* $\iff$ $\forall x \in U$, $\exists \epsilon \in \mathbb{R}^+$, such that $x \in B_\epsilon(x) \subseteq X$.

- $C \subseteq X$ is *closed* $\iff$ $X \setminus C$ is open in X.

- $x \in X$ is a *limit point* of $A \subseteq X$ $\iff$ $\forall \epsilon \in \mathbb{R}^+$, we have $B_\epsilon(x) \cap (A \setminus \{x\}) \neq \emptyset$.

- $C \subseteq X$ is *bounded* if $\exists r > 0$ such that $C \subseteq B_r(x)$ for some $x \in X$.

Proposition 2.1.7, concerning unions and intersections, also carries through in metric spaces. We shall see in Chapter 4 that these properties are key in defining general topological spaces.

Proposition 2.5.3. *Suppose (X, d) is a metric space.*

(a) The empty set, $\emptyset \subseteq X$, is open.

(b) The entire space X is open.

(c) If U and V are open subsets of X, then $U \cap V$ is open.

(d) Suppose for each $k \in \mathcal{I}$, $U_k \subseteq X$ is open. Then $\bigcup_{k \in \mathcal{I}} U_k$ is open.

Proof. See Exercise 2. □

We may even make sense out of *continuous functions* between two metric spaces, in the sense of Definition 2.2.3.

Definition 2.5.4. Suppose (X, d) and (Y, d') are metric spaces. A function $f : X \to Y$ is **continuous** if:

For every $x \in X$, and $\epsilon \in \mathbb{R}^+$, there exists $\delta > 0$, such that $B_{d, \delta}(x) \subseteq f^{-1}[B_{d', \epsilon}(f(x))]$.

Example 35. Let X be any set. Define a metric $d_{\text{disc}} : X \times X \to \mathbb{R}$, called the **discrete metric**, as follows:

$$d(x, y) = d_{\text{disc}}(x, y) = \begin{cases} 0, & x = y \\ 1, & x \neq y \end{cases}.$$

Before moving on, we should prove that d_{disc} is in fact a metric. Axioms 1, 2, and 3 are immediate from the definition. Axiom 4 is proved in two cases.

Case I. Suppose $x = z$. Then $d(x, z) = 0$, which is certainly less than or equal to the sum $d(x, y) + d(y, z)$ for any choice of y.

Case II. Suppose $x \neq z$. Then $d(x, z) = 1$. Now let $y \in X$ be arbitrary. Either $x \neq y$ or $y \neq z$, since if both $x = y$ and $y = z$, then transitivity would give $x = z$. Thus either $d(x, y) = 1$ or $d(y, z) = 1$ (or both), which implies $1 \leq d(x, y) + d(y, z)$, as required.

In the discrete metric, every single point is exactly one unit away from each of its neighbors. If X has only three points, then they could be the vertices of an equilateral triangle. Four such points may be the vertices of a regular tetrahedron. If X is larger (perhaps infinite), then it may be impossible to imagine what X "looks like," though we can still use the properties of d to describe the space. For example, every subset $Y \subseteq X$ must be open; indeed, every singleton set $\{x\} \subseteq X$ is open, since $\{x\} = B_{1/2}(x)$ (more precisely, $\{x\} = B_\epsilon(x)$ for any $0 < \epsilon \leq 1$). Thus $\{x\}$ is a neighborhood of x, and so we may imagine the space X as consisting of entirely isolated points. We shall meet this type of space again in Chapter 4.

Example 36. Define a function d_{taxi} on $\mathbb{R}^2 \times \mathbb{R}^2$ as follows:

$$d_{\mathrm{taxi}}(x, y) = d_{\mathrm{taxi}}((x_1, x_2), (y_1, y_2)) = |x_1 - x_2| + |y_1 - y_2|.$$

In the Exercises, you will show that d_{taxi} is metric. It is called the **taxicab metric** because it models how distance might be calculated by a taxicab negotiating the streets of a city like New York City. The streets are laid out in something like a grid pattern, so in order to travel from the Empire State Building to the United Nations Headquarters, you go roughly 8 blocks east and 9 blocks north, for a total distance of 17 blocks,[12] rather than the shorter diagonal route of length $\sqrt{8^2 + 9^2} \approx 12.04$ "blocks." In New York City, the buildings prevent us from going diagonally. In $\mathbb{R}^2$, there are no "buildings" preventing us from drawing a diagonal line; it's just that when using the metric d_{taxi}, we measure distance as if the path were made up of only vertical and horizontal segments. Thus there is no unique shortest path between two points. The two paths shown in Figure 2.24 each represent a minimal path from $a = (-3, -2)$ to $b = (1, 3)$, each with length $d_{\mathrm{taxi}}(a, b) = |(-3) - 1| + |(-2) - 3| = 9$.

Let's see what a metric ball looks like with respect to the taxicab metric.

$$
\begin{aligned}
B_\epsilon((a, b)) &= \{(x, y) \in \mathbb{R}^2 \mid d_{\mathrm{taxi}}((x, y), (a, b)) < \epsilon\} \\
&= \{(x, y) \in \mathbb{R}^2 \mid |x - a| + |y - b| < \epsilon\}
\end{aligned}
$$

The graph of the *metric circle* $|x - a| + |y - b| = \epsilon$ that bounds this ball has a diamond shape. The ball is interior to the diamond, as shown below.

---

[12]We are ignoring the fact that blocks in New York have different lengths and widths.

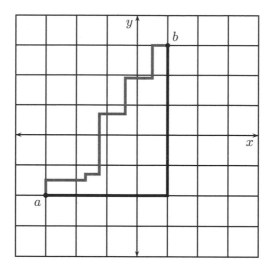

Figure 2.24: Two paths of minimal length from a to b in the taxicab metric.

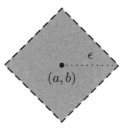

The taxicab metric may be generalized to $\mathbb{R}^n$ in general by

$$d_{\text{taxi}}(x, y) = |x_1 - y_1| + |x_2 - y_2| + \cdots + |x_n - y_n|.$$

In fact, d_{taxi} is one of a whole family of metrics defined in a similar way. Let $p \geq 1$ be a real number. The *p*-**norm** is defined on vectors $x = (x_1, x_2, \ldots, x_n) \in \mathbb{R}^n$ by the formula

$$\|x\|_p = \left(\sum_{k=1}^{n} |x_k|^p \right)^{1/p}.$$

For example,

$$\|x\|_1 = \left(\sum_{k=1}^{n} |x_k|^1 \right)^{1/1} = \sum_{k=1}^{n} |x_k|.$$

The associated *p-norm metric*, d_p, is defined by $d_p(x, y) = \|x - y\|_p$. It is possible to show that d_p is a metric, though it takes careful analysis outside the scope of this text. Note that $d_1 = d_{\text{taxi}}$, and d_2 is the standard Euclidean metric. In the limit as $p \to \infty$, we obtain the **max norm**,

$$\|x\|_\infty = \lim_{p \to \infty} \left(\sum_{k=1}^n |x_k|^p \right)^{1/p} = \max\{|x_k| \mid k \in \{1, 2, \ldots, n\}\},$$

and associated *max norm metric*, $d_\infty(x, y) = \|x - y\|_\infty$.

It turns out that the p-norms ($p \geq 1$) and the max norm on $\mathbb{R}^n$ are all *topologically* equivalent, in the sense that for $p, q \geq 1$, every open d_p-ball contains an open d_q-ball and vice versa. However, differences do arise if we allow the dimension n to increase to infinity.

Metrics on Sequences and Functions

The following examples require some experience in mathematical analysis; feel free to skim or skip. Consider the set of real-number sequences,

$$\mathbb{R}^\infty = \{(x_1, x_2, x_3, \ldots) \mid x_k \in \mathbb{R}\}.$$

The p-norms as defined for finite-dimensional vectors (finite real-number sequences) may not be well defined in $\mathbb{R}^\infty$. Moreover, some sequences have finite p-norms for some p but infinite for others.

Example 37. Let $x_k = 1$ for $k \in \mathbb{N}$; that is, (x_k) is the constant sequence $(1, 1, 1, 1, \ldots)$.

$$\|(x_k)\|_1 = \left(\sum_{k=1}^\infty |1|^1 \right)^{1/1} = 1 + 1 + 1 + 1 + \cdots = \infty$$

In fact, for any finite $p \geq 1$, we have

$$\|(x_k)\|_p = \left(\sum_{k=1}^\infty |1|^p \right)^{1/p} = \left(\lim_{n \to \infty} n \right)^{1/p} = \infty.$$

But the max norm of (x_k) is finite:

$$\|(x_k)\|_\infty = \max\{1, 1, 1, 1, \ldots\} = 1.$$

Let $y_k = \frac{1}{k}$ for $k \in \mathbb{N}$; that is, $(y_k) = (1, \frac{1}{2}, \frac{1}{3}, \ldots)$, which we call the **harmonic sequence**. It is well known that the associated **harmonic series**, $\sum_{k=1}^\infty \frac{1}{k}$, *diverges* to infinity (becomes unbounded); see also Example 40. So

$$\|(x_k)\|_1 = 1 + \frac{1}{2} + \frac{1}{3} + \cdots = \infty.$$

On the other hand, so long as $p > 1$, then the series $\sum_{k=1}^\infty \frac{1}{k^p}$ *converges* to a finite value $\zeta(p)$ (where ζ is called the *Riemann zeta function*), and so

$$\|(x_k)\|_p = \left(\sum_{k=1}^\infty \left(\frac{1}{k} \right)^p \right)^{1/p} = \left(\sum_{k=1}^\infty \frac{1}{k^p} \right)^{1/p} = \zeta(p)^{1/p} < \infty, \qquad \text{if } p > 1.$$

For each $p \geq 1$ and $p = \infty$, define the subset of sequences $\ell^p \subseteq \mathbb{R}^\infty$ by

$$\ell^p = \{(x_k) \in \mathbb{R}^\infty \mid \|(x_k)\|_p < \infty\}.$$

There are certain relationships among the subsets: if $p < q$, then $\ell^p \subseteq \ell^q \subseteq \ell^\infty$. Moreover, each set ℓ^p is a metric space with metric function d_p induced by the p-norm; that is, $d_p((x_k), (y_k)) = \|(x_k) - (y_k)\|_p$.

The family of p-metrics can also be defined on certain classes of functions. Suppose $f : \mathbb{R} \to \mathbb{R}$ is a function, not necessarily continuous, but such that the integral

$$\int_{-\infty}^{\infty} |f(x)|^p \, dx$$

is well defined and finite.[13] Then we say $f \in L^p(\mathbb{R})$, and the p-norm of the function is:

$$\|f\|_p = \left(\int_{-\infty}^{\infty} |f(x)|^p \, dx \right)^{1/p}.$$

Now we may define a distance function,

$$d_p(f, g) = \|f - g\|_p = \left(\int_{-\infty}^{\infty} |f(x) - g(x)|^p \, dx \right)^{1/p}.$$

However, d_p fails to be a metric in general because it fails the Zero-Distance Rule. If two functions f and g differ in value at only a finite number of points,[14] then $d_p(f, g) = 0$. In order to fix this "bug," we say that two functions are equal *almost everywhere* (with respect to the metric d_p) if $d_p(f, g) = 0$.

Metrics in Information Technology

Information is vital to a functioning society, and these days most of that information is digital. According to one study,[15] as of 2013 there were over 4.4 *zetabytes*[16] of data in the digital universe, and that number roughly doubles every two years, reaching an estimated 44 ZB by 2020. Thus analyzing digital data is incredibly important. One way to analyze data is to introduce a notion of distance between strings.

A **string** is a finite sequence of characters, typically bits. A **bit** is a single binary digit, taking on one of the two values, 0 or 1. Thus a string is nothing more than a finite sequence $(a_k)_{k=1}^n$ in the space $\{0, 1\}$ or, equivalently, an element of $\{0, 1\}^n$. When a string is transmitted over the Internet, errors could be introduced. Perhaps one or two bits flip from 0 to 1 because of a random power fluctuation; the error string in that case would would still be very *close*

---

[13]Technically, we require that $|f(x)|^p$ is *Lebesgue integrable* over $\mathbb{R}$.

[14]More generally, if f and g differ on a set of *measure* 0. For more information, see an advanced text on mathematical analysis, such as [Rud87].

[15]ECM Digital Universe Study; see `http://www.emc.com/leadership/digital-universe/`.

[16]One zetabyte (ZB) is equal to $2^{60} \approx 10^{21}$ bytes; each byte is equal to 8 bits.

to the original string. One way to measure this kind of closeness is by the **Hamming distance**.

Define the **XOR (exclusive OR) operation** on two bits a and b by

$$a \oplus b = \begin{cases} 0, & a = b \\ 1, & a \neq b \end{cases}. \tag{2.11}$$

Let $(a_k)_{k=1}^n$ and $(b_k)_{k=1}^n$ be strings. The **Hamming distance** on binary strings is a metric function defined by

$$d_H((a_k), (b_k)) = \sum_{k=1}^n a_k \oplus b_k. \tag{2.12}$$

Note that the sum in (2.12) is to be done with respect to integer addition, not binary digit addition. For example,

$$d_H(01001, 00101) = 0 + 1 + 1 + 0 + 0 = 2.$$

The fact that the Hamming distance is a metric function is not difficult to prove. Axioms 1–3 are obvious. The triangle inequality (axiom 4) follows from a simple argument. Consider three sequences, (a_k), (b_k), and (c_k). **Claim:** For any fixed index k, we have $a_k \oplus b_k \leq a_k \oplus c_k + c_k \oplus b_k$. *Proof of claim:* If $a_k = b_k$, then there is nothing to prove, as $a_k \oplus b_k = 0$ in that case, so let's assume that $a_k \neq b_k$. This implies that $a_k \oplus b_k = 1$. Now since a_k and b_k are different, we know that c_k must be the same as one of them and different from the other. Thus one of $a_k \oplus c_k$ or $c_k \oplus b_k$ is equal to 0 while the other is equal to 1. This proves that $a_k \oplus b_k \leq a_k \oplus c_k + c_k \oplus b_k$. Therefore the sums have the same relationship:

$$
\begin{aligned}
d_H((a_k), (b_k)) &= \sum a_k \oplus b_k \\
&\leq \sum (a_k \oplus c_k + c_k \oplus b_k) \\
&= \sum a_k \oplus c_k + \sum c_k \oplus b_k \\
&= d_H((a_k), (c_k)) + d_H((c_k), (b_k)).
\end{aligned}
$$

Example 38. Consider the metric space $(\{0, 1\}^3, d_H)$, which has eight points:

$$\{0, 1\}^3 = \{000, 001, 010, 011, 100, 101, 110, 111\}.$$

Let's build a model of the metric space. Staring at 000, connect this point to its closest neighbors, and continue until all points are included. The result is shown in the diagram below.

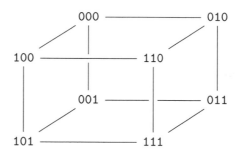

Although the points seem to form the vertices of a unit cube, it's important to realize that the distances between points do not necessarily correspond to Euclidean distances. For example, 110 is two units from 000 (not $\sqrt{2}$), and 111 is three units from 000 (not $\sqrt{3}$). In fact, the distances between vertices coincides with the taxicab metric d_1 in $\mathbb{R}^3$.

An **autocorrect** algorithm is included in the texting and messaging apps on most smartphones. In theory, autocorrect should detect a misspelling and replace it with the word that the user most likely intended. Many autocorrect algorithms work by defining a metric on strings similar to the Hamming distance. When the algorithm sees a string not in its dictionary (i.e., the user has misspelled a word), the algorithm then finds the dictionary word that is *closest* to the misspelled word with respect to its defined metric. This idea is explored further in Exercise 14.

Cauchy Sequences and Completeness

In a metric space, if a sequence converges, then we expect the terms of the sequence to be getting closer to one another. On the other hand, sometimes the terms of a sequence get arbitrarily close to one another and yet do not approach a limiting point in the space. For example, consider the space $\mathbb{Q}$ under the usual Euclidean metric. The sequence $(x_k) = (3, 3.1, 3.14, 3.141, 3.1415, 3.14159, \ldots)$ of decimal approximations to π is a sequence in $\mathbb{Q}$, since any terminating decimal number is rational. Every term past the Nth term is within 10^{-N} of all terms following it. And yet there is no point $x \in \mathbb{Q}$ such that $\lim_{k \to \infty}(x_k) = x$. Of course we know that $\lim_{k \to \infty}(x_k) = \pi \in \mathbb{R}$, but since $\pi \notin \mathbb{Q}$, the sequence (x_k) has no limit (in the space $\mathbb{Q}$). We now define the term *Cauchy* sequence, which captures the idea of terms getting arbitrarily close to one another.

> **Definition 2.5.5.** Suppose $(x_k)_{k \in \mathbb{N}}$ is a sequence in a metric space (X, d). The sequence is called **Cauchy** if, for every $\epsilon \in \mathbb{R}^+$, there is a number $N \in \mathbb{N}$ such that $d(x_k, x_\ell) < \epsilon$ for all $k, \ell \geq N$.

Example 39. The sequence (g_k), whose terms are $g_k = \sum_{m=1}^{k} \frac{1}{2^m}$, that is, the partial sums of the geometric series $\sum_{m=1}^{\infty} \frac{1}{2^m}$, is a Cauchy sequence in $\mathbb{R}$ (or in $\mathbb{Q}$). To verify this, suppose $\epsilon \in \mathbb{R}^+$ is given. Let $N = \log_2(1/\epsilon)$, and suppose $k, \ell \geq N$. Without loss of generality, assume $k \geq \ell$.

$$d(g_k, g_\ell) = \left| \sum_{m=1}^{k} \frac{1}{2^m} - \sum_{m=1}^{\ell} \frac{1}{2^m} \right| = \sum_{m=\ell+1}^{k} \frac{1}{2^m} \qquad (2.13)$$

By factoring out $1/2^{\ell+1}$, reindexing the sum, and by using the well-known geometric sum formula (Exercise 4 in §A.1), we obtain:

$$\sum_{m=\ell+1}^{k} \frac{1}{2^m} = \frac{1}{2^{\ell+1}} \sum_{m=0}^{k-\ell-1} \left(\frac{1}{2} \right)^m = \frac{1}{2^{\ell+1}} \cdot \frac{1 - \left(\frac{1}{2} \right)^{k-\ell}}{1 - \frac{1}{2}}. \qquad (2.14)$$

Then, algebraic simplification and the fact that $1 - y < 1$ when $y > 0$ yields:

$$\frac{1}{2^{\ell+1}} \cdot \frac{1 - \left(\frac{1}{2} \right)^{k-\ell}}{1 - \frac{1}{2}} = \frac{1}{2^\ell} \left(1 - \left(\frac{1}{2} \right)^{k-\ell} \right) < \frac{1}{2^\ell}. \qquad (2.15)$$

Now since $\ell \geq N = \log_2(1/\epsilon)$, we have $2^\ell \geq 1/\epsilon$, or $1/2^\ell \leq \epsilon$. Together with (2.13)–(2.15), this proves that $d(g_k, g_\ell) < \epsilon$, as required.

The Cauchy property is stronger than simply requiring that adjacent terms get arbitrarily close to each other. A sequence (x_k) may fail to be Cauchy even though $d(x_k, x_{k+1}) \to 0$ as $k \to \infty$.

Example 40. The sequence (h_k) whose terms are $h_k = \sum_{m=1}^{k} \frac{1}{m}$, that is, the sequence of partial sums of the harmonic series, is not Cauchy. Suppose $k \geq \ell$ are natural numbers, and mimic the steps shown in (2.13)

$$d(h_k, h_\ell) = \left| \sum_{m=1}^{k} \frac{1}{m} - \sum_{m=1}^{\ell} \frac{1}{m} \right| = \sum_{m=\ell+1}^{k} \frac{1}{m} \qquad (2.16)$$

One way to proceed is by a variant of the *integral test*.

$$\sum_{m=\ell+1}^{k} \frac{1}{m} > \int_{\ell+1}^{k+1} \frac{dx}{x} = \ln(k+1) - \ln(\ell+1) = \ln\left(\frac{k+1}{\ell+1} \right) \qquad (2.17)$$

The last expression of (2.17) is unbounded as k ranges throughout $\mathbb{N}$; therefore (h_k) is not Cauchy. However, note that the distance between *consecutive* terms does tend toward zero.

$$d(h_k, h_{k+1}) = \left| \sum_{m=1}^{k} \frac{1}{m} - \sum_{m=1}^{k+1} \frac{1}{m} \right| = \frac{1}{k+1} \to 0, \text{ as } k \to \infty$$

Cauchy sequences are exactly the ones that converge in $\mathbb{R}^n$. However, some spaces such as $\mathbb{Q}$ have points "missing," which means that some Cauchy sequences may fail to converge. The following definition states exactly what we mean to say that a metric space has "no points missing."

> **Definition 2.5.6.** A metric space X is called **complete** if every Cauchy sequence has a limit in X.

Example 41.   Any closed ball $D_\epsilon(x)$ in $\mathbb{R}^n$ is complete owing to compactness, while an open ball $B_\epsilon(x)$ is not.

Example 42. $\mathbb{Q}$ is not complete, but $\mathbb{R}$ is. The completeness of $\mathbb{R}$ follows from the existence of a *supremum* and *infimum* for every bounded set (see, e.g., Bartle and Sherbert [BS11]). In fact, $\mathbb{R}^n$ is complete for any $n \in \mathbb{N}$. For trivial reasons, $\mathbb{Z}$ is complete (every Cauchy sequence in $\mathbb{Z}$ is eventually constant).

Example 43.   It can be shown that ℓ^p is complete for every $p \geq 1$ and $p = \infty$. Since ℓ^p is also a *normed vector space*, we say that ℓ^p is a **Banach space**. The function spaces $L^p(\mathbb{R})$ are also Banach spaces, as are the the Euclidean spaces $\mathbb{R}^n$.

Exercises

1. Suppose (X, d) is a metric space, and $Y \subseteq X$. Show that (Y, d') is also a metric space, where d' is the function d restricted to $Y \times Y$.

2. Prove all parts of Proposition 2.5.3.

3. Suppose (X, d_X) and (Y, d_Y) are metric spaces. An **isometry** is a function $f : X \to Y$ that preserves the metrics, in the sense that

$$d_Y(f(x_1), f(x_2)) = d_X(x_1, x_2), \quad \forall x_1, x_2 \in X.$$

   (a) Prove that every isometry is continuous.

   (b) Show that translation and rotation in $\mathbb{R}^2$ are isometries with respect to the Euclidean metric (recall §2.2, *Standard Transformations*).

   (c) Is scaling $S_k(x) = kx$ in $\mathbb{R}^2$ an isometry with respect to the Euclidean metric? Note that your answer may depend on k. How does your answer change if $\mathbb{R}^2$ is given the discrete metric?

4. Let X be any set. Suppose d is the discrete metric on X, as defined in Example 35.

   (a) Show that every subset of (X, d) is both open and closed.

   (b) Let $x \in X$ and $\epsilon > 1$. Describe $B_\epsilon(x)$.

5. Prove that the taxicab metric d_{taxi} is indeed a metric on $\mathbb{R}^2$.

6. Show that all p-norm metrics d_p coincide in $\mathbb{R}^1$.

7. Draw the max norm metric ball $B_{d_\infty, 3}(-1, 2) \subseteq \mathbb{R}^2$.

8. Show that d_1 and d_2 are *equivalent* norms on $\mathbb{R}^n$ by completing both of the following steps. (*Hint for both parts:* Draw pictures and find a geometric relationship.)

   (a) For any $x \in \mathbb{R}^n$ and $\epsilon \in \mathbb{R}^+$, show that there is a $\delta > 0$ such that $B_{d_1, \delta}(x) \subseteq B_{d_2, \epsilon}(x)$.

   (b) For any $x \in \mathbb{R}^n$ and $\epsilon \in \mathbb{R}^+$, show that there is a $\delta > 0$ such that $B_{d_2, \delta}(x) \subseteq B_{d_1, \epsilon}(x)$.

9. Using a computer graphing utility, plot the unit metric circles in $\mathbb{R}^2$ centered at the origin $|x|^p + |y|^p = 1$ for various values of p. What happens as p gets larger?

10. Prove that d_1 and d_∞ are equivalent in $\mathbb{R}^n$.

11. Consider the function $d_{\frac{1}{2}} : \mathbb{R}^2 \times \mathbb{R}^2 \to \mathbb{R}$ defined by

$$d_{\frac{1}{2}}((x_1, x_2), (y_1, y_2)) = \left(\sqrt{|x_1 - y_1|} + \sqrt{|x_2 - y_2|} \right)^2.$$

   (a) Draw the set $\{(x, y) \in \mathbb{R}^2 \mid d_{\frac{1}{2}}((x, y), (0, 0)) < 1\}$.

   (b) Show that $d_{\frac{1}{2}}$ is not a metric by finding three points x, y, and z that fail to satisfy the triangle inequality. (*Hint:* The set in part (a) is nonconvex. What direction seems shorter than it *should* be?)

12. Find a sequence $(x_k) \in \mathbb{R}^\infty$ such that $\|(x_k)\|_p = \infty$ for every $p \geq 1$.

13. Let $(\{0, 1\}^4, d_H)$ be the metric space of binary strings of length 4 with the Hamming distance. Determine all of the strings in the metric ball $B_3(1001)$ in this space.

14. Suppose $(a_k)_{k=1}^m$ and $(b_k)_{k=1}^n$ are strings of alphabetic letters. Define a metric function by

$$d((a_k), (b_k)) = \sum_{k=1}^{\min\{m,n\}} \frac{a_k \oplus b_k}{2^k} + \sum_{k=\min\{m,n\}+1}^{\max\{m,n\}} \frac{1}{2^k},$$

where $a_k \oplus b_k$ is the XOR operation (2.11).

(a) Compute the distance between each pair of words in the following list: {CAUCHY, COMPACT, CONNECTED, CONTINUITY, CONTINUOUS}.

(b) List at least five words that are within a distance of 1/8 from WORD.

(c) Find the distance between TEH and THE, and the distance between TEH and TEN. Why would this metric not be the most ideal tool for an autocorrect algorithm?

15. Show that the sequence $(x_k)_{k \in \mathbb{N}}$ defined by $x_k = \frac{1}{k^2}$ is a Cauchy sequence in $\mathbb{R}$. (*Hint:* Use the fact that $\frac{1}{k^2} < \frac{1}{k(k-1)}$ for $k > 1$, and write $\sum \frac{1}{k(k-1)}$ as a *telescoping* sum.)

Supplemental Reading

Metric topology is a standard topic in many topology texts. The list below is by no means exhaustive.

- Bartle and Sherbert [BS11] for background in mathematical analysis.

- Goodman [Goo05], Chapter 1.

- Mendelson [Men90], Chapter 2.

- Munkres [Mun00], Chapter 3 for general Euclidean topology, and §§20–21 for metric spaces in general

- Rudin [Rud76] for further background in mathematical analysis and metric spaces.

- Wall [Wal72], Chapters 1 and 3.

Chapter 3

Vector Fields in the Plane

Imagine being in a little raft out in the middle of the ocean. There is no way to power the raft, so it simply follows the current. The path the raft follows is completely determined by the direction and magnitude of the current at each point (let's assume the currents themselves never change with time). In fact, if we had precise measurements of the direction and magnitude of the flow at every point in the ocean, then we could predict exactly where the raft would end up. These measurements would correspond to a *vector field*, and the particular path the raft took would be a *trajectory* through the field.

This leads to what mathematicians call *dynamical systems* and *differential equations*. However, our purpose is not to solve differential equations explicitly or to provide detailed numerical analysis of dynamical systems. Instead, we find in this chapter how topology may help us describe some aspects of a system. We find where abstraction helps to determine *qualitative* information; for example, where does a trajectory eventually end up if it begins in a certain region? We rely on the language of continuous functions. We make use of the concept of *closeness* from earlier chapters, as trajectories that begin close enough to one another seem to behave about the same.[1] Topology even allows us to ponder questions about all vector fields at once. For example, we can be certain that in *any* vector field on the surface of a sphere, there is at least one point at which nothing moves.

3.1 Trajectories and Phase Portraits

The main objects of study in this chapter are *vector fields* in Euclidean space and, more specifically, those in $\mathbb{R}^1$ and $\mathbb{R}^2$, since these spaces are "small" enough so that the vector fields can easily be visualized. More general situations require extra care. In this chapter we generally follow the convention that boldfaced letters (such as $\mathbf{x}$ and $\mathbf{y}$) refer to points or vectors in $\mathbb{R}^n$.

---

[1] An important exception is when the system experiences *chaos*, in which the slightest changes to an initial condition may result in wildly different outcomes within a short period of time.

Vector Fields and Differential Equations

For now, we content ourselves to work within the familiar setting of Euclidean space $\mathbb{R}^n$. Vector fields may be defined in more general spaces called *differentiable manifolds*, but the mathematical machinery required falls outside the scope of this elementary text.

> **Definition 3.1.1.** A **vector field** on $\mathbb{R}^n$ is a function $\mathbf{V} : \mathbb{R}^n \to \mathbb{R}^n$.

In other words, to each *point (vector)* $\mathbf{x} = (x_1, x_2, \ldots, x_n) \in \mathbb{R}^n$, there is assigned a *vector* $\mathbf{V}(\mathbf{x}) = (v_1, v_2, \ldots, v_n) \in \mathbb{R}^n$. When $n = 2$, for example, we might use either of the following notations,

$$\mathbf{V}(x, y) = (v_1(x, y), v_2(x, y)) = \begin{pmatrix} v_1(x, y) \\ v_2(x, y) \end{pmatrix} = v_1(x, y)\mathbf{i} + v_2(x, y)\mathbf{j},$$

where $\mathbf{i}$ and $\mathbf{j}$ are unit vectors in the direction of the positive x- and y-axis, respectively (recall (1.4) in §1.2). Now in order to apply *topological* reasoning in a vector field, it is necessary to assume the function $\mathbf{V}$ is at least *continuous*, so that all vectors in a small neighborhood of $\mathbf{V}$ have nearly the same length and direction. At each point $\mathbf{x} \in \mathbb{R}^n$, consider the vector $\mathbf{V}(\mathbf{x})$ as indicating the strength and direction of the **flow** through $\mathbf{x}$. In other words, $\mathbf{V}$ is the *velocity field* of the flow. This naturally introduces the dimension of *time* into the picture. Then each component $v_i(\mathbf{x})$ of the vector $\mathbf{V}(\mathbf{x})$ represents the rate of change of x_i with respect to time t at the particular point $\mathbf{x}$. In fact, $\mathbf{V}$ defines a system of n functions in n variables, which we will interpret as a system of **differential equations**.[2]

$$\mathbf{V} : \begin{cases} dx_1/dt & = v_1(x_1, x_2, \ldots, x_n) \\ dx_2/dt & = v_2(x_1, x_2, \ldots, x_n) \\ \quad \vdots \\ dx_n/dt & = v_n(x_1, x_2, \ldots, x_n) \end{cases} \tag{3.1}$$

A solution to the system (3.1) is commonly called a **trajectory** or **orbit**.[3] In this text we often denote a trajectory through $\mathbf{x}$ by $\phi(\mathbf{x}, t)$, where t (representing *time*) varies in $\mathbb{R}$. The trajectory functions $\phi(\mathbf{x}, t)$ for various $\mathbf{x}$ must satisfy certain conditions.

- *Initial condition:* $\phi(\mathbf{x}, 0) = \mathbf{x}$, for any point $\mathbf{x} \in \mathbb{R}^n$.

- *Consistency condition:* $\phi(\phi(\mathbf{x}, t_1), t_2) = \phi(\mathbf{x}, t_1 + t_2)$ for all $\mathbf{x} \in \mathbb{R}^n$ and $t_1, t_2 \in \mathbb{R}$.

---

[2] The field of differential equations is vast. The treatment in this book is necessarily quite incomplete. For instance, here we only work with so-called *autonomous* systems – those that do not explicitly involve the time variable t.

[3] Other common names include **flow line**, **streamline**, and **integral curve**.

The initial condition is straightforward enough, but the purpose of the consistency condition may not be readily apparent. Basically this condition ensures that the solutions found at two different times, t_1 and $t_1 + t_2$, along the same trajectory are related by a time shift in exactly the amount of the difference, t_2. This will be used later in the proof of the Poincaré-Bendixson Theorem (Theorem 3.3.1). For simplicity, we also assume the following, which is guaranteed in the case that each function v_k in (3.1) has continuous partial derivatives with respect to each of the variables x_j:

- *Existence and Uniqueness.* Given any point $\mathbf{x} \in \mathbb{R}^n$, there exists a unique trajectory $\phi(\mathbf{x}, t)$ solving (3.1) for t in some open interval containing $t = 0$.

Uniqueness of solutions implies that if two such curves intersect, then they represent the same trajectory. The points at which the flow is $\mathbf{0}$ are particularly important to analyzing the system.

> **Definition 3.1.2.** Any point $\mathbf{x} \in \mathbb{R}^n$ such that $\mathbf{V}(\mathbf{x}) = \mathbf{0}$ is called a **critical point**. Critical points are also known as **singular points**, **stationary points**, **equilibria**, or **nodes**.

If a dimensionless particle is plopped into the field exactly at a critical point, then it will not move over time. However, as we shall see, flows *near* a critical point could behave in many different ways. Note that, in this text, all critical points will be **isolated**, meaning that there is a neighborhood of the point in which there is no other critical point.

Example 44. Describe the vector field defined by:

$$\begin{cases} dx/dt & = x + y \\ dy/dt & = y - x \end{cases}.$$

Solution: Plot a few sample vectors. Figure 3.1 shows a few sample vectors in the field. Although only finitely many vectors are shown, *each and every point* of the plane actually has a vector associated to it. We have also scaled down the length of each vector so that the picture is not too cluttered. There is a single critical point $(0, 0)$, found by solving the system

$$\begin{cases} 0 = x + y \\ 0 = y - x \end{cases}.$$

Now imagine your raft is placed somewhere in this vector field and it moves according to the direction and magnitude of the vectors. The raft would spiral outward in a clockwise direction (*try tracing the path with your finger*). Each such path is a trajectory in the field. A more accurate representation of the field along with a few representative trajectories (as shown in Figure 3.2) can

$y \setminus x$	-2	-1	0	1	2
2	$\begin{pmatrix}0\\4\end{pmatrix}$	$\begin{pmatrix}1\\3\end{pmatrix}$	$\begin{pmatrix}2\\2\end{pmatrix}$	$\begin{pmatrix}3\\1\end{pmatrix}$	$\begin{pmatrix}4\\0\end{pmatrix}$
1	$\begin{pmatrix}-1\\3\end{pmatrix}$	$\begin{pmatrix}0\\2\end{pmatrix}$	$\begin{pmatrix}1\\1\end{pmatrix}$	$\begin{pmatrix}2\\0\end{pmatrix}$	$\begin{pmatrix}3\\-1\end{pmatrix}$
0	$\begin{pmatrix}-2\\2\end{pmatrix}$	$\begin{pmatrix}-1\\1\end{pmatrix}$	$\begin{pmatrix}0\\0\end{pmatrix}$	$\begin{pmatrix}1\\-1\end{pmatrix}$	$\begin{pmatrix}2\\-2\end{pmatrix}$
-1	$\begin{pmatrix}-3\\1\end{pmatrix}$	$\begin{pmatrix}-2\\0\end{pmatrix}$	$\begin{pmatrix}-1\\-1\end{pmatrix}$	$\begin{pmatrix}0\\-2\end{pmatrix}$	$\begin{pmatrix}1\\-3\end{pmatrix}$
-2	$\begin{pmatrix}-4\\0\end{pmatrix}$	$\begin{pmatrix}-3\\-1\end{pmatrix}$	$\begin{pmatrix}-2\\-2\end{pmatrix}$	$\begin{pmatrix}-1\\-3\end{pmatrix}$	$\begin{pmatrix}0\\-4\end{pmatrix}$

Figure 3.1: *Left*, table of vectors corresponding to $\mathbf{V}(x,y) = (x+y, y-x)$. *Right*, vectors of $\mathbf{V}$ plotted in $\mathbb{R}^2$.

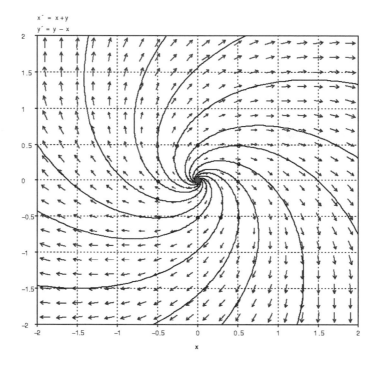

Figure 3.2: Vector field $\mathbf{V} = (x+y, y-x)$. The arrows have been rescaled to a uniform length in this picture; however, it is important to realize that the magnitudes are not constant. Each solid spiral curve is a distinct trajectory in the system. Forward is clockwise; backward is counterclockwise.

be found using Sagemath [Ste], `pplane` [Pol], or other software designed for this purpose.

It is usually quite difficult or even impossible to find a closed-form expression for a trajectory $\phi(\mathbf{x}, t)$ in a given vector field $\mathbf{V}$, so we often estimate the trajectories by sketching them on the vector field or by using numerical analysis techniques for more accurate results. For our purposes, we are more interested in qualitative information about the trajectories, such as how $\phi(\mathbf{x}, t)$ behaves in the future $(t > 0)$ and the past $(t < 0)$, and what happens in the limit when $t \to \infty$ or $t \to -\infty$.

Definition 3.1.3. Let $\phi(\mathbf{x}, t)$ be the trajectory through $\mathbf{x}$ relative to a vector field $\mathbf{V}$.

- The **positive semiorbit** of $\mathbf{x}$ is the set

$$\mathcal{O}^+(\mathbf{x}) = \{\phi(\mathbf{x}, t) \mid t > 0\}.$$

- The **negative semiorbit** of $\mathbf{x}$ is the set

$$\mathcal{O}^-(\mathbf{x}) = \{\phi(\mathbf{x}, t) \mid t < 0\}.$$

Example 45. Let $\mathbf{x} \in \mathbb{R}^2 \setminus \{\mathbf{0}\}$. Describe the positive and negative semiorbits of $\mathbf{x}$ with respect to the vector field of Example 44.

Solution: So long as $\mathbf{x} \neq \mathbf{0}$, the positive semiorbit $\mathcal{O}^+(\mathbf{x})$ is a spiral emanating from $\mathbf{x}$ moving away from the origin forever. The negative semiorbit $\mathcal{O}^-(\mathbf{x})$ is an arc of a spiral between $\mathbf{x}$ and the origin, but the origin is never actually reached.

Phase Portraits

Qualitative information about a vector field can be shown on a *phase portrait*. Figure 3.2 shows a typical phase portrait for a two-dimensional system.

Definition 3.1.4. A **phase portrait** for a vector field is a sketch showing some representative trajectories.

Let us first consider a one-dimensional system, which is defined by a single differential equation, $dx/dt = V(x)$. The critical points are the root(s) of $V(x)$, and the flow can only be in one of two directions: *forward* or *backward* along the x-axis. A phase portrait in this situation simply shows the critical point and the direction of flow between the critical points.

Example 46. Draw phase portraits for each of the following.

(a) $dx/dt = 3x - x^2$ (b) $dx/dt = (x - 1)^2$ (c) $dx/dt = \cos \pi x$

Solution:

(a) Here $0 = 3x - x^2 = x(3 - x) \implies x = 0, 3$ are critical points. The flow is forward (to the right) when $3x - x^2 > 0$, which occurs for $0 < x < 3$. The flow is negative (to the left) when $x < 0$ or $x > 3$.

(b) Here $0 = (x-1)^2 \implies x = 1$ is the only critical point. The flow is forward everywhere else since $(x - 1)^2 > 0$ for $x \neq 1$. However, it is important to realize that the flow does not cross the point $x = 1$. If $a < 1$, then $\mathcal{O}^+(a) = (a, 1)$, with the trajectory never actually reaching $x = 1$ in finite time.

(c) Here $0 = \cos \pi x \implies x = \dots, -\frac{3}{2}, -\frac{1}{2}, \frac{1}{2}, \frac{3}{2}, \dots$ There are infinitely many critical points. The flow alternates between forward and backward depending on the sign of $\cos \pi x$ in each interval.

There are essentially three types of critical point in a one-dimensional system: *sinks*, *sources*, and *semistable nodes*. Sinks and sources are defined in more generality for all systems.

> **Definition 3.1.5.** Suppose $\mathbf{x}_0$ is a critical point. If there is a neighborhood, $U \ni \mathbf{x}_0$ such that for every $\mathbf{x} \in U$,
>
> - $\lim_{t \to \infty} \phi(\mathbf{x}, t) = \mathbf{x}_0$, then $\mathbf{x}_0$ is called a **sink**.
>
> - $\lim_{t \to -\infty} \phi(\mathbf{x}, t) = \mathbf{x}_0$, then $\mathbf{x}_0$ is called a **source**.

An intuitive way to think about these definitions is that the sink *attracts* all nearby orbits while a *source* repels them. In the phase portrait, trajectories seem to enter sinks and exit sources, though we must realize that the sink is never actually reached in finite time, and the source is not the actual starting point of any trajectory but is the limit point as we follow the trajectory in backward time.

Let's get back to the one-dimensional phase portraits. The sign of V on either side of a critical point determines whether it is a sink, source, or neither. Suppose $(c - \delta, c + \delta)$ is a neighborhood of a critical point c containing no other critical point.

- If $v(x) > 0$ on $(c - \delta, c)$ and $v(x) < 0$ on $(c, c + \delta)$, then c is a sink.

- If $v(x) < 0$ on $(c - \delta, c)$ and $v(x) > 0$ on $(c, c + \delta)$, then c is a source.

- If $v(x)$ has the same sign on both $(c - \delta, c)$ and $(c, c + \delta)$, then c is called a **semistable node**.

Flows in a neighborhood of a sink always flow toward it over time, so we say that a sink is **stable**. The points $x = 3$ in Example 46(a) and $x = \ldots, -7/2, -3/2, 1/2, 5/2, \ldots$ in Example 46(c) are stable critical points. By contrast, flows in a neighborhood of a source always flow away from the source, so we say that a source is **unstable**. The points $x = 0$ in Example 46(a) and $x = \ldots, -5/2, -1/2, 3/2, 7/2, \ldots$ in Example 46(c) are unstable. On one side of a semistable node, flows tend toward the node, while on the other side they flow away from the node, and so overall we say that the semistable node is unstable.

Critical Points in Two-Dimensional Systems

It is clear that the critical point at the origin in Figure 3.2 is a source (hence unstable) since all nearby trajectories move away from it. There are many other types of critical point in systems of two or more variables, but for our purposes we will stick to a simple classification.

Definition 3.1.6.

- A critical point x_0 is called **stable** if there is a neighborhood $U \ni x_0$ such that for every point $x \in U$, the positive semiorbit $\mathcal{O}^+(x)$ is a subset of U. Otherwise, x_0 is called **unstable**.

- A critical point x_0 is called **rotational** if every neighborhood of x contains a periodic orbit, and **nonrotational** otherwise.

A stable critical point may be rotational or nonrotational. According to Definition 3.1.5, a sink is the same thing as a nonrotational stable critical point. A source is an example of an unstable critical point, though the unstable critical points come in many others flavors too, as we shall see in later sections. Figure 3.3 shows a sampling of critical points, though there are many more possibilities than those shown here.

Although trajectories never actually reach critical points, those that limit onto critical points (as $t \to \pm\infty$) are of special interest because they can serve as the framework for building an accurate phase portrait.

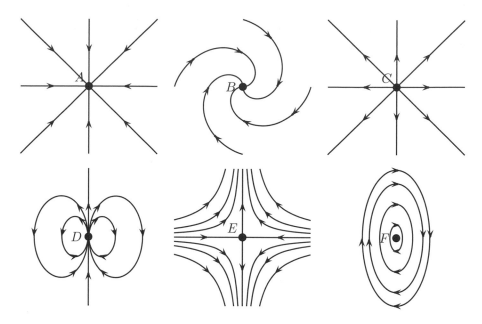

Figure 3.3: A and B are sinks; C is a source; D is called a **dipole**, and E is called a **saddle point**, both of which are unstable. F is a stable rotational critical point, called an **orbital center**.

Definition 3.1.7. Suppose $\phi(\mathbf{x}, t)$ is a trajectory.

- If there is a critical point $\mathbf{x}_0$ such that

$$\lim_{t \to \infty} \phi(\mathbf{x}, t) = \lim_{t \to -\infty} \phi(\mathbf{x}, t) = \mathbf{x}_0,$$

  then ϕ is called a **homoclinic** orbit.

- If there are two distinct critical points $\mathbf{x}_0, \mathbf{x}_1$ such that

$$\lim_{t \to \infty} \phi(\mathbf{x}, t) = \mathbf{x}_1, \qquad \lim_{t \to -\infty} \phi(\mathbf{x}, t) = \mathbf{x}_0,$$

  then ϕ is called a **heteroclinic** orbit.

Example 47. The first picture in Figure 3.4 shows the critical points $\mathbf{x}_1$, $\mathbf{x}_2$, $\mathbf{x}_3$, and a few representative orbits of a vector field. According to the picture, $\mathbf{x}_1$ and $\mathbf{x}_3$ are unstable and nonrotational (but not sources), while $\mathbf{x}_2$ is stable and rotational. A and C are heteroclinic orbits (when occurring as pair like this, it is often called a **heteroclinic cycle**). D is a homoclinic orbit.

B is neither heteroclinic nor homoclinic. Note that in the fleshed-out phase portrait (Figure 3.4, *bottom*), every trajectory drawn within the curve D is also a homoclinic orbit. This is because no other critical points exist, so there cannot be a closed trajectory lying entirely within the region.[4]

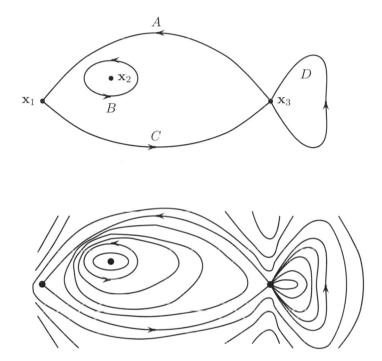

Figure 3.4: *Top*, beginnings of a phase portrait for Example 47. *Bottom*, possible phase portrait for the system.

Exercises

1. Suppose a vector field has a critical point at $\mathbf{x}_0$. What will the trajectory $\phi(\mathbf{x}_0, t)$ look like?

2. Consider the one-dimensional system $V(x) = (x + 1)(x - 1)^2$.

   (a) List all critical points for V, classifying each as a sink, source, or semistable node.

   (b) Determine the following semiorbits in V: $\mathcal{O}^-(-2)$, $\mathcal{O}^+(-2)$, $\mathcal{O}^-(0)$, $\mathcal{O}^+(0)$, $\mathcal{O}^-(2)$, $\mathcal{O}^+(2)$.

---

[4]This is a consquence of Hopf's Theorem (Theorem 3.2.6).

3. Carefully plot a few sample vectors in the field $\mathbf{V}(x,y) = (xy/4, 1)$. Are there any critical points in this system? Describe the positive and negative semiorbit of each of the following points: $(-1,0)$, $(0,0)$, and $(1,0)$.

4. Plot a phase portrait for each system below, either by plotting representative vectors by hand and estimating the flow lines or by using software.

(a) $\begin{cases} dx/dt &= x \\ dy/dt &= y \end{cases}$
$\qquad$
(d) $\begin{cases} dx/dt &= x(y-1) \\ dy/dt &= y(x-2) \end{cases}$

(b) $\begin{cases} dx/dt &= x+y \\ dy/dt &= x-y \end{cases}$
$\qquad$
(e) $\begin{cases} dx/dt &= y \\ dy/dt &= -\sin x \end{cases}$

(c) $\begin{cases} dx/dt &= y \\ dy/dt &= -x \end{cases}$
$\qquad$
(f) $\begin{cases} dx/dt &= xy \\ dy/dt &= y^2 - 1 \end{cases}$

5. For each system from Exercise 4, list all critical points and classify each as stable or unstable, rotational or nonrotational.

6. While in general it is difficult or impossible to solve a system explicitly, it is fairly easy to check whether a given parametrized curve is a solution, simply by taking derivatives. Verify that $(x,y) = (a \sin t, a \cos t)$, where $a \in \mathbb{R}$ is a constant, is a solution to $(dx/dt, dy/dt) = (y, -x)$.

7. Consider the initial value problem $dx/dt = \sqrt{x}$, $x(0) = 0$. Verify that $x = t^2/4$ is a solution for $t \geq 0$. Then verify that for any fixed $t_0 > 0$, the following function is also a solution:[5]

$$x = \begin{cases} 0, & 0 \leq t < t_0, \\ \frac{(t-t_0)^2}{4}, & t \geq t_0. \end{cases}$$

3.2 Index of a Critical Point

A vector field $\mathbf{V}$ as in (3.1) is often called a **dynamical system** because it may describe the motion of a dimensionless particle under the influence of the vectors in the system. While it may be impossible to work out the precise trajectory a particle follows, it is often quite easy to describe its path *qualitatively*. We expect the particle to follow the flow lines of the phase portrait. For example, if we can identify a sink in a system, then we know that a particle nearby will approach the sink over time. In fact, the general behavior of a two-dimensional system may be inferred from the kinds of critical points it has. First we introduce a system of classifying critical points by their *index*.

---

[5]This exercise shows that the differential equation $dx/dt = \sqrt{x}$ fails the *uniqueness* condition.

Index

The behavior of a two-dimensional vector field is largely determined by how flow vectors *turn* near critical points. Consider a small loop in a vector field and track how the flow changes as the loop is traversed once in the counterclockwise sense. **Note:** By *loop* we mean a simple closed curve that is generally *not* a trajectory.

Example 48. Consider the vector field $\mathbf{V} = (-x, -y)$ shown in Figure 3.5. There are two loops drawn in the field: loop A encircling the critical point $(0,0)$, and loop B containing no critical point at all. Note how the direction of flow vectors change as each loop is traversed. As loop A is traversed counterclockwise, the flow vectors on A make a complete counterclockwise turn; while on loop B, the flow vectors vary direction somewhat but do not turn all the way around.

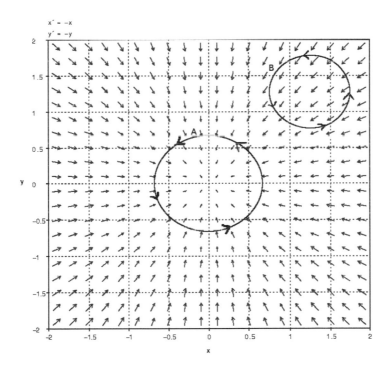

Figure 3.5: Two loops in a field.

Definition 3.2.1. If C is any simple closed curve in a vector field $\mathbf{V}$, then the **index** or **winding number** of C, denoted $\text{Ind}(C)$, is the total number of counterclockwise rotations made by the flow vectors on C as C is traversed once counterclockwise.

The index of a curve can be defined more precisely using integrals, as Exercise 3 indicates. For our purposes, the intuitive definition given above suffices.

Example 49. In Figure 3.5, $\text{Ind}(A) = 1$, and $\text{Ind}(B) = 0$.

Example 50. Suppose $X \subseteq \mathbb{R}^2$ is a subset on which a vector field $\mathbf{V}$ is constant, and let C be any simple closed curve lying within X. Then, since there is zero variation in the angle of vectors throughout X, we have $\text{Ind}(C) = 0$.

Taking Example 50 a step further, suppose a vector field $\mathbf{V}$ on $X \subseteq \mathbb{R}^2$ is not constant but varies only a small amount. Under the right conditions, it follows that $\text{Ind}(C) = 0$ for any simple closed curve lying within X. Suppose that $\mathbf{V}(\mathbf{x}) \neq 0$ for all $\mathbf{x} \in X$. Then for any $\mathbf{x}, \mathbf{y} \in X$, one can define the absolute difference in angle between the vectors $\mathbf{V}(\mathbf{x})$ and $\mathbf{V}(\mathbf{y})$ in the usual way,

$$|\Delta\theta(\mathbf{x}, \mathbf{y})| = \cos^{-1} \frac{\mathbf{V}(\mathbf{x}) \cdot \mathbf{V}(\mathbf{y})}{\|\mathbf{V}(\mathbf{x})\|\|\mathbf{V}(\mathbf{y})\|}.$$

Now if there is a number $\epsilon > 0$ such that $|\Delta\theta(\mathbf{x}, \mathbf{y})| \leq \epsilon < \pi$ for all $\mathbf{x}, \mathbf{y} \in X$, no pair of vectors can be pointing in opposite directions; hence vectors can never turn completely around along any curve within X. Thus $\text{Ind}(C) = 0$ for any simple closed curve within X under such conditions.

Now suppose two curves S_1 and S_2 are contained in this region X (on which $|\Delta\theta| \leq \epsilon < \pi$), both starting at a point $\mathbf{x}$ and ending at a point $\mathbf{y}$, as in Figure 3.6. Let $\Delta\theta_1$ be the total angle swept out in the counterclockwise sense by the flow vectors on S_1 starting at $\mathbf{x}$ and ending at $\mathbf{y}$. Define $\Delta\theta_2$ analogously with respect to S_2. Defining $C = S_1 \cup (-S_2)$, which is shorthand for the curve traversing S_1 forward followed by S_2 backward, and since $\text{Ind}(C) = 0$, it must be the case that $\Delta\theta_1 = \Delta\theta_2$. This implies that if C_1 and C_2 are two loops in the vector field $\mathbf{V}$ (defined on a larger region of $\mathbb{R}^2$) that are exactly the same, except that C_1 contains S_1 and C_2 contains S_2, then it must be the case that $\text{Ind}(C_1) = \text{Ind}(C_2)$.

This observation allows us to prove the following important theorem.

Theorem 3.2.2. *If C_1 and C_2 are two simple closed curves such that one curve lies within the other, and if there are no critical points of $\mathbf{V}$ on either curve or in the region between the two curves, then $\text{Ind}(C_1) = \text{Ind}(C_2)$.*

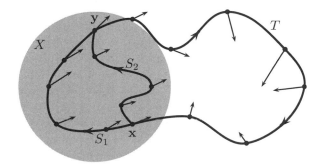

Figure 3.6: Two curves $C_1 = S_1 \cup T$ and $C_2 = S_2 \cup T$ that differ only in a region X. So long as $\mathbf{V} \neq 0$ and the variation of angles of the vector field is small in X, the two curves have the same index.

Proof. Suppose C_2 lies within the curve C_1. Since C_1 is a simple closed curve, its inside is bounded (by the Jordan Curve Theorem 2.4.8). Let A be region between the two curves, including the curves themselves. Since A is closed and bounded, A is compact. Now since $\mathbf{V} \neq 0$ on A (by hypothesis), the angle of each vector (with respect to the positive x-axis, for example) is well defined throughout A. Moreover, since $\mathbf{V}$ is assumed continuous, so is the function giving the angle of $\mathbf{V}(\mathbf{x})$ at each $\mathbf{x} \in A$. Thus, for each point $\mathbf{x} \in A$, there is a small open disk neighborhood $U_{\mathbf{x}} = B_\epsilon(\mathbf{x})$ on which the angle of vectors can be made to vary by as small a value as we please. In other words, any deformation of a curve within a single $U_{\mathbf{x}}$ does not change the index of the curve.

The collection of sets $\{U_{\mathbf{x}} \mid \mathbf{x} \in A\}$ covers A, and since A is compact, there is a finite subcover, say, $\mathscr{U} = \{U_1, U_2, \ldots, U_n\}$. We will deform C_1 to C_2 in finitely many steps. First, if there is any set $U_k \in \mathscr{U}$ such that a segment S of C_1 passes through U_k but no segment of C_2 passes through it, then deform S to S', where S' lies in the intersection of U_k with other open sets. If there are any other segments of C_1 in U_k, then repeat (there cannot be infinitely many such disjoint segments owing to the compactness of the curve C_1). At that point, U_k can be deleted from the set $\mathscr{U}$. After finitely many such deformations of the curve and deletions of open sets, $\mathscr{U}$ will contain only open sets containing a segment of C_2. Figure. 3.7 illustrates this kind of deformation. Now once every open set in $\mathscr{U}$ contains segments of both C_1 and C_2, then each segment of C_1 can be deformed onto the corresponding segment of C_2, entirely within the corresponding open set U_k. Finally, since every deformation happened within a small open set U_k, every intermediate curve has the same index as C_1, and because there could only be finitely many such deformations, the process must terminate, proving that $\mathrm{Ind}(C_1) = \mathrm{Ind}(C_2)$. $\qquad\square$

This result implies that any simple, closed curve surrounding a critical point $\mathbf{x}$ (and no other critical points) has the same index. So the value of the index really only depends on the critical point itself.

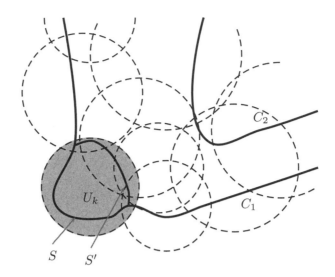

Figure 3.7: Deforming a curve C_1 by pushing a segment S to S' lying entirely within intersections of other open sets. After the deformation, the open set U_k can be discarded, and $\mathrm{Ind}(C) = \mathrm{Ind}(C')$. After finitely many such deformations, C_1 can eventually be deformed onto C_2, proving that $\mathrm{Ind}(C_1) = \mathrm{Ind}(C_2)$.

Definition 3.2.3. If $\mathbf{x} \in \mathbb{R}^2$ is a critical point in a vector field $\mathbf{V}$, then the **index** of $\mathbf{x}$, denoted $\mathrm{Ind}(\mathbf{x})$, is defined by $\mathrm{Ind}(\mathbf{x}) = \mathrm{Ind}(C)$ for any simple closed curve C that encloses $\mathbf{x}$ and no other critical points.

Example 51. Find the index of each of the following kinds of critical points based on their phase portraits shown below.

(a) sink (point A)   (b) source (point C)   (c) saddle point (point E)

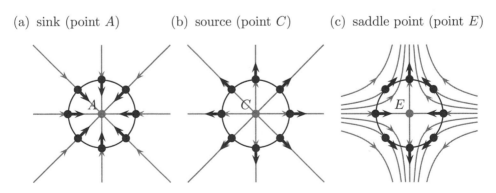

Solution: Draw a circle enclosing each critical point, and draw enough vectors on the circle to determine how many complete counterclockwise turns would occur. A clockwise turn counts as *negative*. (a) $\text{Ind}(A) = 1$; (b) $\text{Ind}(C) = 1$; (c) $\text{Ind}(E) = -1$.

Example 52. Find the index of each critical point in the phase portrait shown below for the system $\mathbf{V} = (-(x-y)(1-x-y), x(2+y))$ shown in Figure 3.8.

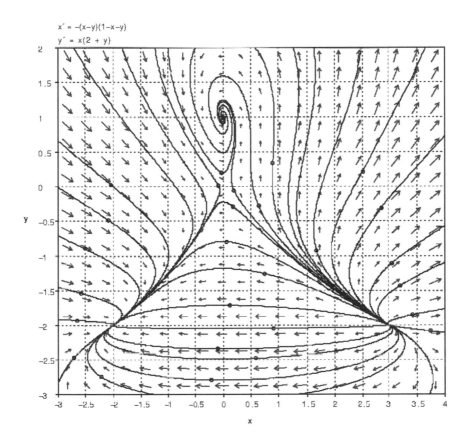

$x' = -(x-y)(1-x-y)$
$y' = x(2+y)$

Figure 3.8: Phase portrait for $\mathbf{V} = (-(x-y)(1-x-y), x(2+y))$.

Solution: First solve $\mathbf{V} = \mathbf{0}$ to find the critical points.

$$-(x-y)(1-x-y) = 0 \implies x = y, \text{ or } x = -y+1$$

When $x = y$ in the second equation, we find: $y(2+y) = 0 \implies y = 0, -2$. Thus two of the critical points are $(0,0)$ and $(-2,-2)$. Similarly, when $x = -y+1$, we find: $(-y+1)(2+y) = 0 \implies y = 1, -2$. This leads to $(0,1)$ and $(3,-2)$. By inspection,

- $(0, 1)$ is a spiral sink. $\text{Ind}((0, 1)) = 1$.

- $(0, 0)$ is a saddle point. $\text{Ind}((0, 0)) = -1$.

- $(-2, -2)$ is a sink. $\text{Ind}((-2, -2)) = 1$.

- $(3, -2)$ is a source. $\text{Ind}((3, -2)) = 1$.

As a corollary to Theorem 3.2.2,

> **Theorem 3.2.4.** *If C is a simple closed curve in a vector field* **V**, *and if there are no critical points on or within C, then* $\text{Ind}(C) = 0$.

Proof. The proof is left as an exercise. $\square$

What can be said about the index of a curve C when there are critical points contained within it? It turns out that there is a simple relationship between $\text{Ind}(C)$ and the index of each critical point within C.

> **Theorem 3.2.5.** *Suppose C is a simple closed curve with no critical points on it, and $\mathbf{x}_1, \mathbf{x}_2, \ldots, \mathbf{x}_n$ are the critical points within C. Then*
> $$\text{Ind}(C) = \sum_{k=1}^{n} \text{Ind}(\mathbf{x}_k).$$

Proof. The proof is by induction on the number of critical points. If there are no critical points, then Theorem 3.2.4 applies, giving $\text{Ind}(C) = 0$, which agrees with the value of the *empty sum*, $\sum_{k \in \emptyset} \text{Ind}(\mathbf{x}_k) = 0$. If $n = 1$, then Theorem 3.2.2 and Definition 3.2.3 yield $\text{Ind}(C) = \sum_{k=1}^{1} \text{Ind}(\mathbf{x}_k) = \text{Ind}(\mathbf{x}_1)$.

Fix $m \geq 1$, and suppose the result is true in all situations for which $n = m$. Consider a simple closed curve C enclosing $m + 1$ critical points, $\mathbf{x}_1, \ldots, \mathbf{x}_m$, $\mathbf{x}_{m+1}$, and let S_1 and S_2 be arcs within the region bounded by C such that S_1 encircles the points $P = \{\mathbf{x}_1, \ldots, \mathbf{x}_m\}$ and $\mathbf{x}_{m+1}$, S_2 encircles x_{m+1}, and two line segments E_1 and E_2 join the curves S_1 and S_2, as shown in Figure 3.9. Let C' by the union of S_1, S_2, E_1, and E_2 with compatible orientations. By Theorem 3.2.2, we have $\text{Ind}(C) = \text{Ind}(C')$.

To compute $\text{Ind}(C')$, we could add up the angle variations around each component, S_1, E_1, S_2, and E_2. Now so long as E_1 and E_2 are close enough to each other, and since these segments are traversed in opposite directions along C', the contribution to angle variation along E_1 (almost) cancels the contribution to angle variation along E_2. Since the vector field is assumed continuous, the total angle variation contributed by $E_1 \cup E_2$ limits to 0 as E_2 approaches E_1.

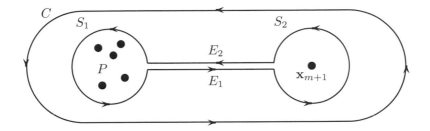

Figure 3.9: Curves C and $C' = S_1 \cup E_1 \cup S_2 \cup E_2$ enclosing the critical points.

Moreover, in the limit when $E_1 = E_2$, the two curves S_1 and S_2 become closed, and we obtain $\mathrm{Ind}(C') = \mathrm{Ind}(S_1) + \mathrm{Ind}(S_2)$. Thus

$$
\begin{aligned}
\mathrm{Ind}(C) = \mathrm{Ind}(C') &= \mathrm{Ind}(S_1) + \mathrm{Ind}(S_2) \\
&= \left[\sum_{k=1}^{m} \mathrm{Ind}(\mathbf{x}_k) \right] + \mathrm{Ind}(\mathbf{x}_{m+1}) \quad \textit{(by inductive hypothesis)} \\
&= \sum_{k=1}^{m+1} \mathrm{Ind}(\mathbf{x}_k).
\end{aligned}
$$

Thus, by induction, the result holds for all $n \in \mathbb{N}$. $\qquad\square$

Finally, we state (without proof) **Hopf's Theorem**, which applies to certain kinds of closed curves: those that are trajectories for the field.

Theorem 3.2.6 (Hopf). *If C is a closed trajectory (orbit) in a vector field, then $\mathrm{Ind}(C) = 1$.*

Theorem 3.2.6 not only implies the existence of at least one critical point within any orbit, but also limits the type of critical points that could be there. For example, if there is only a single critical point $\mathbf{x}$ within the closed trajectory C, then $\mathrm{Ind}(\mathbf{x}) = 1$, which indicates that $\mathbf{x}$ can only be a sink, source, or center (never a saddle point or dipole, etc.). On the other hand, a closed curve C that is not a trajectory can certainly go around any number of critical points of various types.

Example 53. Determine the index of a simple closed curve C in a vector field if C encloses:

(a) Three saddle points.

(b) A dipole and a saddle point.

(c) Two sinks and and a saddle.

In each case, could C possibly be an orbit?

Solution: Simply add the indices of the critical points within the curve. Use Hopf's Theorem to decide whether C could be an orbit.

(a) $\text{Ind}(C) = (-1) + (-1) + (-1) = -3$. C cannot be an orbit.

(b) $\text{Ind}(C) = (2) + (-1) = 1$. C could be an orbit.

(c) $\text{Ind}(C) = (1) + (1) + (-1) = 1$. C could be an orbit.

Sectors

Let us now turn our attention back to trajectories that limit onto a critical point either as $t \to \infty$ or $t \to -\infty$. In a neighborhood of a critical point, these trajectories define regions called **sectors**, which can be used to easily determine the index of the critical point. There are three types of sector. Refer to Figure 3.10.

- A **parabolic** sector is a region in which all trajectories either enter or leave, approaching the critical point as $t \to \infty$ or $t \to -\infty$, respectively.

- An **elliptic** sector is the interior of a homoclinic orbit that reaches the boundary of the neighborhood (recall Definition 3.1.7).

- A **hyperbolic** sector is a region bounded by two trajectories, one approaching the critical point as $t \to \infty$, and the other as $t \to -\infty$, while every other trajectory in the region enters and leaves the neighborhood in finite time.

A critical point surrounded by only parabolic sectors is either a sink or source node, and so the index must be 1. Imagine this is the "base case," and let's see the effect of including an elliptic or hyperbolic sector. Both of these reverse

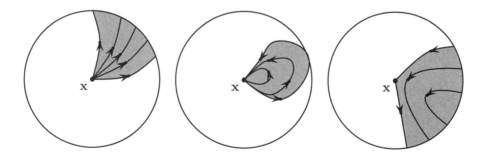

Figure 3.10: *Left*, parabolic sector. *Center*, elliptic sector. *Right*, hyperbolic sector.

the direction of arrows, with the elliptic sector turning the vectors counter-clockwise and the hyperbolic sector turning them clockwise. Thus an elliptic sector contributes $+\frac{1}{2}$ to the index, while a hyperbolic sector contributes $-\frac{1}{2}$. If we let $e(\mathbf{x})$ be the number of elliptic sectors, and $h(\mathbf{x})$ be the number of hyperbolic sectors in a neighborhood of a critical point $\mathbf{x}$, then

$$\mathrm{Ind}(\mathbf{x}) = 1 + \frac{e(\mathbf{x}) - h(\mathbf{x})}{2}. \tag{3.2}$$

Example 54. Referring to Figure 3.3, find the index of points A, C, and D using sectors and (3.2).

(a) A is a source node, which has only parabolic sectors ($e(A) = h(A) = 0$), so $\mathrm{Ind}(A) = 1$.

(b) C is a saddle point, which is surrounded by four hyperbolic sectors ($e(C) = 0$, $h(C) = 4$), so $\mathrm{Ind}(C) = 1 + \frac{0-4}{2} = -1$.

(c) The dipole D has two elliptic sectors (and two parabolic sectors, which do not contribute to the index), so $\mathrm{Ind}(D) = 1 + \frac{2-0}{2} = 2$.

Exercises

1. Find the index of each of the following critical points from Figure 3.3.

   (a) spiral sink (point B)   (c) orbital center (point F)
   (b) dipole (point D)

2. Find the index of each critical point for the vector field $\mathbf{V} = (x, x \mid y^2 \mid y)$, and sketch the phase portrait.

3. Definition 3.2.1 may be stated in more precise mathematical language. Let $\mathbf{V} = (u(x,y), v(x,y))$, and follow the steps outlined below to derive an integral expression that calculates the index of a curve C. Assume there are no critical points on C itself. (*Note:* this exercise requires vector calculus.)

   (a) At each point $\mathbf{x} = (x,y) \in C$, let $\theta = \theta(\mathbf{x})$ represent the angle of the vector $\mathbf{V}(\mathbf{x})$ (measured with respect to the positive x-axis). Express θ in terms of u and v.

   (b) Assume C is parametrized by a smooth function $\mathbf{C} : \mathbb{I} \to \mathbb{R}^2$. Divide $\mathbb{I}$ into n equally spaced subintervals, having endpoints $0 = t_0 < t_1 < t_2 < \cdots < t_{n-1} < t_n = 1$. Let $\Delta\theta_i = \theta(\mathbf{C}(t_i)) - \theta(\mathbf{C}(t_{i-1}))$. Show that the sum $\sum_{i=1}^{n} \frac{\Delta\theta_i}{\Delta t} \cdot \Delta t$ gives the total angle change around $\mathbf{C}$, so long as Δt is small enough.

(c) Using your expression for θ from part (a), and taking the limit of the Riemann sum from part (b), show that

$$\text{Ind}(C) = \frac{1}{2\pi} \int_0^1 \frac{u\frac{dv}{dt} - v\frac{du}{dt}}{u^2 + v^2} \, dt,$$

where $u = u(\mathbf{C}(t))$ and $v = v(\mathbf{C}(t))$.

(d) Let $a > 0$ be a constant, and define $\mathbf{C}(t) = (a\cos 2\pi t, a\sin 2\pi t)$, $t \in \mathbb{I}$. Use the integral from part (c) to find the index of $\mathbf{C}(t)$ with respect to the following vector fields.

(i) $\begin{cases} dx/dt &= x \\ dy/dt &= y \end{cases}$
(ii) $\begin{cases} dx/dt &= x \\ dy/dt &= -y \end{cases}$

4. Prove Theorem 3.2.4.

5. A **quadripole** is a critical point having four elliptic sectors (and no hyperbolic sectors). Compute the index of a quadripole. Then compute the index of an $2n$-pole in general.

6. Show why a *monopole* (one elliptic sector and no hyperbolic sectors) cannot exist in a continuous vector field. What about a *tripole*? What about an *n-pole* where n is an odd number in general?

7. Suppose $\text{Ind}(\mathbf{x}) = 5$. What is the relationship between $e(\mathbf{x})$ and $h(\mathbf{x})$? What is the minimum number of elliptical sectors in any neighborhood of $\mathbf{x}$?

3.3   *Nullclines and Trapping Regions

Stable critical points are especially important in applications because these are the points to which a system can be expected to settle down into a state of motionless equilibrium. Consider a pendulum whose arm can turn around completely, as in Figure 3.11. If the arm is given enough velocity, it may spin around many times before friction finally wins and causes the arm to swing only partway before it comes to a momentary stop, swinging the other way, and then back and forth thereafter. Friction continues to work on the pendulum, causing the arm to swing in smaller and smaller arcs, until finally the swinging is imperceptible, with the arm practically motionless in its lowest position. The final state is a *stable equilibrium* because any small change made to the system (i.e., giving the pendulum a gentle push) will eventually evolve back into that equilibrium state.

There is another equilibrium position, though. If the pendulum is positioned directly above the arm's point of attachment with zero velocity (see Figure 3.12),

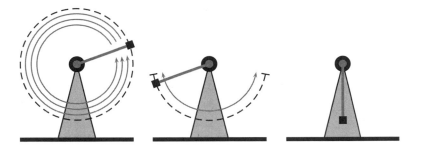

Figure 3.11: *Left*, a pendulum with high velocity, making complete turns over the top. *Center*, the pendulum has lower velocity, and so it swings back and forth but not over the top. *Right*, pendulum in its stable equilibrium position.

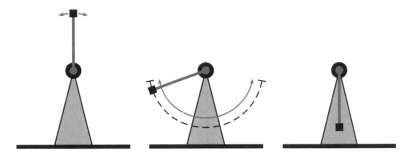

Figure 3.12: *Left*, a pendulum in unstable equilibrium. *Center and right*, a small perturbation of the system takes the pendulum out of this unstable state and into the stable equilibrium state.

then in the absence of any external influences, the pendulum would remain in that position forever. However, our experience in the "real world" reminds us that there are always external influences, from the gentle breeze through the office window, to the random motion of the atoms and molecules making up the pendulum, to quantum fluctuations on the Planck scale, every system must constantly contend with small *perturbations*. Thus the *unstable* equilibrium position can never truly be realized for more than an instant.[6]

The Poincaré-Bendixson Theorem

Now suppose we have a system in which there are no stable equilibria. Then under real-world conditions, the system would never come to rest. Some of these systems, however, exhibit *stable periodic* behavior. In other words, there could be a periodic orbit to which all nearby orbits tend over time. This section

---

[6]Notwithstanding the fact that friction can actually help to make an unstable equilibrium stable with respect to very tiny perturbations.

illustrates how to locate a stable periodic orbit in such a system. The idea is to *trap* it.

Theorem 3.3.1 (Poincaré-Bendixson). *Let* **V** *be a vector field in* $\mathbb{R}^2$. *Suppose there is a compact subset* X *containing no critical points and such that for every* $\mathbf{x} \in X$, *we have* $\mathcal{O}^+(\mathbf{x}) \subseteq X$. *Then there exists a stable periodic orbit in* X.

A compact set X as described in Theorem 3.3.1 is called a **trapping region**. The proof below is adapted from [Goo05].

Proof. Let $\mathbf{x} \in X$ be arbitrary. Let $\omega(\mathbf{x})$ be the set of all points $\mathbf{y} \in \mathbb{R}^2$ that are limit points of sequences of the form $(\phi(\mathbf{x}, t_n))$, where $t_n \to \infty$ (the so-called **omega limit set** of $\mathbf{x}$). You will show in Exercise 4 that $\omega(\mathbf{x})$ is closed.

Claim: If there is a point $\mathbf{y} \in \omega(\mathbf{x})$, then the entire orbit $\phi(\mathbf{y}, t)$ is also contained in $\omega(\mathbf{x})$. *Proof of claim:* Suppose $\mathbf{y}_0 = \phi(\mathbf{y}, t_0)$ for some $t_0 \in \mathbb{R}$ (so that $\mathbf{y}_0$ is on the orbit of $\mathbf{y}$). Let $(\mathbf{x}_n = \phi(\mathbf{x}, t_n))_{n \in \mathbb{N}}$ be the sequence converging to $\mathbf{y}$, so we have $\lim_{n \to \infty} \mathbf{x}_n = \mathbf{y}$. Consider a new sequence, $(\phi(\mathbf{x}, t_n + t_0))_{n \in \mathbb{N}}$. By definition, $\lim_{n \to \infty} \phi(\mathbf{x}, t_n + t_0) \in \omega(\mathbf{x})$, but also by the consistency condition, we have for each $n \in \mathbb{N}$, $\phi(\mathbf{x}, t_n + t_0) = \phi(\phi(\mathbf{x}, t_n), t_0) = \phi(\mathbf{x}_n, t_0)$; by continuity, $\lim_{n \to \infty} \phi(\mathbf{x}_n, t_0) = \phi(\lim_{n \to \infty} \mathbf{x}_n, t_0) = \phi(\mathbf{y}, t_0) = \mathbf{y}_0$. Thus $\mathbf{y}_0 \in \omega(\mathbf{x})$, and since $\mathbf{y}_0$ was arbitrary in the orbit of $\mathbf{y}$, we've shown that the entire orbit $\phi(\mathbf{y}, t)$ is contained within $\omega(\mathbf{x})$.

Compactness of X and the assumption that $\mathcal{O}^+(\mathbf{x}) \subseteq X$ guarantee that all limit points of $\mathcal{O}^+(\mathbf{x})$ are contained within X. Moreover, there must be at least one such limit point $\mathbf{y} \in X$ (for any $\mathbf{x} \in X$, the sequence $(\phi(\mathbf{x}, n))_{n \in \mathbb{N}}$ must have a convergent subsequence by the Bolzano-Weierstrass Theorem, Theorem 2.3.9). This shows that $\omega(\mathbf{x}) \subseteq X$, and $\omega(\mathbf{x})$ is nonempty. Let $\mathbf{y} \in \omega(\mathbf{x})$. By what we have shown above, the entire orbit $\phi(\mathbf{y}, t)$ is also in $\omega(\mathbf{x})$. This orbit is attracting, hence stable, by definition of $\omega(\mathbf{x})$.

Now this orbit $\phi(\mathbf{y}, t)$ cannot be a single point, because we have assumed there are no critical points in X. We claim that the only possibility is that $\phi(\mathbf{y}, t)$ is a periodic orbit. Consider an arc $A \subseteq X$ that is never tangent to any vector in **V**. In particular, the flow across A must always be in the same direction (recall that there are also no critical numbers in X). Suppose there are times $t_1 < t_2$ such that the points $\mathbf{y}_1 = \phi(\mathbf{y}, t_1)$ and $\mathbf{y}_2 = \phi(\mathbf{y}, t_2)$ are on A. Let C be the simple closed curve consisting of the part of the trajectory $\phi(\mathbf{y}, t)$ for which $t_1 \leq t \leq t_2$ and the segment of the arc A between $\mathbf{y}_1$ and $\mathbf{y}_2$, as illustrated in Figure 3.13. By the Jordan Curve Theorem (Theorem 2.4.8), C separates the plane into an inside and outside. In particular, either the positive or negative semiorbit of $\mathbf{y}$ must eventually be contained inside of C; without loss of generality, we may assume that $\mathcal{O}^+(\mathbf{y}_2)$ is on the inside of C (simply "reverse time" to cover the other case).

However, now there is a huge problem: the trajectory can never cross A again. If it did cross at a point $\mathbf{y}_3 = \phi(\mathbf{y}, t_3)$ for some time $t_3 > t_2$, then the flow would be in the opposite direction at $\mathbf{y}_3 \in A$. This implies any arc A in

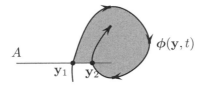

Figure 3.13: The simple closed curve C is bounded by a segment of A and an arc of the trajectory between $\mathbf{y}_1$ and $\mathbf{y}_2$. The positive semiorbit $\mathcal{O}^+(\mathbf{y}_2)$ is contained in the shaded region bounded by C.

X that is never tangent to $\mathbf{V}$ can only intersect our trajectory $\phi(\mathbf{y}, t)$ at most once, so that in fact $\mathbf{y}_2 = \mathbf{y}_1$, and $\phi(\mathbf{y}, t)$ is periodic. □

As stated, the Poincaré-Bendixson Theorem only applies to vector fields in $\mathbb{R}^2$. What about $\mathbb{R}^3$ or higher-dimensional spaces? Everything in the proof above applies to higher-dimensional fields *except* the Jordan Curve Theorem. In fact, if $\mathbf{V}$ is a vector field in $\mathbb{R}^n$ for $n \geq 3$, and there is a trapping region X (compact, with no critical points, and containing $\mathcal{O}^+(\mathbf{x})$ for every $\mathbf{x} \in X$), then there is guaranteed to be an *attractor* in X (a subset to which nearby trajectories approach over time) – it just might not be periodic.

Figure 3.14 shows the **Lorenz attractor**, which is a stable trajectory in the three-dimensional system (for appropriate values of the parameters σ, ρ, β):

$$\begin{cases} dx/dt & = \sigma(y - x), \\ dy/dt & = x(\rho - z) - y, \\ dz/dt & = xy - \beta z. \end{cases}$$

Just because the orbit is stable doesn't mean it's well behaved, though. This strange orbit experiences **chaos**, meaning that a small perturbation in intial conditions may cause a dramatic change in the orbit over time. This is commonly called the *butterfly effect*.[7] Indeed, the Lorenz attractor comes from a model that describes atmospheric air flow, a major component of weather prediction.

Nullclines

In order to find a trapping region, we must be able to construct a subset of the plane from which no trajectory ever leaves. To do this, we will plot some useful guides called the *nullclines* of the system.

[7]The butterfly effect states that something as trivial as the movement of a butterfly wing could ultimately cause a tornado halfway across the globe, and is not a reference to the butterfly shape of the Lorenz attractor. According to Lorenz [Lor63] himself, "One meteorologist remarked that if the theory were correct, one flap of a sea gull's wings would be enough to alter the course of the weather forever. The controversy has not yet been settled, but the most recent evidence seems to favor the sea gulls."

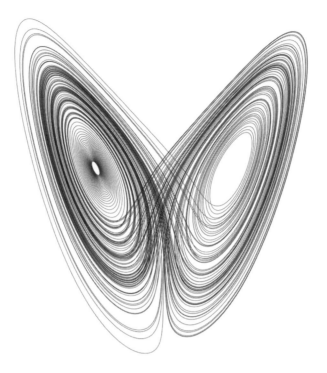

Figure 3.14: Lorenz attractor ($\sigma = 10$, $\rho = 28$, $\beta = 8/3$).

Definition 3.3.2. A **nullcline** in a vector field $\mathbf{V}$ is the set of points at which one component of the flow is 0: the x_i-nullcline is the set of points satisfying $dx_i/dt = 0$.

When $\mathbf{V}$ is two dimensional, there is an x-nullcline and a y-nullcline, each of which consists of a curve or set of curve(s) in the plane. The intersection points of the x- and y-nullclines are (by definition) the critical points of the system.

Example 55. Find and plot the nullclines of the system from Example 52:

$$\mathbf{V} = \begin{cases} dx/dt & = -(x - y)(1 - x - y), \\ dy/dt & = x(2 + y). \end{cases}$$

Solution: The x-nullcline is found by setting $-(x-y)(1-x-y) = 0$, which yields two intersecting lines, $y = x$ and $y = -x + 1$. Similarly, the y-nullcline is found by setting $x(2 + y) = 0$, resulting in the pair of lines, $x = 0$ and $y = -2$.

Autocatalytic Chemical Reactions

As a case study, we explore a class of chemical reactions, called **autocatalytic** reactions, that exhibit spontaneous periodic, or oscillating, behavior. We assume that the reaction is governed by a given system of differential equations. While we cannot solve the system explicitly, we will see how index theory and the concept of limit sets help to identify when there must be a periodic stable solution, which corresponds to observable periodic behavior in the chemical reaction.

An autocatalytic chemical reaction is one in which one or more of the reactants is also a product. The process forms a feedback loop, which causes the reaction rates to oscillate, resulting in periodic changes to the amounts of chemicals in the reaction. If some of those chemicals are colored, then the oscillatory behavior can be observed as periodic changes in color or even more complicated ripple patterns.[8]

Boris Belousov [Bel85] discovered a mixture that exhibited stable oscillations in reactant amounts in the 1950s. The ingredients were potassium bromate, cerium(IV) sulfate, propanedioic acid, and citric acid. When mixed, the color of the solution oscillated in regular time intervals between pale yellow and colorless. Belousov could not publish the results because the editors believed this reaction violated the laws of thermodynamics, but the same reaction was rediscovered by Anatol Zhabotinsky [Zha64] in the early 1960s. Today, reactions of this type are called Belousov-Zhabotinsky (BZ) reactions, and there have been many more discovered or theorized to exist, including the *Oregonator*, *Edelstein reaction*, *Feinberg-Horn-Jackson reaction*, and *Briggs-Rauscher reaction*. In what follows, we will explore the *Brusselator*, which is a theoretic reaction that illustrates how oscillatory behavior arises in complicated systems.

Using the theory of *chemical kinetics*, one can produce a system of differential equations describing the rates of change of all reactants and products in a chemical reaction. It is not important to understand the details of the theory in order to analyze the resulting system. In the Brusselator, we will be most interested in two species of reactant, whose concentrations in the solution are labeled x and y. Other species may be present in abundance, so their concentrations, a, b, etc., are effectively constant. After some simplifying assumptions, the Brusselator may be described by the system of differential equations

$$\begin{cases} dx/dt & = 1 - (b+1)x + x^2y \\ dy/dt & = bx - x^2y, \end{cases} \qquad \text{where } b > 0 \text{ is constant.} \qquad (3.3)$$

Set $dx/dt = 0$ to find the x-nullcline:

$$0 = 1 - (b+1)x + x^2y \quad \implies \quad y = \frac{(b+1)x - 1}{x^2}.$$

---

[8]At the time of this writing, there are many videos of this phenomenon available online, e.g., https://www.youtube.com/watch?v=3JAqrRnKFHo.

Set $dy/dt = 0$ to find the y-nullcline:

$$0 = bx - x^2y = x(b - xy) \quad \Longrightarrow \quad xy = b \quad \text{or} \quad x = 0.$$

The nullclines intersect in a single critical point $(1, b)$. Let us choose a particular value $b = 2.5$ (different values of b exhibit different behavior). Thus the system is:

$$\begin{cases} dx/dt & = 1 - 3.5x + x^2y \\ dy/dt & = x(2.5 - xy), \end{cases} \tag{3.4}$$

with x-nullcline $y = \frac{3.5x-1}{x^2}$ and y-nullcline consisting of the vertical line $x = 0$ and the hyperbola $y = \frac{2.5}{x}$. Determine the flow by sampling on the nullclines, which in turn determines the general flow in each region. See Figure 3.15 for a plot of the nullclines on the vector field.

The precise methods used to analyze the vector flow are beyond the scope of this textbook; however, we will sketch a general idea. In courses such as Differential Equations and Dynamics, you will discover a fuller set of tools used to work with these systems.

First, we determine that the flow near the critical point $(1, 2.5)$ is *repelling*. In practice, we use the *Jacobian*[9] of the system to determine this, but you may verify that the flow is generally outward from the critical point in a small neighborhood containing the point. Second, we build a trapping region $\mathscr{R}$ containing the critical point. Then the existence of a stable periodic orbit would be guaranteed by the Poincaré-Bendixson Theorem (Theorem 3.3.1).

In may help to refer to Figure 3.16 as you read over the following construction. The line $y = 0$ is a natural choice for the bottom boundary of $\mathscr{R}$, since the flow is up/left there (Region IV). Following this line left to the x-nullcline, turn at right angles at the x-intercept of the x-nullcline, which is $x = 1/3.5 \approx 0.2857$. The line $x = 1/3.5$ forms the left boundary of $\mathscr{R}$. Observe that the flow on this line is up/right (Region I). Then turn at right angles when the line $x = 1/3.5$ intersects the y-nullcline; this happens at $y = \frac{2.5}{(1/3.5)} = 8.75$. Now to the right, on the line $y = 8.75$, we know the flow is down/right (since this is part of Region II); however, there is not another nullcline to the right that we could use to make another turn. Instead, we will make a 45° turn at some point. It matters a great deal where that turn is made: the diagonal segment of the boundary, which we will call L, must also be *trapping*. Careful analysis indicates that we could begin turning at the point $(2, 8.75)$ – we just have to show that the flow, which is generally down/right on L (since L is in Region II), has larger *down* component than *right* component. In other words, we must show that every flow vector on L has slope $m < -1$. This is an exercise left to the reader (see Exercise 2), but graphical evidence may be found by carefully studying the phase plane. Finally, the rightmost segment of the boundary is located where L meets the x-nullcline.

---

[9]Specifically, we compute $J(x, y) = \dfrac{\partial(v_1, v_2)}{\partial(x, y)} = \begin{vmatrix} -3.5 + 2xy & x^2 \\ 2.5 - 2xy & -x^2 \end{vmatrix} = x^2$. Since $J(1, 2.5) = 1 > 0$, the critical point $(1, 2.5)$ is repelling.

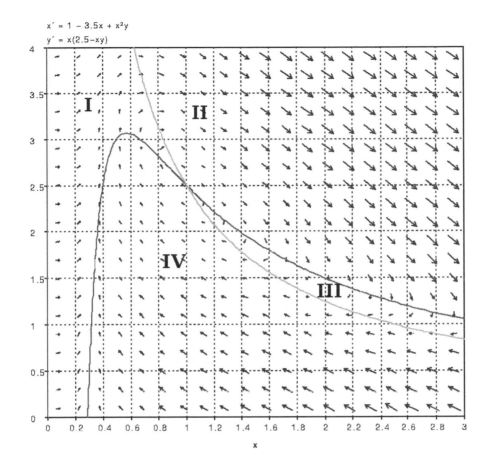

Figure 3.15: Nullclines and general flow. Region I: up/right. Region II: down/right. Region III: down/left. Region IV: up/left.

If we parametrize the line L by $(x, y) = (2 + t, 8.75 - t)$, then we have to solve the following equation for t:

$$y = \frac{3.5x - 1}{x^2} \implies 8.75 - t = \frac{3.5(2 + t) - 1}{(2 + t)^2} \implies t \approx 8.4234.$$

Therefore $x = 2 + (8.4234) = 10.4234$ forms the right boundary line. Flow on the vertical line is always to the left (down/left in Region II; up/left in Region IV), so the rightmost segment of the boundary is also trapping.

Now that we have shown that region $\mathcal{R}$ is trapping, and the only critical point in $\mathcal{R}$ is a repellor, the Poincaré-Bendixson Theorem implies that the system has a stable periodic solution. In other words, the Brusselator reaction exhibits a self-sustaining oscillatory behavior. Using an ODE-solver, the orbit

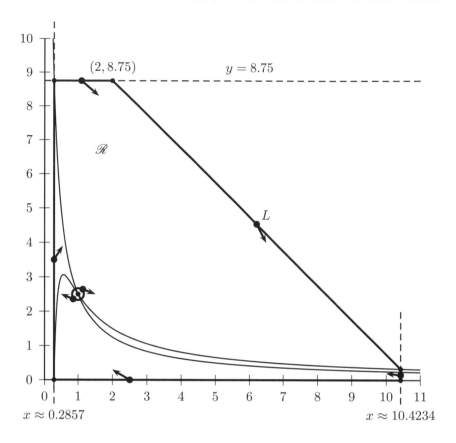

Figure 3.16: Trapping region $\mathscr{R}$ for the Brusselator system.

can be found, as shown in Figure 3.17. (**ODE** stands for *ordinary differential equations*.) Trajectories beginning within the region will be bounded by the orbit spiral outward to approach it; trajectories starting on the exterior will spiral in to approach it.

Exercises

1. For each system from §3.1, Exercise 4, find the x- and y-nullclines.

2. Finish the argument that the region $\mathscr{R}$ in Figure 3.16 is *trapping*, by verifying that all flow vectors on line L have slope less than -1. The following steps should help you do so.

   (a) Find the equation of line L, in $y = mx + b$ form, and then substitute this expression for y into (3.4).

   (b) Use your results from (a) to show that every vector on L has slope less than -1. (*Hint:* Consider the sum $\frac{dx}{dt} + \frac{dy}{dt}$, which will be less

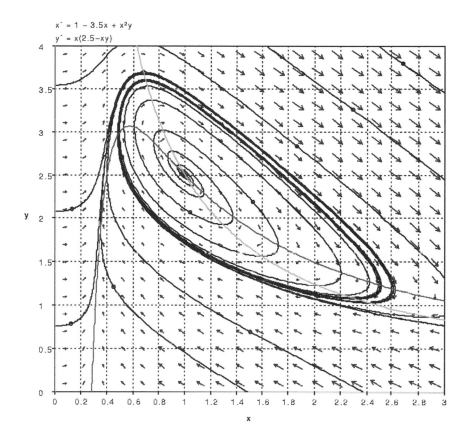

$x' = 1 - 3.5x + x^2y$

$y' = x(2.5-xy)$

Figure 3.17: Phase portrait for the Brusselator system. All nearby trajectories approach the stable periodic orbit (which is implied by the close grouping of flows).

> than 0 if the negative contribution from $\frac{dy}{dt}$ is greater than the positive
> contribution from $\frac{dx}{dt}$.)

3. Analyze the Brusselator system (3.3) with the following values of the parameter b. Are there always periodic orbits?

   (a) $b = 1$ (b) $b = 5$

4. Finish the details in Theorem 3.3.1 by showing that $\omega(\mathbf{x})$ is a closed subset of X. (*Hint:* Use the definition of limit point in terms of open sets to show that there is an open neighborhood around any point $\mathbf{y} \notin \omega(\mathbf{x})$).

Supplemental Reading

- Ault and Holmgreen [AH]. Unpublished analysis of the Brusselator.

- Boyce and DiPrime [BD08], Chapters 2, 7, and 9. A more analytic treatment of phase portraits and systems of differential equations.

- Devaney [Dev03], Part 2. Detailed treatment of chaos in dynamical systems.

- Goodman [Goo05], Chapter 5.

- Henle [Hen79], Chapter 2.

- Prasolov [Pra95], Chapter 6.

- Strogatz [Str94]. Analysis of nonlinear dynamical systems such as those used to describe the Brusselator.

Part II

Abstract Topology with Applications

Chapter 4

Abstract Point-Set Topology

What is topology, anyway? In Chapter 1, we state that topology is the *study of the qualities of a space that are preserved under invertible continuous maps*, without clearly defining the terms "space" or "continuous map." Then in Chapter 2, we developed continuity (among other things) in the context of familiar Euclidean spaces, and finally in general metric spaces. In this chapter, we introduce a concept of *space* that not only encompasses metric spaces but also allows us to talk about much more exotic structures. A topological space is simply a *set with additional structure* (to be defined in §4.1). The elements of the set are called the *points* of the space, so we often use the term **point-set** for a topological space – hence the traditional moniker for the subject, **point-set topology**. Using the language of point-set topology, the real power and elegance of the theory become apparent in what topology abstracts from Euclidean geometry and how it may be applied to unrelated – often very non-geometric – problems.

4.1 The Definition of a Topology

The language of topology is based in the language of set theory. Let X be a set – not necessarily a subset of $\mathbb{R}^n$. We will call X a **point-set** if we want to alert the reader that we consider the elements of X as individual points. Just as a set of points in the plane may have a certain *geometry* (i.e., a particular geometric structure defining distances and angles), X may have a **topology**. In what may seem like a circular definition, a topology on a point-set X is a mathematical description of the essential topological properties of X. But what properties of X are most *essential*?

- A way to determine whether a subset $U \subseteq X$ is open (or closed, or neither, or both).

- A way to define limit points, and thus to define $\lim\limits_{n \to \infty} x_n$.

- Connectedness, compactness, etc.

- A criterion to determine whether a function $f : X \to Y$ is continuous or not.

What is *not* important (*topologically* speaking)?

- Any particular representation of the space (e.g., recall Figures 1.6, 1.7, 1.9, and 1.16).

- Length, area, volume, angle, and similar measures of geometric quantity. Even the distinction between *bounded* and *unbounded* is meaningless. For example, we have seen in §2.2 that there is a homeomorphism $(a, b) \approx (-\infty, \infty)$.

Indeed, the whole idea of a metric d and the ϵ-balls defined in Chapter 2 are somehow irrelevant, vestigial concepts, convenient for defining certain topological concepts in $\mathbb{R}^n$ (and other metric spaces), but too restrictive to be of general use in abstract topology. But then how could we hope to define the *open* sets in a space X without referrence to a metric on X? The trick is to make *open*ness part of the very structure of a topology on X. In fact, all that is *essential* (topologically) is knowing which subsets of the point-set are considered to be open, since the concepts of limit points, connectedness, compactness, and continuity can be defined in relation to open sets.[1] On the other hand, we must be careful not to allow too much freedom. The sets that we designate as *open* should satisfy some general properties that we find useful. These defining properties are called the **axioms** for a topology, and over the years mathematicians have a agreed upon a list of four.[2]

Definition 4.1.1. A **topology** $\mathscr{T}$ on a point-set X is a subset of the power set $\mathcal{P}(X)$ satisfying the following four axioms:

1. $\emptyset \in \mathscr{T}$.

2. $X \in \mathscr{T}$.

3. If $U, V \in \mathscr{T}$, then $U \cap V \in \mathscr{T}$.

4. If $\{U_k\}_{k \in \mathcal{I}}$ is a collection of sets $U_k \in \mathscr{T}$, then $\bigcup\limits_{k \in \mathcal{I}} U_k \in \mathscr{T}$.

The notation $(X, \mathscr{T})$ means that X is a point-set with a particular topology $\mathscr{T}$, and then we may call X a (**topological**) **space**.

---

[1] In fact, once you see the definitions in terms of open sets, you may realize that these definitions are more natural and elegant than those found in Chapter 2.

[2] Felix Hausdorff's original definition in 1914 included a fifth axiom that we call nowadays the *Hausdorff condition* [Hau14] (see Definition 4.2.6).

To say that $\mathscr{T}$ is a subset of $\mathcal{P}(X)$ means that every element of $\mathscr{T}$ is a particular subset of X. That is, $U \in \mathscr{T} \implies U \subseteq X$ (but not the converse). Typically, not *every* subset of X is in a given topology $\mathscr{T}$ on X. So what does Definition 4.1.1 have to do with open sets? Each set $U \in \mathscr{T}$ is *considered to be open* with respect to the topology $\mathscr{T}$ on X. Think of $\mathscr{T}$ as the master list of absolutely every subset of X that we wish to label "open."

Definition 4.1.2. Suppose $(X, \mathscr{T})$ is a topological space. A subset $U \subseteq X$ is **open** (with respect to $\mathscr{T}$) if and only if $U \in \mathscr{T}$.

Axioms 1–4 of Definition 4.1.1 may be expressed roughly as follows:

1. The empty set is open.
2. The whole set X is open.
3. The intersection of two open sets is open.
4. The union of arbitrarily many open sets is again open.

Note that each of the four axioms is a *property* of open sets in any metric space (recall Proposition 2.5.3 and Exercise 2 in §2.5), so every metric space (X, d) is a topological space with topology $\mathscr{T} = \mathscr{T}_{(X,d)}$ defined to be the set of all open subsets of X, where "open" is defined in terms of metric balls.

Definition 4.1.3. If (X, d) is any metric space, then the **metric topology** on X is the set $\mathscr{T}_{(X,d)} \subseteq \mathcal{P}(X)$ defined by

$$\mathscr{T}_{(X,d)} = \{U \subseteq X \mid \forall x \in U, \exists \epsilon > 0 \text{ such that } x \in B_{d,\epsilon}(x) \subseteq U\}.$$

Since Euclidean space is a particular kind of metric space, $\mathbb{R}^n$ has a topology $\mathscr{T}_E \subseteq \mathcal{P}(\mathbb{R}^n)$ called the **Euclidean topology**, in which $\mathscr{T}_E$ consists of all sets $U \subseteq \mathbb{R}^n$ that are open according to Definition 2.1.5. Metric spaces are important examples of topological spaces; however, not all topological spaces are metric spaces.

Example 56. Let $X = \mathbb{N}$. For each $n \in \mathbb{N}$, let $U_n = \{x \in \mathbb{N} \mid x > n\}$. Let $\mathscr{T} = \{\emptyset\} \cup \{U_n \mid n \in N\}$. Show that $\mathscr{T}$ is a topology on X.

We must verify the four axioms. Before doing so, it helps to get a better sense for the sets U_n. Observe that $U_1 \supseteq U_2 \supseteq U_3 \supseteq \cdots$.

$$
\begin{aligned}
U_1 &= \{x \in \mathbb{N} \mid x \geq 1\} = \{1, 2, 3, 4, 5, 6, \ldots\} \\
U_2 &= \{x \in \mathbb{N} \mid x \geq 2\} = \{2, 3, 4, 5, 6, 7, \ldots\} \\
U_3 &= \{x \in \mathbb{N} \mid x \geq 3\} = \{3, 4, 5, 6, 7, 8, \ldots\} \\
U_4 &= \{x \in \mathbb{N} \mid x \geq 4\} = \{4, 5, 6, 7, 8, 9, \ldots\} \\
&\;\;\vdots \qquad \vdots
\end{aligned}
$$

1. Clearly, $\emptyset \in \mathscr{T}$.

2. Since $U_1 = \mathbb{N}$, we see that $\mathbb{N} = X \in \mathscr{T}$.

3. Let $U, V \in \mathscr{T}$. If either of U or V is $\emptyset$, then $U \cap V = \emptyset \in \mathscr{T}$. (Note that in the future, we will not verify Axiom 3 when one of the two sets is empty, as this check is trivial.) Now suppose $U = U_n$ and $V = U_m$ for some $n, m \in \mathbb{N}$. If $n \leq m$, then $U_n \cap U_m = U_m$, which proves $U \cap V \in \mathscr{T}$. We need not check $n \geq m$, as $U \cap V = V \cap U$. Thus the intersection of two open sets is open.

4. Let $\{U_{n_k}\}_{k \in \mathcal{I}}$ be an arbitrary set of sets in $\mathscr{T}$. Let $W = \bigcup_{k \in \mathcal{I}} U_{n_k}$. For any $x \in \mathbb{N}$, we see that $x \in W$ if and only if $x \geq n_k$ for some $k \in \mathcal{I}$. That is, $x \in W$ if and only if $x \geq \min\{n_k \mid k \in \mathcal{I}\}$. Now in any set of natural numbers, there must be a least element, n_0 (by the *well-ordering principle*). Thus $x \in W$ if and only if $x \geq n_0$, which shows that $W = U_{n_0}$. But $U_{n_0} \in \mathscr{T}$; hence the arbitrary union of open sets is open.

It is easy to show, using induction, that Axiom 3 of the definition of a topology extends to finitely many intersections (see Exercise 2). But what about infinite intersections? Nothing in the axioms implies that an intersection of infinitely many open sets should be open, and for good reason – this property does not hold with respect to Euclidean open sets in $\mathbb{R}^n$, as Exercise 4 of §2.1 illustrates.

Now that we have a topological definition for open sets, we may define closed sets. Compare to the definition given in Definition 2.1.8.

> **Definition 4.1.4.** Suppose $\mathscr{T}$ is a topology on X. A subset $C \subseteq X$ is **closed** if and only if $X \setminus C \in \mathscr{T}$. In other words, C is closed if and only if its complement in X is open with respect to the topology.

Example 57. Describe the closed sets of $\mathbb{N}$ with respect to the topology $\mathscr{T}$ defined in Example 56.

The open sets are, by definition, the elements of $\mathscr{T}$: $\emptyset$, and $U_n = \{x \in \mathbb{N} \mid x \geq n\}$, for each $n \in \mathbb{N}$. Therefore the closed sets, which are the complements of these, are:

- The whole space, $\mathbb{N}$.

- The sets $\{x \in \mathbb{N} \mid x < n\} = \{1, 2, 3, \ldots, n-1\}$ for each $n \in \mathbb{N}$ (which includes $\emptyset$, when $n = 1$).

We say that N is an **neighborhood** of a point $x \in X$ if there is some $U \in \mathscr{T}$ such that $x \in U \subseteq N$.

Discrete and Indiscrete Spaces

Consider the *lattice points* $\mathbb{Z}^2$ in the plane (i.e., points $(x, y) \in \mathbb{R}^2$ such that both x and y are integers), as shown in Figure 4.1. Each point is separated from all others, in the sense that you could draw a circle around the point containing none others. In fact, as a subset of $\mathbb{R}^2$, each singleton set $\{(x, y)\} \subseteq \mathbb{Z}^2$ is considered open in $\mathbb{Z}^2$ (of course, $\{(x, y)\}$ is *not* open in $\mathbb{R}^2$). Indeed, since arbitrary unions of open sets are open, *every* subset of $\mathbb{Z}^2$ is open. This is an example of a *discrete space*. In a discrete space, every point belongs to a neighborhood consisting of *only* that point.

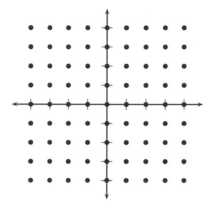

Figure 4.1: $\mathbb{Z}^2 \subseteq \mathbb{R}^2$ is a discrete space.

Definition 4.1.5. The **discrete topology** on a point-set X is the topology $\mathcal{T}_D$ defined by $\mathcal{T}_D = \mathcal{P}(X)$. In other words, *every* subset of X is open.

Implicit in Definition 4.1.5 is the fact that $\mathcal{T}_D$ is indeed a topology on X. In fact, $\mathcal{T}_D$ is the metric topology corresponding to the *discrete metric*, d_{disc}, on X (see Example 35 of §2.5). Of course the discrete topology represents something of an extreme in the sense that no other topology on X can have more open sets than the discrete topology does. The following two terms help us to compare topologies in general.

Definition 4.1.6. Fix a point-set X, and let $\mathcal{T}_1$ and $\mathcal{T}_2$ be two topologies defined on X. If $\mathcal{T}_1 \subseteq \mathcal{T}_2$, then we say that $\mathcal{T}_1$ is **coarser** than $\mathcal{T}_2$ and that $\mathcal{T}_2$ is **finer** than $\mathcal{T}_1$.

Thus, according to Definition 4.1.6, the discrete topology $\mathcal{T}_D$ is the finest topology that can be defined on a point-set X. Any other topology $\mathcal{T}$ on X

must be coarser than $\mathscr{T}_D$. What is the coarsest topology possible on X? Looking through the axioms, we see that at least X and $\emptyset$ must be included, and if no other subset of X is included, then axioms 3 and 4 are automatically satisfied. We call this extreme case the *indiscrete topology* on X; it's the coarsest topology on X.

> **Definition 4.1.7.** The **indiscrete topology** on a point-set X is the topology $\mathscr{T}_I$ defined by $\mathscr{T}_I = \{\emptyset, X\}$.

In the space $(X, \mathscr{T}_I)$, the only neighborhood of a given point $x \in X$ is the whole space itself. Think of X as being completely amorphous – all of the points are indistinguishable and even interchangeable (see Example 65).

Topologies on Finite Sets

In general, if X is an infinite set and $\mathscr{T}$ a topology on X, it may be impossible to list all of the open sets of X (unless of course $\mathscr{T}$ is the indiscrete topology on X). On the other hand, if X is finite, then $\mathscr{T} \subseteq \mathcal{P}(X)$ must be finite, and it becomes a straightforward exercise to list all the open and closed sets of X with respect to $\mathscr{T}$. Though the study of finite topologies has very little practical application, it can still be useful for sharpening our abstract topological reasoning.[3]

Example 58. Let $X = \{a, b, c, d\}$, and let

$$\mathscr{T} = \{\emptyset, \{a\}, \{d\}, \{a, b\}, \{a, d\}, \{a, b, d\}, \{a, b, c, d\}\}.$$

(a) Show that $\mathscr{T}$ is a topology on X.   (c) List all the closed sets of X.

(b) List all the open sets of X.

Solution:

(a) We find both $\emptyset$ and X in $\mathscr{T}$. Next consider all intersections of pairs of sets in $\mathscr{T}$. We may omit any situation in which $U \subseteq V$, since then $U \cap V = U$. In particular, we never have to check any intersection involving $\emptyset$ or X itself.

$$\begin{aligned}
\{a\} \cap \{d\} &= \emptyset \in \mathscr{T} \\
\{d\} \cap \{a, b\} &= \emptyset \in \mathscr{T} \\
\{a, b\} \cap \{a, d\} &= \{a\} \in \mathscr{T}
\end{aligned}$$

Next consider all unions of two or more sets in $\mathscr{T}$ (omitting cases in which one set is a subset of another; for example, we need not check whether

---

[3]William Thurston (1946-2012), a prominent and influential topologist, wrote that finite topology is "an oddball topic that can lend good insight to a variety of questions" [Thu94].

$\{a\} \cup \{d\} \cup \{a,b\} \in \mathscr{T}$ since $\{a\} \cup \{a,b\} = \{a,b\}$, and this reduces to checking that $\{d\} \cup \{a,b\}$ is in the topology).

$$\{a\} \cup \{d\} = \{a,d\} \in \mathscr{T}$$
$$\{d\} \cup \{a,b\} = \{a,b,d\} \in \mathscr{T}$$
$$\{a,b\} \cup \{a,d\} = \{a,b,d\} \in \mathscr{T}$$

Thus $\mathscr{T}$ is a topology on X.

(b) By definition, the open sets are all the elements of $\mathscr{T}$:

$$\emptyset, \{a\}, \{d\}, \{a,b\}, \{a,d\}, \{a,b,d\}, \{a,b,c,d\}.$$

(c) By definition, the closed sets are the complements of open sets in X:

$$X \setminus \emptyset, X \setminus \{a\}, X \setminus \{d\}, X \setminus \{a,b\}, X \setminus \{a,d\}, X \setminus \{a,b,d\}, X \setminus \{a,b,c,d\},$$

Thus the closed sets are: $\{a,b,c,d\}, \{b,c,d\}, \{a,b,c\}, \{c,d\}, \{b,c\}, \{c\}, \emptyset$.

For small sets X, a diagram in which the open sets are circled may help in visualizing the topology.

Example 59. Below is a diagram for the topology $\mathscr{T}$ on $\{a,b,c,d\}$ from Example 58.

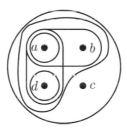

Base of a Topology

In linear algebra, we learn about the space of all vectors in $\mathbb{R}^n$. Clearly there are infinitely many such vectors,[4] all having the form

$$\mathbf{x} = \begin{pmatrix} x_1 \\ x_2 \\ \vdots \\ x_n \end{pmatrix}$$

---

[4]Indeed, there are *uncountably* many vectors.

for real numbers x_i. On the other hand, each vector $\mathbf{x}$ may be thought of as a *linear combination* of a finite set of *basis* vectors

$$\mathbf{x} = x_1\mathbf{e}_1 + x_2\mathbf{e}_2 + \cdots + x_n\mathbf{e}_n,$$

where $\mathbf{e}_i$ is the ith unit vector of $\mathbb{R}^n$. Thus the finite set $B = \{\mathbf{e}_i \mid i = 1, 2, \ldots, n\}$ is sufficient to describe all of the vectors $\mathbf{x} \in \mathbb{R}^n$, using operations that are essential to linear algebra, such as addition and scalar multiplication.[5]

In a similar way, a topology $\mathscr{T}$ on a point-set X may be described using a much smaller subset $\mathscr{B} \subseteq \mathscr{T}$, such that every open set $U \in \mathscr{T}$ can be built up from sets in $\mathscr{B}$ using operations that are essential to topology, such as unions. The motivating observation is that every open set in a metric space is equal to a union of (perhaps infinitely many) open metric balls. This idea can be abstracted to any topological space.

Definition 4.1.8. A subset $\mathscr{B} \subseteq \mathcal{P}(X)$ is a **base** on X if $\mathscr{B}$ satisfies:

1. For each $x \in X$, there is at least one $B \in \mathscr{B}$ such that $x \in B$.

2. If $B_1, B_2 \in \mathscr{B}$ and $B_1 \cap B_2 \neq \emptyset$, then for every $x \in B_1 \cap B_2$, there is a set $B \in \mathscr{B}$ such that $x \in B \subseteq B_1 \cap B_2$.

Condition 1 of Definition 4.1.8 ensures the sets of $\mathscr{B}$ *cover* all of X; in other words, every point has an open neighborhood. Condition 2 specifies enough structure so that the topology on X can be built up from $\mathscr{B}$ by taking unions. See Figure 4.2 for an illustration of the two conditions. The following lemma shows how to build the topology $\mathscr{T}$ from a given base.

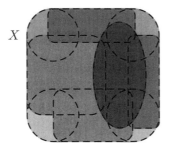 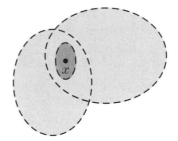

Figure 4.2: *Left*, condition 1: A base must cover X. *Right*, condition 2: Any point in the intersection of base sets must be contained within a base set that is itself a subset of the intersection.

---

[5]See §B.2 for further details about vector spaces.

Lemma 4.1.9. *If $\mathscr{B}$ is a base on X, then the collection $\mathscr{T}$ of arbitrary unions of sets from $\mathscr{B}$ is a topology on X.*

Proof. The empty union of base sets (i.e., the union of no sets at all) is $\emptyset$, so Axiom 1 is satisfied. Now for each $x \in X$, let $B_x \in \mathscr{B}$ be a set containing x. Clearly $\bigcup_{x \in X} B_x = X$, satisfying Axiom 2. Let's establish Axiom 4 next (since we will need it to show Axiom 3). Suppose $U_k \in \mathscr{T}$ for all k in some indexing set $\mathcal{I}$. By definition, each U_k is an arbitrary union of sets from $\mathscr{B}$, so we may write

$$U_k = \bigcup_{B \in \mathscr{B}_k} B,$$

where $\mathscr{B}_k$ is the collection of the open sets whose union is U_k. Let U be the union of the sets U_k.

$$U = \bigcup_{k \in \mathcal{I}} U_k = \bigcup_{k \in \mathcal{I}} \bigcup_{B \in \mathscr{B}_k} B$$

This shows that U is simply a union of all of the various base sets B from all of the collections $\mathscr{B}_k$ as k ranges throughout $\mathcal{I}$; by definition, $U \in \mathscr{T}$.

Finally, to verify Axiom 3, consider two sets $U, V \in \mathscr{T}$. Express $U = \bigcup_{B_1 \in \mathscr{B}_1} B_1$ and $V = \bigcup_{B_2 \subset \mathscr{B}_2} B_2$, for some collections $\mathscr{B}_1, \mathscr{B}_2 \subset B$. Consider $W = U \cap V$.

$$W = \left(\bigcup_{B_1 \in \mathscr{B}_1} B_1 \right) \cap \left(\bigcup_{B_2 \in \mathscr{B}_2} B_2 \right) = \bigcup_{B_1 \in \mathscr{B}_1, B_2 \in \mathscr{B}} (B_1 \cap B_2)$$

By Condition 2 of a base, for each $x \in B_1 \cap B_2$, we can find a set $B_x \in \mathscr{B}$ such that $x \in B_x \subseteq B_1 \cap B_2$. Thus

$$B_1 \cap B_2 = \bigcup_{x \subset B_1 \cap B_2} B_x,$$

which shows that $B_1 \cap B_2 \in \mathscr{T}$ since it is an arbitrary union of base sets (which is open by Axiom 4). Then, since W is an arbitrary union of open sets, we have shown that $W \in \mathscr{T}$ (again invoking Axiom 4). $\qquad \square$

We say that $\mathscr{T}$ in Lemma 4.1.9 is the topology **generated** by $\mathscr{B}$.

Example 60. Show that for any point-set X, the collection of singleton sets, $\mathscr{B} = \{\{x\} \mid x \in X\}$, is a base for the discrete topology on X.

Solution: First we verify that $\mathscr{B}$ is a base. Clearly condition 1 is satisfied. Condition 2 is vacuously true, since there are no nonempty intersections. Now let's figure out the topology $\mathscr{T}$ generated by $\mathscr{B}$. By definition, every $U \in \mathscr{T}$ is an arbitrary union of base sets. In particular, if $U \subseteq X$ is *any* subset, then we have $U = \bigcup_{x \in U} \{x\} \in \mathscr{T}$. Thus $\mathscr{B}$ generates the topology $\mathscr{T} = \mathcal{P}(X)$, which is the discrete topology $\mathscr{T}_D$ on X.

Example 61. The collection $\mathscr{B} = \{B_\epsilon(x) \mid \epsilon > 0, x \in \mathbb{R}^n\}$ is a base for the metric topology on $\mathbb{R}^n$. In fact, there is a *countable* base for this topology:

$$\mathscr{B}_0 = \{B_\epsilon(x) \mid \epsilon \in \mathbb{Q}^+, x \in \mathbb{Q}^n\}. \tag{4.1}$$

The details are left to the reader (see the Exercises).

To say that a base $\mathscr{B}$ is **minimal**, we mean that no sets can be removed from $\mathscr{B}$ without changing the topology generated.

Example 62. Find a minimal base for the topology $\mathscr{T}$ on $\{a, b, c, d\}$ from Example 58.

Solution: Any open set that is the union of smaller open sets need not be a part of the base (because such unions will be generated from the base sets). You can verify that $\mathscr{B} = \{\{a\}, \{d\}, \{a, b\}, \{a, b, c, d\}\}$ is a base for $\mathscr{T}$, and no smaller subset suffices. The diagram below shows only the base sets, but it is understood that the topology generated by this diagram also includes the unions $\{a, d\}$ and $\{a, b, d\}$.

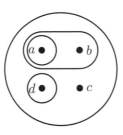

Exercises

1. Show that X and $\emptyset$ are always both open and closed no matter what topology $\mathscr{T}$ is defined on X.

2. Show that Axiom 3 of Definition 4.1.1 generalizes to finitely many intersections of open sets. That is, if $\mathscr{T}$ is a topology on X, and if $U_1, U_2, \ldots, U_n \in \mathscr{T}$, then $\bigcap_{i=1}^{n} U_i \in \mathscr{T}$.

3. Use the definition of a topology to prove that the union of two closed sets is closed. Show by example that the union of arbitrarily many closed sets may not be closed.

4. Show that an arbitrary intersection of closed sets is closed.

5. If X has the discrete topology, show that every subset is closed (hence every subset is both open and closed in a discrete space).

6. How many closed sets are there in an indiscrete space? What are they?

7. Determine all topological spaces that are simultaneously discrete and indiscrete.

8. List all of the possible topologies that can be defined on the two-point set, $X = \{a, b\}$.

9. Let $X = \{a, b, c, d, e\}$, and let $\mathscr{T}$ be the following subset of $\mathcal{P}(X)$,

$$\{\{a, b, c, d, e\}, \{a, b, c, e\}, \{a, b, d, e\}, \{a, b, e\}, \{c, d, e\}, \{c, e\}, \{d, e\}, \{e\}, \emptyset\}.$$

   (a) Verify that $\mathscr{T}$ is a topology on X.
   (b) List all open sets.
   (c) List all closed sets.
   (d) List all sets that are both closed and open.
   (e) List all sets that are neither open nor closed.

10. Let $X = \{a, b, c, d, e, f\}$, and let $\mathscr{B} = \{\{a\}, \{a, c\}, \{a, d\}, \{b, e\}, \{b, e, f\}\}$.

   (a) Verify that $\mathscr{B}$ is a base.
   (b) List all open sets in the topology $\mathscr{T}$ generated by $\mathscr{B}$.
   (c) List all closed sets.
   (d) List all sets that are both closed and open.
   (e) List all sets that are neither open nor closed.

11. Suppose that (X, d) is a metric space. Let $\mathscr{B} = \{B_\epsilon(x) \mid \epsilon > 0, x \in X\}$.

   (a) Prove that $\mathscr{B}$ is a base.
   (b) Show that the topology generated by $\mathscr{B}$ is the metric topology on X. (*Hint:* show that a subset $U \subseteq X$ is open in (X, d) if and only if U is equal to a union of open metric balls.)

12. Show that the topology generated by $\mathscr{B}_0$ as defined in (4.1) is the metric topology on $\mathbb{R}^n$.

13. Let $\mathscr{B} = \{[a, b) \mid a, b \in \mathbb{R}, \text{ with } a < b\}$.

   (a) Show that $\mathscr{B}$ is a base.
   (b) The topology generated by $\mathscr{B}$ is called the *lower-limit topology* on $\mathbb{R}$. Prove that the lower-limit topology is finer than the Euclidean topology on $\mathbb{R}$.

14. Suppose that $\mathscr{B}_1$ and $\mathscr{B}_2$ are bases on a set X, and suppose that $\mathscr{B}_1 \subseteq \mathscr{B}_2$. Prove that the topology $\mathscr{T}_1$ generated by $\mathscr{B}_1$ is either the same or coarser than than the topology $\mathscr{T}_2$ generated by $\mathscr{B}_2$.

4.2   Continuity and Limits

This section shows how the definition of continuity from §2.2 may be abstracted to functions of topological spaces.

Continuous Functions

When you first learned about graphing functions, $y = f(x)$, perhaps you were told that f is *continuous* if its graph can be drawn without lifting your pencil. This "definition" leaves a lot to be desired. Then, in calculus, you finally encountered a precise definition in terms of *limits*.

> Let $D \subseteq \mathbb{R}$ be an open subset of the domain of f. A real-variable function $f : D \to \mathbb{R}$ is **continuous** on D if for every $c \in D$, we have $\lim_{x \to c} f(x) = f(c)$.

However, the definition is useless unless we know how to compute limits. Limits are all about *closeness*: the notation $\lim_{x \to c} f(x) = L$ means, intuitively, that "when x approaches a value c, then the values of $f(x)$ approach a limit value L." But how does x *approach* c anyway? And if the values of $f(x)$ approach L but never actually reach L, then how can we claim the "limit" is L? Certainly, for a limit to exist, we require the values of f to get closer and closer to L, but how close is close enough? The quandary is quite deep. Not only do these questions cause immeasurable grief to anyone studying calculus, but they also gave mathematicians quite a headache in the past as they wrestled with the foundational concepts behind what makes calculus tick. Over time, mathematicians[6] refined the concept of limit, arriving at what we often call the "ϵ-δ defintion."

> The limit equation $\lim_{x \to c} f(x) = L$ means that for every $\epsilon > 0$, there exists a value $\delta > 0$ (whose value depends on ϵ) such that if $0 < |x - c| < \delta$, then $|f(x) - L| < \epsilon$.

Putting this together with the definition of continuity above, we obtain the precise definition of continuity that we have already encountered in Definition 2.2.1, which is reproduced below in a more concise form.

> Let $f : D \to \mathbb{R}$, where $D \subseteq \mathbb{R}$ is an open subset of the domain of f. We say that f is **continuous** on D if:
>
> $\forall c \in D, \forall \epsilon > 0, \exists \delta > 0$ such that
>
> $$(x \in D \text{ and } |x - c| < \delta) \implies |f(x) - f(c)| < \epsilon.$$

---

[6]Among them was Bernard Bolzano, who is often credited with formulating the ϵ-δ definition in 1817.

Now this definition was generalized in Definitions 2.2.2 and 2.2.3 to arbitrary functions in Euclidean spaces using the machinery of open balls. Recall that underneath all of the details there was a simple nugget: a function is continuous if and only if the inverse image of any open set (in the form of open ϵ-balls) in the codomain is an open set in the domain. Presently, we abstract the definition even further.

Definition 4.2.1. Suppose $f : X \to Y$ is a function from one set to another, and suppose X has a topology $\mathscr{T}_X$, and Y has a topology $\mathscr{T}_Y$. Then f is **continuous** (with respect to the topologies on X and Y) if

$$f^{-1}[V] \in \mathscr{T}_X \text{ for every } V \in \mathscr{T}_Y.$$

Definition 4.2.1 states that for a function f to be continuous, all that is required is that the preimage of every open set of Y is some open set of X. Notice that there is no explicit mention of the individual points of X or Y. Topologically, individual points are not as important as whole collections of points that are close to one another (open sets).

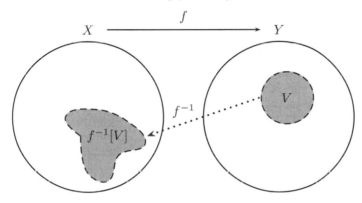

Although the diagram shows "f^{-1}," we cannot assume that the inverse function for f exists. Just think of f^{-1} as an operation that can be applied to a set, yielding a set as output (see Definition A.2.8 of §A.2). Another word of caution: The definition for continuity does **not** imply that $f[U]$ is open for all open $U \subseteq X$. If a continuous function f does satisfy the condition $U \in \mathscr{T}_X \implies f[U] \in \mathscr{T}_Y$, then we call f an **open map**.

Example 63. Let X and Y be nonempty topological spaces, and fix a point $y_0 \in Y$. Let $f : X \to Y$ be the map that sends $x \mapsto y_0$ for every $x \in X$ (this kind of map is called a **constant map**). Prove that f is continuous. Under what conditions would f be an open map?

Solution: We simply check that the preimage of every open set of Y is open in X. Even though we do not explicitly know the open sets of X or Y in this example, we have just enough information to proceed. If $V \in \mathscr{T}_Y$, then either $y_0 \in V$ or $y_0 \notin V$. If $y_0 \in V$, then $f^{-1}[V] = X$ (since $f(x) = y_0$ for all $x \in X$). If $y_0 \notin V$, then $f^{-1}[V] = \emptyset$. By the definition of a topology, we know both $\emptyset \in \mathscr{T}_X$ and $X \in \mathscr{T}_X$. Thus $f^{-1}[V]$ is open in all cases, proving that f is continuous.

Since $f[U] = \{y_0\}$ for every open set U, we find that f would be an open map if and only if the singleton set $\{y_0\}$ is open in Y. (Here f would not be open, for example, if $Y = \mathbb{R}^n$ for $n \geq 1$, since a single point is closed, and not open, in Euclidean space. On the other hand, if $Y = \mathbb{Z}$ with the discrete topology, then f must be an open map.)

We shall see the real power of abstraction, as we can now prove the following important composition theorem in a few lines (compare Exercise 1 from §2.2).

Theorem 4.2.2. *The composition of continuous functions is continuous. More precisely, if X, Y, and Z are topological spaces, and we have continuous functions $f : X \to Y$ and $g : Y \to Z$, then the composite function $g \circ f : X \to Z$ is continuous.*

Proof. Let V be any open set of Z. We must show that $(g \circ f)^{-1}[V] = f^{-1}\left[g^{-1}[V]\right]$ is open in X. Since g is continuous, $g^{-1}[V]$ is open in Y. Then, since f in continuous, $f^{-1}\left[g^{-1}[V]\right]$ is open in X. $\qquad\square$

Continuity may also be defined in terms of closed sets and closures of sets.

Theorem 4.2.3. *Let X and Y be topological spaces, and let $f : X \to Y$. Then the following are equivalent:*

(i) f is continuous (for every open set $V \subseteq Y$, $f^{-1}[V] \subseteq X$ is open).

(ii) for every closed set $C \subseteq Y$, $f^{-1}[C] \subset X$ is closed.

(iii) f preserves closure; that is, for every $A \subseteq X$, $f[\overline{A}] \subseteq \overline{f[A]}$.

Our plan for this this proof is to show that (i) $\Longleftrightarrow$ (ii) and (ii) $\Longleftrightarrow$ (iii).

Proof.   ((i) $\Longrightarrow$ (ii)): Suppose $f : X \to Y$ is continuous, and let $C \subseteq Y$ be closed. Then, by definition, $C = Y \setminus V$ for some open set V. There is a set identity, $f^{-1}[C] = f^{-1}[Y \setminus V] = f^{-1}[Y] \setminus f^{-1}[V]$ (see §A.2, Exercise 7(d)). Observe that $f^{-1}[Y] = X$, and $U = f^{-1}[V]$ is an open set in X, and so we have shown that $f^{-1}[C] = X \setminus U$ is closed.

The converse ((ii) $\Longrightarrow$ (i)) is left to the reader.

((ii) $\Longrightarrow$ (iii)): Assume that the inverse image of any closed set is closed. Let $A \subseteq X$ be arbitrary. Since $f[A] \subseteq \overline{f[A]}$, we have $f^{-1}[f[A]] \subseteq f^{-1}\left[\overline{f[A]}\right]$. Moreover, since $A \subseteq f^{-1}[f[A]]$ (see Proposition A.2.10), we have by transitivity $A \subseteq f^{-1}\left[\overline{f[A]}\right]$. Now since $\overline{f[A]}$ is closed, so is $f^{-1}\left[\overline{f[A]}\right]$, but then the previous set inclusion implies that $\overline{A} \subseteq f^{-1}\left[\overline{f[A]}\right]$ because the closure $\overline{A}$ is the *smallest* closed set containing the set A (Exercise 12, §2.1). Applying the forward image to both sides, we find $f[\overline{A}] \subseteq f\left[f^{-1}\left[\overline{f[A]}\right]\right] \subseteq \overline{f[A]}$.

((iii) $\implies$ (ii)): Assume that $f[\overline{A}] \subseteq \overline{f[A]}$ for any subset $A \subseteq X$. Let $C \subseteq Y$ be a closed set, and let $A = f^{-1}[C]$. Our goal is to show that $A = \overline{A}$, which would imply that $A \subseteq X$ is closed. Observe that $f[A] = f\left[f^{-1}[C]\right] \subseteq C$. But by assumption, we also have $f[\overline{A}] \subseteq \overline{f[A]}$, so that $f[\overline{A}] \subseteq \overline{C} = C$. Now $\overline{A} \subseteq f^{-1}\left[f[\overline{A}]\right] \subseteq f^{-1}[C] = A$. Finally, since $A \subseteq \overline{A}$ and $\overline{A} \subseteq A$, we have $A = \overline{A}$, proving $A = f^{-1}[C]$ closed.

$\square$

Homeomorphism

Suppose X and Y are two topological spaces. A function $f : X \to Y$ is called a **homeomorphism** if it is both *bijective* and *bicontinuous* (recall Definition 2.2.4 of Chapter 2). Equivalently, f is a homeomorphism if it is a bijective continuous open map, since first bijectivity implies that there exists a function $f^{-1} : Y \to X$, and then openness of f implies that f^{-1} is continuous since $(f^{-1})^{-1}[U] = f[U]$ is open in Y for every open $U \subseteq X$.

If a homeomorphism exists between X and Y, then we say that X and Y are **homeomorphic**, or **topologically equivalent**, and we write $X \approx Y$. To a topologist, the two spaces X and Y are indistinguishable, even if they "look" very different on the page. For example, as discrete spaces, $\mathbb{Z} \approx \mathbb{Z}^2$. To prove this claim, we need to define a homeomorphism $f : \mathbb{Z} \to \mathbb{Z}^2$. First note that every map f from one discrete space to another is both continuous and open (see Exercise 1). So long as there is a bijection $f : \mathbb{Z} \to \mathbb{Z}^2$, then f is guaranteed to be a homeomorphism. The tricky part is to construct a bijection.

The method shown here is a variant of that used by Georg Cantor[7] to show that there is a bijection $\mathbb{N} \to \mathbb{Q}$. All that we have to do is construct an embedded copy of $\mathbb{Z}$ in $\mathbb{Z}^2$ that hits every point. Figure 4.3 shows the construction.[8]

---

[7] A more famous result, known as *Cantor's Diagonalization Argument*, shows that there is no bijection $f : \mathbb{Z} \to \mathbb{R}$.

[8] You may be worried that we have not expressed the function $f : \mathbb{Z} \to \mathbb{Z}^2$ in a standard or explicit way. This is a valid concern, as oftentimes a visual "proof" has subtle flaws that invalidate it. But as you might imagine, the function f would be quite hard to write down in terms of operations involving the input variable n, and even after writing it down it would be extremely tedious to verify that the expression defines an injective and surjective map anyway. In this case, the "proof by picture" in Figure 4.3 is in fact explicit enough to prove the point.

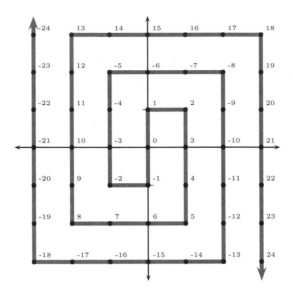

Figure 4.3: A visual bijection $f : \mathbb{Z} \to \mathbb{Z}^2$. For each $n \in \mathbb{Z}$, $f(n)$ is defined as the point in $\mathbb{Z}^2$ whose label is n. For example, $f(0) = (0,0)$, $f(5) = (1,-2)$, $f(-11) = (2,-1)$, etc.

Example 64. Let $n \in \mathbb{N}$. Prove that $\mathbb{R}^n \approx B_\epsilon(\mathbf{0})$ for any $\epsilon > 0$.

Solution: It suffices to prove the homeomorphism for a specific value of ϵ, since every ϵ-ball (for various $\epsilon > 0$) is homeomorphic to the open unit ball via scaling. The function (4.2) is a homeomorphism $\mathbb{R}^n \to B_{\pi/2}(\mathbf{0})$.

$$f(\mathbf{x}) = \begin{cases} \mathbf{0} & \mathbf{x} = \mathbf{0} \\ \frac{\tan^{-1}\|\mathbf{x}\|}{\|\mathbf{x}\|}\mathbf{x} & \mathbf{x} \neq \mathbf{0} \end{cases} \tag{4.2}$$

The idea behind (4.2) is to extend Example 21 from Chapter 2 to higher dimensions. Every line through the origin is sent by f to a finite open segment of length π centered at $\mathbf{0}$, just as the single-variable function $y = \tan^{-1} x$ sends the entire real number line $(-\infty, \infty)$ to the open segment $(-\pi/2, \pi/2)$.

We note that f is continuous away from $\mathbf{0}$ because it is the composition of continuous functions, including the inverse tangent function and the norm function. The fact that f is continuous at $\mathbf{x} = \mathbf{0}$ can be established by analyzing the norm of $f(\mathbf{x})$ itself. We simply have to show that $\|f(\mathbf{x})\|$ can be made arbitrarily small by taking $\mathbf{x}$ close enough to $\mathbf{0}$. When $\mathbf{x} \neq \mathbf{0}$, we have

$$\|f(\mathbf{x})\| = \frac{\tan^{-1}\|\mathbf{x}\|}{\|\mathbf{x}\|}\|\mathbf{x}\| = \tan^{-1}\|\mathbf{x}\|. \tag{4.3}$$

Now since $\tan^{-1} x$ is a continuous function, and since $\tan^{-1} 0 = 0$, there exists a small neighborhood U of $\mathbf{0}$ such that if $\mathbf{x} \in U$, then $\tan^{-1}\|\mathbf{x}\| < \epsilon$ for any prescribed $\epsilon > 0$. Thus f is continuous on all of $\mathbb{R}^n$.

It should be clear that f is injective and that the image is all of $B_{\pi/2}(\mathbf{0})$. We leave it to you to verify that f is an open map, which is the final ingredient needed to prove that $f : \mathbb{R}^n \to B_{\pi/2}(\mathbf{0})$ is a homeomorphism.

Limits of Sequences

We've already encountered sequences and limits in $\mathbb{R}^n$ (see §2.3). But to a topologist, $\mathbb{R}^n$ is simply one out of an infinite variety of different spaces, and so we extend Definition 2.3.6 to cover sequences of points in an arbitrary topological space X.

> **Definition 4.2.4.** A **sequence** in a topological space X is a function $x : \mathbb{N} \to X$.

As in Chapter 2, the usual notation for a sequence is:

$$(x_k) = (x_k)_{k \in \mathbb{N}} = (x_1, x_2, x_3, \ldots), \quad x_k \in X, \forall k.$$

Definition 2.3.7 also extends easily to arbitrary spaces X.

> **Definition 4.2.5.** A point $x \in X$ is a **limit** of a sequence (x_k) if for every open set U containing x, there is a number $N \in \mathbb{N}$ such that $x_k \in U$ for all $k \geq N$. If such a point exists, then we call the sequence **convergent**.

We must be careful though, because in the most general case, we cannot expect a convergent sequence to converge to a unique limit, as the next two examples illustrate.

Example 65. Let $\mathscr{T}_I$ be the indiscrete topology on $\mathbb{R}$. If (x_k) is any sequence in $\mathbb{R}$, then *every* point $x \in \mathbb{R}$ is a limit of the sequence. Because the only nonempty set in $\mathscr{T}_I$ is the whole space $\mathbb{R}$, Definition 4.2.5 is automatically satisfied, regardless of the sequence *or* the limit point.

Example 66. Let X be a point-set. The **cofinite topology** $\mathscr{T}_C$ on X is defined by

$$\mathscr{T}_C = \{\emptyset\} \cup \{U \subseteq X \mid X \setminus U \text{ is finite}\}.$$

Consider the real number line $\mathbb{R}^1$ with cofinite topology. Let (x_k) be a sequence of *distinct* terms in $\mathbb{R}$. Then the sequence (x_k) converges to any point $x \in \mathbb{R}$. This is because every nonempty open set U has the property that $X \setminus U$ is finite, and so only finitely many of the terms of (x_k) can be in $X \setminus U$. If x_K is the highest-index term in $X \setminus U$, then $x_k \in U$ for all $k \geq K + 1$. *Question: What could happen if the terms of (x_k) were not required to be distinct?*

On the other hand, if the topology on X is what is called *Hausdorff*, then every sequence has at most one limit.

> **Definition 4.2.6.** A space X is called **Hausdorff** if for each pair of distinct points $x, y \in X$ there exist disjoint open sets $U, V \subseteq X$ such that $x \in U$ and $y \in V$.

Definition 4.2.6 is an example of a *separation condition*[9] – any two points of a Hausdorff space can be separated by a pair of open sets. Let's see how this leads to unique limits.

Suppose $(x_k)_{k \in \mathbb{N}}$ is a sequence in a Hausdorff space X, and suppose that x and y are both limits of (x_k). If $x \neq y$, then there are disjoint open sets $U, V \subseteq X$ such that $x \in U$ and $y \in V$ (see Figure 4.4). Since U is open containing the limit x, there is a number $N \in \mathbb{N}$ such that if $k \geq N$, then $x_k \in U$. But V is also open containing a limit y, so there is a number $M \in \mathbb{N}$ such that if $k \geq M$, then $x_k \in V$. Thus, for $k \geq \max\{M, N\}$, we have $x_k \in U \cap V$, contradicting $U \cap V = \emptyset$.

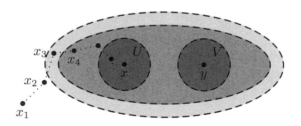

Figure 4.4: If $x \neq y$, then there are disjoint open sets U and V that separate x and y. This leads to unique limits.

Intrinsic Definitions

At this point, we have adapted many of the definitions from Euclidean or metric topology into the more abstract setting of point-set topology. The definitions

---

[9]There is a rich hierarchy of separation conditions for topological spaces. These conditions are classified $T_0, R_0, T_1, R_1, T_2, T_{2\frac{1}{2}}, T_3$, etc., or by name, *Kolmogorov, Fréchet, Hausdorff, regular, normal*, etc. See Schechter [Sch97] or Steen and Seebach [SS95] for more details.

are *intrinsic* in that they depend only on the topology $\mathscr{T}$ on a given point-set X, not on any particular representation of the space. The following table summarizes what we have defined so far as well as a few additional definitions.

Suppose $(X, \mathscr{T}_X)$ and $(Y, \mathscr{T}_Y)$ are topological spaces.

- $U \subseteq X$ is *open* $\iff$ $U \in \mathscr{T}_X$

- $C \subseteq X$ is *closed* $\iff$ $X \setminus C \in \mathscr{T}_X$

- $f : X \to Y$ is *continuous* $\iff$ $\forall V \in \mathscr{T}_Y, f^{-1}[V] \in \mathscr{T}_X$

- $x \in X$ is a *limit point* of $A \subseteq X$ $\iff$ $\forall U \in \mathscr{T}_X$ such that $x \in U$, we have $U \cap (A \setminus \{x\}) \neq \emptyset$

- $x \in X$ is an *interior point* of $A \subseteq X$ $\iff$ $\exists U \in \mathscr{T}_X$ such that $x \in U \subseteq A$

- $x \in X$ is an *exterior point* of $A \subseteq X$ $\iff$ $\exists U \in \mathscr{T}_X$ such that $x \in U \subseteq X \setminus A$

The *interior* and *closure* of $A \subseteq X$ have particularly nice interpretations with respect to only open and closed sets (rather than individual points). The interior of A is the union of all open sets contained in A, and the closure is the intersection all all closed sets containing A. In the language of set theory,

$$\text{int}(A) = \bigcup \{U \mid (U \in \mathscr{T}_X) \text{ and } (U \subseteq A)\}, \text{ and}$$

$$\overline{A} = \bigcap \{C \mid (C = X \setminus U \text{ for some } U \in \mathscr{T}_X) \text{ and } (C \supseteq A)\}.$$

Exercises

1. Let X and Y be arbitrary topological spaces. Prove the following.

   (a) If X is a discrete space, then every function $f : X \to Y$ is continuous.

   (b) If Y is a discrete space, then every function $f : X \to Y$ is an open map.

   (c) If Y is an indiscrete space, then every surjective function $f : X \to Y$ is continuous.

   (d) If X is an indiscrete space and Y is discrete, then the only continuous functions $f : X \to Y$ are the constant functions.

2. Suppose that X is a discrete space. Show that the only convergent sequences are those that are eventually constant. By definition, a sequence is **eventually constant** if there is a constant $x \in X$ and an index $K \in \mathbb{N}$ such that $x_k = x$ for all $k \geq K$.

3. Finish the proof of Theorem 4.2.3 by showing that (ii) $\implies$ (i).

4. Suppose $X = (\mathbb{R}, \mathscr{T}_{\text{lower}})$ where $\mathscr{T}_{\text{lower}}$ is the lower-limit topology as defined in Exercise 13 of §4.1, and let $Y = (\mathbb{R}, \mathscr{T}_E)$ be the real line with standard Euclidean topology. Show why the *step function* $f : X \to Y$ defined by $f(x) = 0$ if $x < 0$, and $f(x) = 1$ if $x \geq 0$, is continuous.

5. Let X be any point-set, and let $\mathscr{T}_C$ be the cofinite topology as defined in Example 66.

   (a) Prove that $\mathscr{T}_C$ is a topology on X.

   (b) Describe the closed sets in the cofinite topology on X.

   (c) Prove that the cofinite topology on a space X is the same as the discrete topology on X if and only if X is finite.

6. Prove that if $A \subseteq B$, then $\text{int}(A) \subseteq \text{int}(B)$ and $\overline{A} \subseteq \overline{B}$.

7. Prove that every metric space is Hausdorff.

8. Determine which of the following spaces is Hausdorff.

   (a) An indiscrete space having at least two points.

   (b) The space $\mathbb{N}$ having topology $\mathscr{T}$ given in Example 56.

   (c) The finite topological space from Example 58.

   (d) $(\mathbb{R}, \mathscr{T}_{\text{lower}})$, where $\mathscr{T}_{\text{lower}}$ is the lower-limit topology as defined in Exercise 13 of §4.1.

   (e) $(\mathbb{R}, \mathscr{T}_C)$, where $\mathscr{T}_C$ is the cofinite topology as defined in Example 66.

4.3 Subspace Topology and Quotient Topology

Suppose X is a topological space, and A is a subset of X. It seems reasonable to expect a topology on A related to that of X. In this section, we explore the concept of the *subspace topology* on a subset $A \subseteq X$. We also consider a natural way to define a topology on equivalence classes of a space, the so-called *quotient topology*.

Subspace Topology

In any metric space X (including the Euclidean spaces), any subset $A \subseteq X$ inherits the metric by restriction of domain, and so has a topology *induced* from the topology on X. In a similar way, if X is a point-set with topology $\mathscr{T}$, and $A \subseteq X$, then there is a natural way to induce a topology on A by "restricting" $\mathscr{T}$.

Definition 4.3.1. A subset A of a topological space $(X, \mathscr{T})$ may be given the **subspace topology** $\mathscr{T}|_A$, defined by:

$$V \in \mathscr{T}|_A \iff V = A \cap U, \quad \text{for some } U \in \mathscr{T}.$$

If $V \in \mathscr{T}|_A$, we say V is **open in** A. Similarly, C is **closed in** A if $A \setminus C \in \mathscr{T}|_A$.

Caution: A set Y may be open in $A \subseteq X$, but not open in X, and Y may be closed in A, but not closed in X.

Example 67. For any $m < n$, the Euclidean topology on $\mathbb{R}^m$ is the subspace topology induced from $\mathbb{R}^n$. Indeed, any open ball in $\mathbb{R}^m$,

$$B = \{y \in \mathbb{R}^m \mid d(x, y) < \epsilon\},$$

is simply the intersection of an open ball in $\mathbb{R}^n$ and the subspace $\mathbb{R}^m \times \mathbf{0} \subseteq \mathbb{R}^n$, where $\mathbf{0} \in \mathbb{R}^{n-m}$ (see Example 141):

$$B = \{y \in \mathbb{R}^n \mid d(x, y) < \epsilon\} \cap (\mathbb{R}^m \times \mathbf{0}).$$

Example 68. Consider the sphere $\mathbb{S}^2 \subseteq \mathbb{R}^3$ with respect to the subspace topology. An open ball $B_\epsilon(\mathbf{x})$ that intersects the sphere does so in an open "lens," as shown in Figure 4.5. All open sets of $\mathbb{S}^2$ are made up of unions of these open lenses.

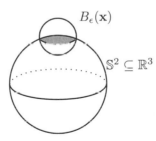

Figure 4.5: Open sets in the sphere $\mathbb{S}^2$ with respect to the subspace topology.

Example 69. Let A be the x-axis considered as a subspace of $\mathbb{R}^2$, and let $U = \{(x, 0) \mid 1 < x < 3\} \subseteq A$, as illustrated in Figure 4.6. Note that U is not open in the larger space $\mathbb{R}^2$ (neither is U closed in $\mathbb{R}^2$ – *why?*), but U is open in A because $U = A \cap B_1((2, 0))$. Intuitively, we may ignore the y-direction and identify $A \approx \mathbb{R}^1$. Then $U \approx (1, 3) = B_1(2)$ is an open interval of the real line.

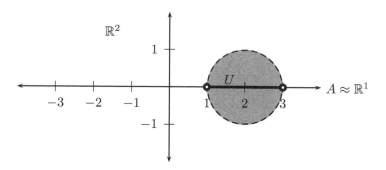

Figure 4.6: The set U is open in $A \approx \mathbb{R}^1$, but not open in $\mathbb{R}^2$.

Theorem 4.3.2. *Let A be a subspace of X.*

- *If U is open in A and A is open in X, then U is open in X.*

- *If U is closed in A and A is closed in X, then U is closed in X.*

Proof. Suppose U is open in A and A is open in X. Since U is open in A, and A is a subspace of X, there is an open set $W \subseteq X$ such that $U = W \cap A$. But since A is an open set of X, the intersection $W \cap A$ is also open in X. The proof of the second point is left to the reader. □

One important implication of Definition 4.3.1 is that the **inclusion** map from a subpace to its containing space is always continuous with respect to the subspace topology. If $A \subseteq X$, define the following.

$$\begin{aligned} i \;:\; & A \to X \\ i(x) \;=\; & x \end{aligned}$$

The inclusion map may seem quite trivial; after all, it's a map that seems to do *nothing* (input and output are the same: x). But the importance of this map is that it allows us to interpret the subset relationship (set theory) in terms of a continuous function (topology). In fact, we could have defined the subspace topology on A as the coarsest topology on A such that the inclusion $i : A \to X$ is continuous.

Theorem 4.3.2 does not address the situations in which a subset U is open in a *closed* subset C of X, or vice versa. In fact, Example 69 demonstrates a set U that is open in $\mathbb{R}^1$, but not open in $\mathbb{R}^2$. However, any *closed* subset of the real number line, such as $[0, 1]$ must be closed in $\mathbb{R}^2$ as well, by Theorem 4.3.2.

Example 70. Let X be the point-set with topology $\mathscr{T}$ from Examples 58 and 59. Let $A = \{b, c\}$. Determine $\mathscr{T}|_A$.

Solution: Intersect each set in $\mathscr{T}$ with A to find the open sets of $\mathscr{T}|_A$:

$$
\begin{aligned}
A \cap \emptyset &= \emptyset \\
A \cap \{a\} &= \emptyset \\
A \cap \{d\} &= \emptyset \\
A \cap \{a, b\} &= \{b\} \\
A \cap \{a, d\} &= \emptyset \\
A \cap \{a, b, d\} &= \{b\} \\
A \cap \{a, b, c, d\} &= \{b, c\}.
\end{aligned}
$$

Thus $\mathscr{T}|_A = \{\emptyset, \{b\}, \{b, c\}\}$. Observe that $\{b\}$ is open in A but not open in X. Similarly, $\{c\}$ is closed in A, since $A \setminus \{c\} = \{b\} \in \mathscr{T}|_A$, and $\{c\}$ happens to be closed in X as well.

Our work in Example 70 could have been reduced by only considering those sets $U \in \mathscr{T}$ that have elements in common with A (since we already know $\emptyset \in \mathscr{T}|_A$). Another way to be more efficient, especially when the topology is not finite, is to work with the bases of the topologies. It turns out that a base for the subspace topology can be found by taking intersections of a base of the larger space.

Theorem 4.3.3. *If $\mathscr{B}$ is a base for a topology $\mathscr{T}$ on X, and $A \subset X$, then a base for $\mathscr{T}|_A$ is*

$$
\mathscr{B}|_A = \{B \cap A \mid B \in \mathscr{B}\}.
$$

Example 71. Consider the unit square

$$
\mathbb{I}^2 = \mathbb{I} \times \mathbb{I} = \{(x, y) \in \mathbb{R}^2 \mid 0 \le x \le 1 \text{ and } 0 \le y \le 1\}
$$

as a subspace of $\mathbb{R}^2$. We know that $\mathscr{B} = \{B_\epsilon(x, y) \mid \epsilon > 0, (x, y) \in \mathbb{R}^2\}$ is a base for $\mathbb{R}^2$. By Theorem 4.3.3, the set of all intersections of these open balls with the square forms a base for $\mathbb{I}^2$. To avoid trivial intersections, we may specify that the center of each open ball must be a member of $\mathbb{I}^2$.

$$
\mathscr{B}|_{\mathbb{I}^2} = \{B_\epsilon(x, y) \cap \mathbb{I}^2 \mid \epsilon > 0, x, y \in \mathbb{I}\}
$$

Notice that some sets in $\mathscr{B}|_{\mathbb{I}^2}$ would not be considered open in $\mathbb{R}^2$. For example, the semicircular region $U = B_{1/4}(0, 1/2) \cap \mathbb{I}^2$ shown below is open in $\mathbb{I}^2$, but U is not an open subset of $\mathbb{R}^2$.

Note that since $\mathbb{I}^2$ is not open in $\mathbb{R}^2$, Theorem 4.3.2 cannot be used to "prove" U is open in $\mathbb{R}^2$.

Example 72. Describe the topologies on $\mathbb{Z}$ and $\mathbb{Q}$ as subspaces of $\mathbb{R}$.

Solution: $\mathbb{Z} = \{\ldots, -3, -2, -1, 0, 1, 2, 3, \ldots\}$ has the discrete topology as a subspace of $\mathbb{R}$, since each open ball $B_1(k) = (k - 1, k + 1) \subseteq \mathbb{R}$ intersects $\mathbb{Z}$ in the single point $\{k\}$.

The subspace topology on $\mathbb{Q}$ is generated by the rational intervals $(a, b) \cap \mathbb{Q}$. In contrast to $\mathbb{Z}$, the space topology on $\mathbb{Q}$ is not discrete. However, $\mathbb{Q}$ is a **totally disconnected** space, meaning that the only connected subsets of $\mathbb{Q}$ are the singleton sets $\{x\}$ (see Example 78).

Suppose now that $f : X \to Y$ is a continuous function. Is the restriction $f : X \to f[X]$ still continuous with respect to the subspace topology on $f[X] \subseteq Y$? Consider an open set $U \subseteq f[X]$. By definition, there is an open set $W \subseteq Y$ such that $U = W \cap f[X]$. Then $f^{-1}[U] = f^{-1}[W \cap f[X]]$. The points in $W \setminus f[X]$ do not contribute to $f^{-1}[U]$, so we may write $f^{-1}[U] = f^{-1}[W]$. Finally, since f is continuous $X \to Y$, the inverse image $f^{-1}[W]$ $(= f^{-1}[U])$ is open in X, proving that $f : X \to f[X]$ is also continuous.

Quotient Topology

Consider a length of string, a good model for the interval $\mathbb{I} = [0, 1]$. We know $\mathbb{I} \not\simeq \mathbb{S}^1$; however, if the two ends of the string are tied or glued together, then the resulting loop is a circle. The process of *gluing* together parts of one space to make a new space is called **identification**, and it is more than just sticking pieces together. To say that two points x and y in a space X should be *identified* or *glued* means that the points now act as a single point. That is, if we identify the endpoints 0 and 1 of the interval $\mathbb{I}$, then this means that 0 is *the same point* as 1 in the result. We will use the concept of equivalence relations and equivalence classes to make this precise (see §A.2). It's important to realize that the space may be topologically different after the identification is made (*identification* is not a type of *deformation*). In the above example, the new space, which we denote by $\mathbb{I}/\{0 \sim 1\}$, is in fact homeomorphic[10] to the circle $\mathbb{S}^1$ (see Figure 4.7).

If X is a set with an equivalence relation $\sim$ defined on it, then there is a natural way to define a surjective function $X \to X/\sim$, sending each point to its corresponding equivalence class. We call this function the **quotient map** q, and it is defined as follows:

$$q \ : \ X \to X/\sim$$
$$q(x) \ = \ [x].$$

---

[10]Note that we haven't actually shown that the result is a topological space at all, let alone a circle; read further for more details.

Figure 4.7: Identifying $0 \sim 1$ in the interval $\mathbb{I}$ results in a space homeomorphic to $\mathbb{S}^1$.

The quotient map is not injective, unless the equivalence relation is simply equality. Indeed, if $x, y \in X$ are distinct elements and if $x \sim y$, then $q(x) = q(y)$ in $X/\sim$. We define a topology on $X/\sim$ using the map q and the topology of X itself.

> **Definition 4.3.4.** If X has a topology $\mathscr{T}$, then $X/\sim$ has an induced topology called the **quotient topology** $\mathscr{T}/\sim$, defined by:
>
> $$U \in \mathscr{T}/\sim \iff q^{-1}[U] \in \mathscr{T}.$$
>
> Then $X/\sim$ is called a **quotient space** or **identification space** of X with respect to the relation $\sim$.

You will be asked in the Exercises to prove that the quotient topology is indeed a topology on the quotient set $X/\sim$. By definition, the quotient map $q : X \to X/\sim$ is continuous ($q^{-1}[U]$ is open in X for every U open in $X/\sim$).

Example 73. Let's revisit the quotient space $\mathbb{I}/\sim$ in which the equivalence relation is defined[11] by $0 \sim 1$ (i.e., $[0] = [1]$ in $\mathbb{I}/\sim$). The *points* of $\mathbb{I}/\sim$ are in fact certain subsets of $\mathbb{I}^2$: the equivalence classes of $\mathbb{I}$ under $\sim$, which are, in this case, the singleton sets, $[t] = \{t\}$ for $t \subset (0,1)$, and the set of two identified points, $[0] = \{0,1\}$.

$$\mathbb{I}/\sim \; = \{[t] \mid t \in (0,1)\} \cup \{[0]\}$$

What are the open sets of $\mathbb{I}/\sim$? For any $0 \le a < b \le 1$, define the sets $V_{a,b} = \{[t] \in \mathbb{I}/\sim \; \mid a < t < b\} \subseteq \mathbb{I}/\sim$. Let $q : \mathbb{I} \to \mathbb{I}/\sim$ be the projection map. Then $q^{-1}[V_{a,b}] = (a,b)$ is open in $\mathbb{I}$. Therefore $V_{a,b}$ is open in $\mathbb{I}/\sim$. Now define a related set $W_{a,b} = \{[t] \in \mathbb{I}/\sim \; \mid t < a \text{ or } t > b\}$. Since $q^{-1}[W_{a,b}] = [0,a) \cup (b,1] \subseteq \mathbb{I}$, the sets $W_{a,b}$ are open in $\mathbb{I}/\sim$. These two collections of open sets $V_{a,b}$ and $W_{a,b}$ (for various a, b) form a base for the topology on $\mathbb{I}/\sim$.

You may be wondering what happened to intervals containing just one endpoint. After all, $[0,a)$ is open in $\mathbb{I}$. However, the analogous set in $\mathbb{I}/\sim$ is *not*

---

[11]When we say an equivalence relation $\sim$ on X is **defined by** a set of specific relations, we mean that there is a minimal subset $E \subseteq X \times X$ such that $(x,y) \in E$ for all given relations $x \sim y$, and such that E satisfies Definition A.2.2. In particular, it is understood that $x \sim x$ for all $x \in X$.

open. For $0 < a < 1$, let $Y_a = \{[t] \in \mathbb{I}/\sim \;|\; 0 \le t < a\}$. Since $[0] = [1]$, we have $q^{-1}[Y_a] = [0, a) \cup \{1\} \subseteq \mathbb{I}$, which is not open in $\mathbb{I}$. This shows, by definition, that Y_a is not an open set of $\mathbb{I}/\sim$. In fact, any open set containing $[0]$ must contain points $[t]$ such that $t > 0$ (near 0) *and* points $[u]$ such that $u < 1$ (near 1).

Our next goal is to prove that $\mathbb{I}/\sim \;\approx\; \mathbb{S}^1$. The easiest way to do so is to set up another map,

$$f \;:\; \mathbb{I} \to \mathbb{S}^1$$
$$f(t) \;=\; (\cos 2\pi t, \sin 2\pi t).$$

Here we are using a particular model for $\mathbb{S}^1$, namely, the *unit circle*,

$$\mathbb{S}^1 = \{(x, y) \in \mathbb{R}^2 \mid x = \cos 2\pi t, y = \sin 2\pi t, t \in \mathbb{R}\}.$$

Note that $f(0) = (\cos 0, \sin 0) = (\cos 2\pi, \sin 2\pi) = f(1)$, so we consider related function $g : \mathbb{I}/\sim \;\to \mathbb{S}^1$ defined by $g([t]) = f(t)$ for all $t \in \mathbb{I}$. We have defined g on equivalence classes, while f is defined on the actual points of $\mathbb{I}$. Observe that both f and g are injective when restricted to $(0, 1)$; only the behavior at the endpoints is distinct. While f is not injective at the endpoints ($f(0) = f(1)$ but $0 \ne 1$ in $\mathbb{I}$), g *is* injective on $\mathbb{I}/\sim$, since even though $g([0]) = g([1])$, the classes $[0]$ and $[1]$ are not distinct elements of $\mathbb{I}/\sim$. The map g is also surjective to $\mathbb{S}^1$, since every point of the unit circle corresponds to a particular angle $\theta = 2\pi t$.

Thus there is a bijection between $\mathbb{S}^1$ and $\mathbb{I}/\sim$. All that remains is to prove that the maps f and f^{-1} are continuous, but we leave this for the reader to verify.

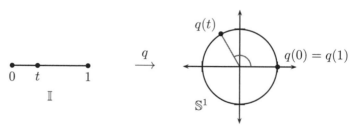

Example 74. Consider the unit square $\mathbb{I}^2$. We will make a number of identification spaces from $\mathbb{I}^2$. In what follows, we often drop the equivalence class notation.[12] In other words, for $x \in X$, we may write x instead of $[x]$ for the corresponding equivalence class in $X/\sim$.

(a) If identifications are made along the left and right edges so that $(0, y) \sim (1, y)$ for all $y \in \mathbb{I}$, then we get a **cylinder** (having no top or bottom surface). You can easily construct it by bending a sheet of paper it until the two side edges meet, and gluing or taping those two edges together, as shown below.

---

[12]This is what typically happens in practice, but it requires the reader to be extra vigilant in using context to determine whether an element is simply a point or represents an equivalence class.

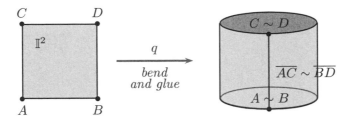

(b) Now suppose the identifications are made with a *twist*. That is, $(0, y) \sim (1, 1-y)$ for all $y \in \mathbb{I}$. What results is called a **Möbius strip** (see Figure 4.8). The Möbius strip differs from a cylinder in that there is only one side and one edge; we say it is *one sided*, while the cylinder is *two sided*. Suppose an ant begins a journey at point $(0, 0)$ and moves with increasing x-value. First the ant traverses the "bottom" edge, going through $(1/2, 0)$, $(0.9, 0)$, etc. Now, arriving at $(1, 0)$, the ant suddenly finds itself on the "top" edge, because $(1, 0) \sim (0, 1)$ (note that on the Möbius strip, there is no "top" and "bottom" edge, and as far as the ant is concerned there was no perceptible change).

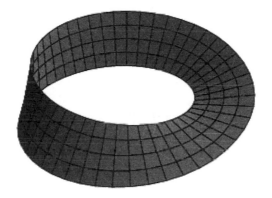

Figure 4.8: A representation of the Möbius strip.

Example 75. In the previous examples, identifications were only made along two of the edges of the square $\mathbb{I}^2$. To define the next space, we will identify points along the top and bottom as well. Define the **torus** as the quotient space,

$$\mathbb{T} = \mathbb{I}^2 / \{(0, y) \sim (1, y), \ (x, 0) \sim (x, 1)\}.$$

In the above notation, we mean that $\sim$ is the equivalence relation generated by all pairwise identifications $(0, y) \sim (1, y)$ for $y \in \mathbb{I}$, and $(x, 0) \sim (x, 1)$ for $x \in \mathbb{I}$.

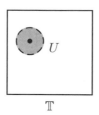 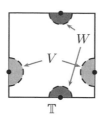 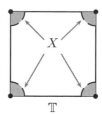

Figure 4.9: Open neighborhoods of various points in the torus. U, V, W, and X are each connected open disk neighborhoods.

Note that these identifications produce some two-element equivalence classes, $[(x,0)] = \{(x,0),(x,1)\}$ for $x \neq 0,1$, $[(0,y)] = \{(0,y),(1,y)\}$ for $y \neq 0,1$, and a four-element class,

$$[(0,0)] = \{(0,0),(1,0),(0,1),(1,1)\}.$$

We've already encountered the torus in Chapter 1 – see Figure 1.3 for a representation of the torus as a subset of $\mathbb{R}^3$. What are the neighborhoods of points of $\mathbb{T}$? Since the identifications happen only on the edges of $\mathbb{I}^2$, the subspace,

$$\mathbb{T} \setminus (\{(0,y) \mid y \in \mathbb{I}\} \cup \{(x,0) \mid x \in \mathbb{I}\}),$$

is topologically identical to the interior of $\mathbb{I}^2$ (note that we must be careful not to use the term "interior" of $\mathbb{T}$, because what might at first seem like boundary points have been identified to points on the opposite side, and so are no longer really on the edges). Now suppose $x \neq 0,1$. The point $(x,0) \in \mathbb{T}$ represents two identified points of $\mathbb{I}^2$, namely, $(x,0)$ and $(x,1)$. Therefore any neighborhood of $(x,0)$ must contain points on both "sides," as Figure 4.9 illustrates. A similar remark applies to $(0,y) \in \mathbb{T}$ for $y \neq 0,1$. Finally, a neighborhood of $(0,0)$ must contain points near all four members of the equivalence class, $(0,0)$, $(1,0)$, $(0,1)$, and $(1,1)$.

Now let's see how the identified square can actually be formed into the more familiar doughnut shape that we call a torus. Find a flexible rubber or fabric square – paper won't quite work. Make some marks on the square as in Figure 4.10. Now the identifications $(0,y) \sim (1,y)$ imply that the left and right sides of the square should be joined together, making sure to match the markings correctly. This forms a cylinder. Then the identifications $(x,0) \sim (x,1)$ imply that the top circle must be matched to the bottom circle. In order to match the markings in the correct order, you'll have to bend the cylinder around and meet the "top" to the "bottom" to form a doughnut shape. Note how all four of the vertices of the square (point a) meet to form a single point in the torus.

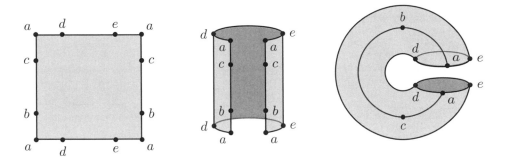

Figure 4.10: A representation of a torus.

Example 76. Consider $(\mathbb{R}^2 \setminus \{0\})/\sim$, where $(x_1, x_2) \sim (y_1, y_2)$ if and only if there is a nonzero $\lambda \in \mathbb{R}$ such that $(y_1, y_2) = \lambda(x_1, x_2) = (\lambda x_1, \lambda x_2)$. In the exercises, you will prove that this is an equivalence relation. The equivalence class of a point $x = (x_1, x_2)$ is the set of points on a line through the origin and x, but not including the origin itself. See Figure 4.11.

$$[x] = \{(\lambda x_1, \lambda x_2) \mid \lambda \in \mathbb{R} \setminus \{0\}\}$$

One may choose a *representative* in each equivalence class. Since each line hits the unit circle in two points, let's choose the point on the unit circle in the upper half-plane, except when the line is horizontal: then choose $(1, 0)$. Topologically, the representatives form what looks like a semicircle; however, since $(-1, 0) \sim (1, 0)$, the space is actually homeomorphic to $\mathbb{S}^1$. This space, which may be denoted by $\mathbb{P}^1$, is an example of a *projective space*; higher-dimensional projective spaces are defined analogously (see Exercise 11).

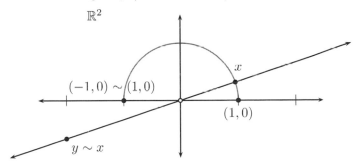

Figure 4.11: Projective space $\mathbb{P}^1 = (\mathbb{R}^2 \setminus \{0\})/\sim$. Choosing representatives on the upper half of the unit circle, with the identification $(-1, 0) \sim (1, 0)$, we see that $\mathbb{P}^1 \approx \mathbb{S}^1$.

Exercises

1. Define a subset $H \subseteq \mathbb{R}^2$ called the **Hawaiian earring** as follows:

$$H = \bigcup_{k=1}^{\infty} C_k,$$

   where C_k is the circle with center $\left(\frac{1}{2k}, 0\right)$ and radius $r = \frac{1}{2k}$. Let H have the subspace topology $\mathscr{T}|_H$ with respect to $\mathbb{R}^2$. Describe a base for $\mathscr{T}|_H$. What does a small neighborhood of $(0,0)$ look like?

2. Prove the second point of Theorem 4.3.2.

3. Let $A = (0, \infty)$, considered as a subspace of $\mathbb{R}^1$. Classify the following subsets of A as *open, closed*, or *neither* in A and *open, closed*, or *neither* in $\mathbb{R}^1$. Assume that $0 < a < b$ are real numbers.

   (a) $[a, b]$ (c) (a, b) (e) $(0, b]$

   (b) $[a, b)$ (d) $\{a, b\}$ (f) $\{1/k \mid k \in \mathbb{N}\}$

4. Considering $\mathbb{R}^1$ as a subspace of $\mathbb{R}^2$ as in Example 69, classify the following subsets of $\mathbb{R}^1$ as *open, closed*, or *neither* in $\mathbb{R}^1$ and *open, closed*, or *neither* in $\mathbb{R}^2$. Assume that $a < b$ are real numbers.

   (a) $[a, b]$ (c) (a, b) (e) $\mathbb{Z}$

   (b) $[a, b)$ (d) $\{a\}$ (f) $\mathbb{R}$

5. Show that every subspace of a discrete space is discrete, and every subspace of an indiscrete space is indiscrete.

6. Let $(X, \mathscr{T})$ be a topological space, and suppose that $\sim$ is an equivalence relation on X. Prove that the quotient topology $\mathscr{T}/\sim$ as defined in Definition 4.3.4 is indeed a topology on $X/\sim$.

7. Consider the Möbius strip M as defined in Example 74(b). Determine whether the following sets are open in M (note that sets are written in terms of $\mathbb{I}^2$, but it should be understood that the points are representative of equivalence classes in M).

   (a) $A = \{(x, y) \mid \frac{1}{3} < x < \frac{2}{3}, \frac{1}{3} < y < \frac{2}{3}\}$
   (b) $B = \{(x, y) \mid 0 \le x < \frac{2}{3}, \frac{1}{3} < y < \frac{2}{3}\}$
   (c) $C = \{(x, y) \mid \frac{1}{3} < x < \frac{2}{3}, 0 \le y < \frac{2}{3}\}$
   (d) $D = \{(x, y) \mid 0 \le x < \frac{1}{4}, \frac{1}{4} < y < \frac{1}{2}\} \cup \{(x, y) \mid \frac{3}{4} < x \le 1, \frac{1}{4} < y < \frac{1}{2}\}$

(e) $E = \{(x, y) \mid 0 \leq x < \frac{1}{4}, \frac{1}{4} < y < \frac{1}{2}\} \cup \{(x, y) \mid \frac{3}{4} < x \leq 1, \frac{1}{2} < y < \frac{3}{4}\}$

8. Use a computer to graph the following embeddings of $\mathbb{T}$ in $\mathbb{R}^3$. Choose various values of the parameters $b > a > 0$. (Note that b is the distance from the center of the hole to the center of the torus tube, and a is the radius of the tube.)

   (a) As an implicitly defined equation, $\left(b - \sqrt{x^2 + y^2}\right)^2 + z^2 = a^2$.

   (b) As parametric equations,

   $$\begin{cases} x &= (b + a \cos s) \cos t, \\ y &= (b + a \cos s) \sin t, \\ z &= a \sin s. \end{cases}$$

9. Show how the parametric equations from Exercise 8(b) provides a homeomorphism from $\mathbf{T} = \mathbb{I}^2 / \sim$ to the graph of the torus embedded in $\mathbb{R}^3$. (*Hint:* Consider (s, t) as points in the square $[0, 2\pi] \times [0, 2\pi]$, and examine how the boundary edges of the square are identified.)

10. Define a space called the **Klein bottle** by

    $$\mathbb{K} = \mathbb{I}^2 / \{(x, 0) \sim (x, 1), \ (0, y) \sim (1, 1 - y)\}.$$

    (a) Find all elements in each of the following equivalence classes:
    $[(\frac{1}{3}, \frac{1}{3})], \ [(\frac{1}{3}, 0)], \ [(0, \frac{1}{3})], \ [(0, 0)]$.

    (b) Draw neighborhoods around each point of $\mathbb{K}$ from part (a).

    (c) The torus can be constructed by physically identifying the edge points. Try to construct the Klein bottle in a similar way, using a flexible material, like a thin rubber sheet or fabric. What, if anything, prevents you from making a model of the Klein bottle? (Note that the righthand picture in Figure 1.3 represents what a Klein bottle might look like as a subset of $\mathbb{R}^3$, except that there should be no *self-intersection* where the "handle" enters the body of the bottle.)

11. Let $n \geq 1$ be a natural number. **Projective space**, $\mathbb{P}^n$, is defined as a quotient space by $\mathbb{P}^n = (\mathbb{R}^{n+1} \setminus \{\mathbf{0}\}) / \sim$, where $x \sim y$ if and only if there is a nonzero $\lambda \in \mathbb{R}$ such that $y = \lambda x$.

    (a) Prove that $\sim$ is an equivalence relation.

    (b) Draw a picture of $\mathbb{P}^2$ as an identification space in $\mathbb{R}^3$. By choosing representatives appropriately, explain how $\mathbb{P}^2$ can be visualized as a hemisphere with certain identifications along the boundary circle.

12. Define a quotient space of the unit cube $\mathbb{I}^3$ by

$$\mathbb{T}^3 = \mathbb{I}^3 / \{(0, y, z) \sim (1, y, z), \ (x, 0, z) \sim (x, 1, z), \ (x, y, 0) \sim (x, y, 1)\}.$$

The space $\mathbb{T}^3$ is called a **three-dimensional torus**. Taking the classroom as a model for $\mathbb{I}^3$, explain how the walls, floor, and ceiling of the room would be identified. What would you see if you looked straight ahead? Up? Down? Into one corner of the room?

13. Consider the space X defined by

$$X = ((\mathbb{R} \times \{0\}) \cup (\mathbb{R} \times \{1\})) / \sim, \quad \text{where } (x, 0) \sim (x, 1) \iff x \neq 0.$$

Since $(x, 0) \sim (x, 1)$ for all nonzero x, we may identify the equivalence class of $[(x, 0)]$ simply by x (for $x \neq 0$). Let $0 = (0, 0)$ and $0' = (0, 1)$.

(a) Show that X is not Hausdorff.

(b) Consider the sequence $(1, \frac{1}{2}, \frac{1}{3}, \frac{1}{4}, \frac{1}{5}, \ldots)$ in X. Prove that both 0 and $0'$ are limits of the sequence.

(c) List an open set containing 0 but not $0'$. Then list an open set containing $0'$ but not 0. (This type of separation makes X into a so-called T_1 space.)

4.4   Compactness and Connectedness

The topological definitions of compactness and connectedness are exactly the same as in §2.3 and §2.4, since those definitions only refer to open sets. However, there is no analog to the Heine-Borel Theorem (Theorem 2.3.5) in general. That is, even if we have a notion of *boundedness* in a topological space (which we do not in general), then it still does not follow that every bounded closed subset is compact. First let's recall the definition of open cover and compact.

> **Definition 4.4.1.** An **open cover** of a topological space (or subspace) X is a collection of open sets $\mathscr{U}$ such that
>
> $$X \subseteq \bigcup_{U \in \mathscr{U}} U.$$

> **Definition 4.4.2.** A space (or subspace) X is **compact** if *every* open cover $\mathscr{U}$ has a *finite* subcover.

Compact spaces are quite important in topology precisely because of the *finiteness* condition on open covers. A typical mathematical argument involving a compact space X usually goes something like this:

Let $\mathscr{U}$ be an arbitrary open cover of the compact space X (perhaps $\mathscr{U}$ consists of open sets satisfying some useful property). Let $\mathscr{U}' \subseteq \mathscr{U}$ be a finite subcover of X. Then perform some mathematical procedure or check some property within each set $U \in \mathscr{U}'$ (which is possible to do because there are only finitely many such open sets).

For example, the proof of Theorem 2.3.3 follows this paradigm. In fact, since the proof involves only open sets, the result generalizes to arbitrary topological spaces.

> **Theorem 4.4.3.** *Any infinite subset of a compact space X must have a limit point in X.*

Proof. Exercise 3. □

Any topological space in which every infinite subset must have a limit point is called **limit point compact**. So Theorem 4.4.3 states that every compact space is also limit point compact. The converse is not true in general.[13]

Example 77. Consider the set $X = \mathbb{N} \times \{0, 1\}$, and let

$$\mathscr{B} = \{U_n = \{(n, 0), (n, 1)\} \mid n \in \mathbb{N}\}.$$

The topology $\mathscr{T}$ generated by the base $\mathscr{B}$ is non-Hausdorff (*why?*). We will show that *every* nonempty subset of X has a limit point, not just the infinite subsets. Let $A \subseteq X$ be nonempty. Suppose $(n, 0) \in A$; then $(n, 1) \in X$ is a limit point of A since every open set U containing $(n, 1)$ must contain the base set U_n, which intersects A nontrivially. Similarly, if there is a point $(n, 1)$ in A, then $(n, 0) \in X$ is a limit point of A. This proves that X is limit point compact. However, X is not compact.

Compactness is also important because it is one of those rare topological properties that is passed down to (certain) subspaces and images of maps. Let's develop these ideas more rigorously.

> **Theorem 4.4.4.** *Suppose X is a compact topological space and $Y \subseteq X$ has the subspace topology. If Y is closed in X, then Y is also compact.*

Proof. Let $\mathscr{U}$ be an arbitrary open cover of Y. Now since Y is closed in X, the set $X \setminus Y$ is open (in X). Hence the collection $\mathscr{U} \cup \{X \setminus Y\}$ in an open cover of X. Since X is compact, there is a finite subscover $\mathscr{U}' \subseteq \mathscr{U} \cup \{X \setminus Y\}$. If $X \setminus Y \in \mathscr{U}'$, then remove $X \setminus Y$. The resulting collection is a finite subset of $\mathscr{U}$ of open sets covering Y. Thus Y is compact. □

---

[13] However, if the space is a metric space, or if a metric function can be found that induces the topology on the space, then limit point compactness is equivalent to compactness [Mun00].

> **Theorem 4.4.5.** *If $f : X \to Y$ is a continuous function of topological spaces, and if X is compact, then so is the image of f, $f[X] \subseteq Y$.*

Proof. Let $\{V_\alpha\}_{\alpha \in J}$ be an open cover for $f[X]$. Let $U_\alpha = f^{-1}(V_\alpha)$ for each $\alpha \in J$, and observe that by continuity of f, each U_α is open. Note that for any $x \in X$, we have $f(x) \in f[X]$, so there is at least one α such that $f(x) \in V_\alpha$ (since the sets V_α cover $f[X]$), which implies $x \in U_\alpha$ (for at least one α). This shows that $\{U_\alpha\}_{\alpha \in J}$ is an open cover for X. But now, since X is compact, there must exist a *finite* subcover, $X \subseteq U_{\alpha_1} \cup U_{\alpha_2} \cup \cdots \cup U_{\alpha_n}$. Applying the function f, we have

$$
\begin{aligned}
f[X] &\subseteq f[U_{\alpha_1} \cup U_{\alpha_2} \cup \cdots \cup U_{\alpha_n}] \\
&= f[U_{\alpha_1}] \cup f[U_{\alpha_2}] \cup \cdots \cup f[U_{\alpha_n}] \\
&= f\left[f^{-1}[V_{\alpha_1}]\right] \cup f\left[f^{-1}[V_{\alpha_2}]\right] \cup \cdots \cup f\left[f^{-1}[V_{\alpha_n}]\right] \\
&\subseteq V_{\alpha_1} \cup V_{\alpha_2} \cup \cdots \cup V_{\alpha_n}.
\end{aligned}
$$

The last line shows that there exists a finite subcover for $f[X]$ within the arbitrary open cover $\{V_\alpha\}_{\alpha \in J}$; hence $f[X]$ is compact. $\square$

We cannot expect compact subspaces to be closed in general, unless the space is Hausdorff.

> **Theorem 4.4.6.** *Every compact subspace of a Hausdorff space is closed.*

Proof. Suppose X is a Hausdorff space, and let $C \subseteq X$ be compact. To show that C is closed, we show that $X \setminus C$ is open. Let $x \in X \setminus C$. For each point $y \in C$, let U_y and V_y be disjoint open sets such that $x \in V_y$ and $y \in U_y$ (possible since X is Hausdorff). Now $\mathscr{U} = \{U_y \mid y \in C\}$ is an open cover of C, so by compactness there exists a finite subcover $\mathscr{U}' = \{U_{y_1}, U_{y_2}, \ldots, U_{y_n}\}$. Let $V = V_{y_1} \cap V_{y_2} \cap \cdots \cap V_{y_n}$. Observe that V is open (being the intersection of only finitely many open sets). We also have $V \cap U_{y_k}$ for each $k = 1, 2, \ldots, n$, so V is disjoint from $\bigcup_{k=1}^{n} U_{y_k}$, and since C is contained in this union, we also see that $V \cap C = \emptyset$. Thus we have found an open set $V \ni x$ within $X \setminus C$. Since x was arbitrary, $X \setminus C$ is open, and hence C is closed. $\square$

Compact subsets of a given space are typically rare, but in some special cases they may be incredibly common. If the point-set X is finite, then every subset of X is automatically compact, since any topology $\mathscr{T}_X$ on X must necessarily have only finitely many open sets. Similarly, any topological space $(X, \mathscr{T}_X)$ for which $\mathscr{T}_X$ is finite has the property that every subset is compact. For example, in the indiscrete topology, every open cover of any nonempty subset consists of exactly one open set. At the other end of the extreme, the only compact subsets of a discrete space are finite subsets. A good rule of thumb is: *The finer the topology, the rarer the compact subsets.*

Sequential Compactness

Compactness has a lot to do with the existence of limits. In Euclidean spaces, a sequence (x_k) within a compact space is guaranteed to have a convergent subsequence (see Theorem 2.3.9); however, this is no longer true in more general topological spaces. For this reason, we make a separate definition.

Definition 4.4.7. A topological space (or subspace) X is called **sequentially compact** if every sequence of points (x_k) in X has a convergent subsequence.

It is challenging to produce a space that is compact but not sequentially compact or a space that is sequentially compact but not compact, but such exotic spaces do exist. For example, the uncountable product $\mathbb{I}^{\mathbb{I}} = \prod_{k \in \mathbb{I}} \mathbb{I}$ is compact (since it's the product of compact sets; see §4.5), but not sequentially compact. Consider a sequence $(x_k)_{k \in \mathbb{N}}$ in $\mathbb{I}^{\mathbb{I}}$ defined as follows: Each $x_k \in \mathbb{I}^{\mathbb{I}}$ is the function $\mathbb{I} \to \mathbb{I}$ that sends $t \in \mathbb{I}$ to the kth digit of the binary expansion of t (see §4.5 for an explanation of how an element of a product can be interpreted as a function). Thus we have a sequence of functions.

$$
\begin{aligned}
x_1(t) &= \text{1st digit of } t \\
x_2(t) &= \text{2nd digit of } t \\
x_3(t) &= \text{3rd digit of } t \\
&\vdots
\end{aligned}
$$

The sequence (x_k) cannot have a convergent subsequence, for suppose $(x_{k_1}, x_{k_2}, x_{k_3}, \ldots)$ converges to some $x \in \mathbb{I}^{\mathbb{I}}$. Then for every $t \in \mathbb{I}$, the sequence of values $(x_{k_1}(t), x_{k_2}(t), x_{k_3}(t), \ldots)$ must converge (in $\mathbb{I}$) to $x(t)$. But consider a real number $t \in \mathbb{I}$ whose binary representation has k_j^{th} digit equal to 1 if j is odd and 0 if j is even (with other digits being chosen at random). For this value of t, the sequence $(x_{k_j}(t))$ does not converge. This argument is a variation on Cantor's Diagonalization argument.

Connectedness

The topological definitions of *separation* and *connectedness* are essentially the same as the ones given in Definitions 2.4.4 and 2.4.5 (recall §2.4).

Definition 4.4.8.

- A **separation** of a topological space $(X, \mathscr{T})$ is a disjoint pair of sets $U, V \in \mathscr{T}$, neither of which is empty, such that $X = U \cup V$.

- If X has a separation, then X is called **disconnected**; otherwise, X is **connected**.

- The **connected components** of X are the *maximally* connected subspaces of X.

A couple of points should be addressed concerning these definitions. First, even though the definition is given in terms of a space X, it applies equally well to any subset $A \subseteq X$ with the understanding that A has the subspace topology inherited from X. In this way, Definition 4.4.8 coincides precisely with the concepts of separation and connectedness already defined for subsets of Euclidean space in §2.4. Second, we must be precise about that word *maximally*. We say that a subspace $A \subseteq X$ is **maximal** with respect to a property $\mathscr{Q}$ if, whenever there is another subspace $B \supseteq A$ having property $\mathscr{Q}$, then it must be the case that $B = A$. (Analogously, A is **minimal** with respect to $\mathscr{Q}$ if, whenever $B \subseteq A$ has property $\mathscr{Q}$, then $B = A$.) Moreover, the properties of separations and connectedness given in Proposition 2.4.7 carry over to the general setting. Some additional important results are given in the following proposition.

Proposition 4.4.9. *Let X be a topological space.*

1. *The set of connected components of X forms a partition of X.*

2. *Each connected component of X is closed.*

Proof. Suppose X has connected components $\{X_k\}_{k \in \mathcal{I}}$. For each point $x \in X$, the singleton set $\{x\}$ is connected, and so $\{x\} \subseteq X_k$ for some k. Thus the connected components *cover* X. Next, suppose X_{k_1} and X_{k_2} are any two distinct connected components. If $X_{k_1} \cap X_{k_2} \neq \emptyset$, then $X_{k_1} \cup X_{k_2}$ would be connected, contradicting maximality of both X_{k_1} and X_{k_2} with respect to connectedness. Thus all of the connected components are mutually disjoint. This proves that the set of connected components forms a partition of X.

Let X_k be one of the connected components of X and consider its closure, $\overline{X_k}$. Suppose U, V is a separation of $\overline{X_k}$ (we'll arrive at a contradiction). Since X_k is connected and $X_k \subseteq \overline{X_k}$, then either $X_k \subseteq U$ or $X_k \subseteq V$. Without loss of generality, we may assume that $X_k \subseteq U$. It follows that $\overline{X_k} \subseteq \overline{U}$. Now since $\overline{U} = U$ (see Exercise 2), this implies $V = \emptyset$, a contradiction. Thus $\overline{X_k}$ is connected, but since X_k is *maximally* connected, it must be that $\overline{X_k} = X_k$, proving that X_k is closed. $\qquad\square$

When there are only finitely many connected components in a space X, then each is also open; however, this may no longer hold when there are infinitely many components, as the next example illustrates.

Example 78. Describe the connected components of $\mathbb{Q}$. Show that none of the components is open in $\mathbb{Q}$.

Solution: Let $C \subseteq \mathbb{Q}$ be a connected subset, and suppose $x_1 < x_2$ are elements of C. It is well known that between any two rational numbers there exists an irrational number $z \in \mathbb{R}$. Let $\widetilde{U} = (-\infty, z) \subseteq \mathbb{R}$, and $\widetilde{V} = (z, \infty) \subseteq \mathbb{R}$. In the subspace topology on $\mathbb{Q} \subseteq \mathbb{R}$, the sets $U - \widetilde{U} \cap \mathbb{Q}$ and $V = \widetilde{V} \cap \mathbb{Q}$ are open. We have $x_1 \in U$ and $x_2 \in V$ (so that $U \cap C$ and $V \cap C$ are nonempty), and $U \cap V = \emptyset$, so U, V is a separation of C. This implies there can be no two distinct points in C; hence every connected component of $\mathbb{Q}$ is a singleton set $\{x\}$ where $x \in \mathbb{Q}$. Now in $\mathbb{Q}$, points are not open, as no finite-length open interval (a, b) contains just a single rational number.

Theorem 4.4.10. *If $f : X \to Y$ is a continuous function of topological spaces, and if X is connected, then so is the image of f, $f[X] \subseteq Y$.*

Proof. Recall that the restriction $f : X \to f[X]$ is also continuous, so we may assume $Y = f[X]$. Suppose to the contrary that Y is disconnected, and let U, V be a separation of Y. Since U, V are both open, and since f is continuous, we have $f^{-1}[U]$ and $f^{-1}[V]$, both open, nonempty, and disjoint; that is, these two sets form a separation of X. This contradicts the hypothesis that X is connected and implies that Y must be connected. $\square$

Local Properties

A space may fail to be compact and yet a point in that space may have a compact neighborhood. For example, Euclidean space $\mathbb{R}^n$ (for a fixed $n \in \mathbb{N}$) is not compact, but every point $x \in \mathbb{R}^n$ lies within a compact disk neighborhood. Similarly, a space may fail to be connected (or arc-connected) and yet a point in that space may have a connected (or arc-connected) neighborhood. These local properties are especially useful in reducing certain arguments on more general spaces to arguments about a compact or connected subspace.

Definition 4.4.11. Let X be a topological space and $x \in X$. X is called **locally compact**, **locally connected**, or **locally arc-connected** at x if x has, respectively, a compact, connected, or arc-connected neighborhood. If every $x \in X$ has this property, then the whole space X is said to be locally compact, locally connected, or locally arc-connected, respectively.

Every compact space is automatically locally compact, and similarly for connected and arc-connected spaces. Every discrete space X is locally compact, connected, and arc-connected even though X fails to be compact if it has infinitely many points, and fails to be connected or arc-connected if it has more than one point.

Example 79. $\mathbb{Q}$ is neither locally compact, locally connected, nor locally arc-connected at any point x. A neighborhood N of a point $x \in \mathbb{Q}$ must contain some open rational interval, $J = (a, b) \cap \mathbb{Q}$. We may take a, b to be irrational (*why?*), so then $\overline{J} = J$ with respect to the subspace topology on $\mathbb{Q}$. If N were compact, then the closed set $J \subseteq N$ would also be compact, but if J is compact, so is $i[J] \subseteq \mathbb{R}$, where $i : \mathbb{Q} \to \mathbb{R}$ is the inclusion map. But as a subset of $\mathbb{R}$, the set $i[J] = J$ is not closed, and hence not compact.

J is also disconnected, as Example 78 shows, and a separation of J would imply a separation of N as well. Finally, since N is not connected, it cannot be arc-connected.

Topological Invariants

Theorem 4.4.5 implies that compactness is a topological invariant, and Theorem 4.4.10 does the same for connectedness. In fact, we have defined quite a few properties of spaces that are invariants, which we state presently without further proof.

> **Proposition 4.4.12.** *The properties of compactness, limit point compactness, sequential compactness, connectedness, arc-connectedness, and local versions of these properties are all topological invariants.*

Exercises

1. Suppose X has the *cofinite* topology, as defined in Example 66. Prove that every nonempty subset $A \subseteq X$ is compact, and if X is infinite, then A is also connected.

2. Let X have a separation U, V. Prove that both U and V are closed sets.

3. Prove Theorem 4.4.3.

4. Consider the space $X = \mathbb{N} \times \{0, 1\}$ as defined in Example 77.

   (a) Show that X is not Hausdorff.

   (b) List a few open sets in X. Explain why we might say that X is a "discrete" set of "indiscrete" subsets.

   (c) Show that X is not compact.

5. Suppose that C_1 and C_2 are compact subspaces of X.

   (a) Prove that $C_1 \cup C_2$ is compact.

   (b) Prove that if X is Hausdorff, then $C_1 \cap C_2$ is compact.

   (c) Consider the space $X = \mathbb{Z} \cup \{a, b\}$, whose topology is generated by the base $\mathscr{B} = \{\{n\} \mid n \in \mathbb{Z}\} \cup \{U_a, U_b\}$, where $U_a = \mathbb{Z} \cup \{a\}$ and $U_b = \mathbb{Z} \cup \{b\}$. Show that both U_a and U_b are compact, but $U_a \cap U_b$ is not.

6. Prove that both the torus $\mathbb{T}$ and Klein bottle $\mathbb{K}$ (§4.3, Exercise 10) are connected and compact. (*Hint:* Use the fact that each of these spaces is a quotient of the unit square.)

7. Let $n \geq 1$. Prove that projective space $\mathbb{P}^n$ (as defined in §4.3, Exercise 11) is connected and compact.

8. Suppose X is any point-set. Determine under what conditions X must be compact, connected, locally compact, or locally connected if X has:

   (a) the indiscrete topology $\mathscr{T}_I$.

   (b) the discrete topology $\mathscr{T}_D$.

4.5 Product and Function Spaces

In this section, we present a way to define a topology on the product of two topological spaces. Then we explore spaces in which the "points" are functions. Far from being useless oddities, the so-called *function spaces* play a major role in the theory of partial differential equations, and hence are essential in understanding many complex phenomena in physics.

Product Topology

Recall that the Cartesian product of two sets, X and Y, is the set of all ordered pairs,
$$X \times Y = \{(x, y) \mid x \in X, \ y \in Y\}.$$

What if X and Y are topological spaces? How could one define the open sets of $X \times Y$? Perhaps the most obvious way to start is to insist that if $U \subseteq X$ is open, and $V \subseteq Y$ is open, then $U \times V \subseteq X \times Y$ should also be open.

> **Definition 4.5.1.** If $(X, \mathscr{T}_X)$ and $(Y, \mathscr{T}_Y)$ are topological spaces, then the product $X \times Y$ may be given the **box product topology** $\mathscr{T}_{\text{box}}$, which has as its base the set of all products of open sets from X and Y:
>
> $$\mathscr{B}_{X \times Y} = \mathscr{T}_X \times \mathscr{T}_Y = \{U \times V \mid U \in \mathscr{T}_X, \ V \in \mathscr{T}_Y\}.$$

It must be proven that the set $\mathscr{B}_{X \times Y}$ in Definition 4.5.1 does indeed define a base for a topology. Condition 1 of Definition 4.1.8 is trivially satisfied, since $X \times Y \in \mathscr{B}_{X \times Y}$. Now suppose $B_1, B_2 \in \mathscr{B}_{X \times Y}$ and $B_1 \cap B_2 \neq \emptyset$. Let $z \in B_1 \cap B_2$. By definition of the box product topology, $B_1 = U_1 \times V_1$ and $B_2 = U_2 \times V_2$ for open sets $U_1, U_2 \in X$ and $V_1, V_2 \in Y$. Let $U = U_1 \cap U_2$ (open in X), and $V = V_1 \cap V_2$ (open in Y). Note that $z = (x, y)$ for some $x \in X$ and $y \in Y$. Moreover, since $z \in B_1$, we have $x \in U_1$ and $y \in V_1$. Similarly, since $z \in B_2$, we have $x \in U_2$ and $y \in V_2$. Thus $z = (x, y) \in U \times V \subseteq B_1 \cap B_2$, satisfying condition 2 of Definition 4.1.8.

Now that we have shown that the box product topology is well defined, let's see an example of what it looks like.

Example 80.   Consider two open intervals of $\mathbb{R}^1$, $X = (a, b)$ and $Y = (c, d)$ (where $a < b$ and $c < d$). The product $X \times Y$ is a rectangular region in $\mathbb{R}^2$ not including the boundary edges. A basic open set in $X \times Y$ containing a point (x, y) is shown by the dotted box in the diagram below; this box is the Cartesian product $(x - \delta, x + \delta) \times (y - \epsilon, y + \epsilon)$ for small positive δ and ϵ. (This is why the topology is called the *box* product topology.)

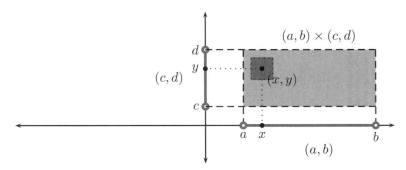

The box product topology on $X \times Y$ is equivalent to the subspace topology inherited from $\mathbb{R}^2$, because within each "box" neighborhood $(x - \delta, x + \delta) \times (y - \epsilon, y + \epsilon)$, there is an open ball (just take the radius to be the minimum of ϵ and δ), and within each open ball $B_\epsilon((x, y))$, there is a small box neighborhood (try to define the dimensions of an appropriate box in terms of ϵ).

Example 81.   Describe $\mathbb{S}^1 \times \mathbb{I}$ as a topological space with the box product topology.

Solution: A good model for $\mathbb{S}^1$ is the unit circle, $\{(x, y) \in \mathbb{R}^2 \mid x^2 + y^2 = 1\}$, while $\mathbb{I} = [0, 1]$ is the unit closed interval in $\mathbb{R}^1$. By embedding our picture in $\mathbb{R}^3$, we find that $\mathbb{S}^1 \times \mathbb{I} = \{(x, y, z) \mid x^2 + y^2 = 1, \ 0 \leq z \leq 1\}$ is a **cylinder**. The bottom edge, $\mathbb{S}^1 \times \{0\}$, and top edge, $\mathbb{S}^1 \times \{1\}$, are both copies of the circle; indeed, for any $z \in \mathbb{I}$, the set $\mathbb{S}^1 \times \{z\}$ is a circle. On the other hand, for every point $(x, y) \in \mathbb{S}^1$, there is a copy of the unit segment, $\{(x, y)\} \times \mathbb{I}$. In this way, the cylinder is simultaneously a *segment of circles* and a *circle of segments*.

Basic open subsets in $\mathbb{S}^1 \times \mathbb{I}$ are of the form $W = C \times (a, b)$, where C is an open arc of the circle, and $(a, b) \subseteq [0, 1]$ is an open interval. It can be proven that this model of cylinder is homeomorphic to the quotient space $\mathbb{I}^2 / \{(0, y) \sim (1, y),\ y \in \mathbb{I}\}$ given in Example 74(a).

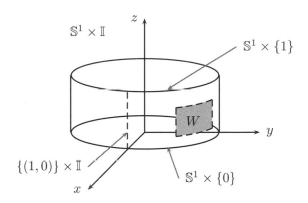

Given a product of sets, $X \times Y$, there are two important set functions called the **projection** maps, which send an ordered pair $(x, y) \in X \times Y$ to each of its components.

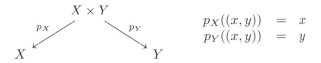

$$p_X((x, y)) = x$$
$$p_Y((x, y)) = y$$

The projection maps are continuous in the box topology (Exercise 3). Suppose now that $(X_1, \mathscr{T}_1), (X_2, \mathscr{T}_2), \ldots, (X_m, \mathscr{T}_m)$ are topological spaces. The box topology on $\prod_{k=1}^{m} X_k$ has as its base the product of the topologies, $\prod_{k=1}^{m} \mathscr{T}_k$, and there are associated projection maps,

$$p_j \ : \ \prod_{k=1}^{m} X_k \to X_j, \quad \text{for each } j \in \{1, 2, \ldots, m\}$$
$$p_j(x_1, x_2, \ldots, x_j, \ldots, x_m) = x_j.$$

In fact, the box topology is the coarsest topology on $\prod_{k=1}^{m} X_k$ such that each projection map p_j is continuous. Unfortunately, this statement no longer holds true in the box topology on arbitrary products, $\prod_{k \in \mathcal{I}} X_k$, in which the indexing set $\mathcal{I}$ is infinite. There is another way to define a product topology, which turns out to have more useful properties than the box topology. Because of its importance, it is simply called the **product topology**.

> **Definition 4.5.2.** Let $\mathcal{I}$ be an indexing set, and let $(X_k, \mathscr{T}_k)$ be a topological space for each $k \in \mathcal{I}$. The **product topology** $\mathscr{T}_{\text{prod}}$ on $X = \prod_{k \in \mathcal{I}} X_k$ has as its base the set of all products of the form $\prod_{k \in \mathcal{I}} U_k$ such that
>
> - $U_k \in \mathscr{T}_k \ \forall k \in \mathcal{I}$, and
>
> - $U_k = X_k$ for all but finitely many $k \in \mathcal{I}$.

The product topology is equivalent to the box product topology when $\mathcal{I}$ is finite; however, only the product topology has the property that it is the coarsest topology on an arbitrary product such that every projection map is continuous.

It is often useful to regard the component spaces X and Y as somehow existing as subspaces within the product $X \times Y$, just as the coordinate axes are copies of $\mathbb{R}^1$ that exist within the Euclidean plane $\mathbb{R}^2 = \mathbb{R}^1 \times \mathbb{R}^1$, but often there is no natural choice for the identification of X or Y within the product. Instead, if we choose ahead of time specific points $x_0 \in X$ and $y_0 \in Y$, then we may define the **injection** maps, i_X and i_Y according to the diagram below.

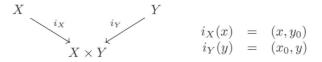

$$i_X(x) = (x, y_0)$$
$$i_Y(y) = (x_0, y)$$

If $X = \prod_{k \in \mathcal{I}} X_k$ is an arbitrary product with either the box or product topology, then each injection map, $i_k : X_k \to X$, is continuous, and there is a homeomorphism, $X_k \approx i_k[X_k] \subseteq X$. For example, in the case of two spaces, X and Y, we have $X \approx X \times \{y_0\} \subseteq X \times Y$, and $Y \approx \{x_0\} \times Y \subseteq X \times Y$.

Topological Properties of Products

Products preserve some important properties.

> **Theorem 4.5.3.** *If X and Y are both connected, then so is $X \times Y$.*

Proof. Suppose there is a separation U, V for the product $X \times Y$. Let $(x_0, y_0) \in U$ (which is possible since U is supposed to be nonempty). Then the $\{x_0\} \times Y$ must also be in U since $Y \approx \{x_0\} \times Y$ is assumed to be connected. Now for each $y \in Y$, we have $(x_0, y) \in U$, but by connectedness of X, this implies the entire set $X \times \{y\}$ is also in U. But this means that every point $(x, y) \in X \times Y$ is in U, leaving V completely empty. Thus U, V could not be a separation in the first place, and $X \times Y$ must be connected. $\qquad \square$

Theorem 4.5.4. *If X and Y are both compact, then so is $X \times Y$.*

Proof. Suppose $\mathcal{U}$ is an open cover of $X \times Y$. Without loss of generality, we may assume every open set in $\mathcal{U}$ is a basic open set $U \times V$ where $U \in \mathcal{T}_X$ and $V \in \mathcal{T}_Y$ *(why?)*. Let $x_0 \in X$ be arbitrary, and let $\mathcal{U}_0 \subseteq \mathcal{U}$ be a set of open sets that cover $\{x_0\} \times Y \subseteq X \times Y$ and that contain no set disjoint from $\{x_0\} \times Y$. Now $\mathcal{U}_0$ is an open cover of the compact set $\{x_0\} \times Y \approx Y$, so there is a finite subcover, $\mathcal{U}_0' \subseteq \mathcal{U}_0$. Label the sets in this subcover:

$$\mathcal{U}_0' = \{U_1 \times V_1, U_2 \times V_2, \ldots, U_m \times V_m\}.$$

Now consider the set $W_0 = \bigcap_{k=1}^m U_k$. Since each U_k is an open set in X, and since there are only finitely many of them, $W_0 \subseteq X$ is an open set containing the point x_0 (see Figure 4.12).

Claim: $\mathcal{U}_0'$ covers $W_0 \times Y$. *Proof of claim:* Let $(x, y) \in W_0 \times Y$ be arbitrary. The point $(x_0, y) \in \{x_0\} \times Y$ belongs to $U_k \times V_k \in \mathcal{U}_0'$ for some $k \in \{1, 2, \ldots m\}$, so $y \in V_k$. On the other hand, since $W_0 \subseteq X_k$ for every k, we also have $x \in U_k$; hence $(x, y) \in U_k \times V_k$.

Since $x_0 \in X$ was chosen arbitrarily, every $x \in X$ is contained in an open neighborhood $W \subseteq X$ such that the strip $W \times Y$ can be covered by finitely many sets chosen from $\mathcal{U}$. By compactness of X, only finitely many such sets W cover all of X, say, $W_1, W_2, \ldots, W_r$. Clearly $X \times Y$ is covered by the (finitely many) strips, $W_1 \times Y, W_2 \times Y, \ldots, W_r \times Y$, each of which can in turn be covered by finitely many open sets chosen from the original cover $\mathcal{U}$. Thus a finite open subcover for $X \times Y$ can be constructed.[14] $\qquad \square$

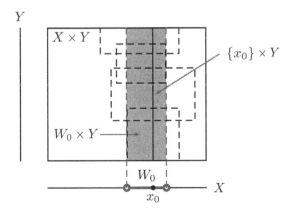

Figure 4.12: $X \times Y$. $\mathcal{U}_0'$ (dashed boxes) covers the strip $W_0 \times Y$.

---

[14]Proof adapted from Munkres [Mun00].

Example 82. The n-fold product of the unit interval, $\mathbb{I}^n = \mathbb{I} \times \mathbb{I} \times \cdots \times \mathbb{I}$ (also called an n-dimensional **hypercube**), is compact since $\mathbb{I}$ is compact.

Example 83. Recall the Heine-Borel Theorem (Theorem 2.3.5 from §2.3). We showed that every compact subset of $\mathbb{R}^n$ is both closed and bounded and claimed that the converse also holds. However, we only proved that closed bounded intervals in $\mathbb{R}^1$ are compact. Now that we have more tools at our disposal, it will be relatively easy to extend the proof.

> If a subset $X \subseteq \mathbb{R}^n$ is closed and bounded (with respect to the Euclidean metric), then X is compact.

Proof. Since X is bounded, there exists $r \in \mathbb{R}^+$ such that $X \subseteq B_r(\mathbf{0})$. The coordinates of any point $\mathbf{x} = (x_1, x_2, \ldots, x_n)$ in the sphere of radius r must necessarily satisfy $-r \leq x_i \leq r$; therefore $B_r(\mathbf{0}) \subseteq [-r, r]^n$. Now $[-r, r]^n$ is compact because the closed bounded interval $[-r, r] \subseteq \mathbb{R}^1$ is compact. By Theorem 4.4.4, since X is a closed subspace of a compact space $[-r, r]^n$, it follows that X is compact. $\square$

In fact, an arbitrary product of compact spaces is compact – a result known as the **Tychonoff Theorem** – so long as the topology is the product topology given in Definition 4.5.2. However, the proof of this important theorem is beyond the scope of this textbook.

The following theorem shows that the product of subspaces is equivalent to a subspace of the product. The proof of this key result boils down to an easy exercise in set theory.

> **Theorem 4.5.5.** *If $A \subseteq X$ and $B \subseteq Y$ are subspaces, then the product topology on $A \times B$ is the same as the subspace topology that $A \times B$ inherits from $X \times Y$.*

Proof. Let $\mathscr{T}_X$ and $\mathscr{T}_Y$ be the topologies on X and Y, respectively. The subspace topologies on A and B are, respectively, $\mathscr{T}_X|_A = \{U \cap A \mid U \in \mathscr{T}_X\}$ and $\mathscr{T}_Y|_B = \{V \cap B \mid V \in \mathscr{T}_Y\}$. So the product topology on $A \times B$ is generated by

$$\mathscr{T}_X|_A \times \mathscr{T}_Y|_B = \{(U \cap A) \times (V \cap B) \mid U \in \mathscr{T}_X \text{ and } V \in \mathscr{T}_Y\}. \qquad (4.4)$$

On the other hand, $(U \cap A) \times (V \cap B) = (U \times V) \cap (A \times B)$ (see §A.1, Exercise 14), and so (4.4) also constitutes a base for the subspace topology on $A \times B \subseteq X \times Y$:

$$\mathscr{B}_{X \times Y}|_{A \times B} = \{(U \times V) \cap (A \times B) \mid U \in \mathscr{T}_X \text{ and } V \in \mathscr{T}_Y\}.$$

$\square$

Function Spaces

What does it mean for a function to be "close" to another function? In calculus and analysis, we learn that certain kinds of functions can be approximated by polynomials (*Taylor polynomials*), sums of trigonometric functions (*Fourier approximation*), and so on. What we really mean here is that there is some topology $\mathscr{T}$ defined on a set of functions $\mathcal{F}$, and under certain circumstances, a *sequence* of functions $(f_n)_{n \in \mathbb{N}}$ in $\mathcal{F}$ may *converge* to another function $f \in \mathcal{F}$ with respect to that topology $\mathscr{T}$.

Certain function spaces may be identified with set-theoretic products, and vice versa. Suppose $\mathcal{F}$ is the set of all functions whose domain is X and codomain Y (with no requirement for continuity). There is a bijection $\mathcal{F} \to Y^X$, where Y^X is the product of Y with itself over the indexing set X.

$$\mathcal{F} = \{f : X \to Y\} \quad \longrightarrow \quad Y^X$$
$$f \quad \mapsto \quad (f(x))_{x \in X}$$

Example 84. Fix $n \geq N$. Euclidean space $\mathbb{R}^n$ can be identified with the n-fold product of $\mathbb{R}$ with itself or, in other words, with the product $\mathbb{R}^{\{1,2,\ldots,n\}}$. This in turn can be identified with the set of all functions,

$$x : \{1, 2, \ldots, n\} \to \mathbb{R}.$$

Given any such function x, the n outputs are simply the n components of a vector $x \in \mathbb{R}^n$.

$$x = (x(1), x(2), \ldots, x(n)) = (x_1, x_2, \ldots, x_n)$$

The space $\mathbb{R}^\infty = \mathbb{R}^{\mathbb{N}}$ may be regarded as the space of functions $\mathbb{N} \to \mathbb{R}$, for example, sequences of real numbers $(x_k)_{k \in \mathbb{N}}$.

With this identitification, we may give $\mathcal{F} = \{f : X \to Y\} \approx Y^X$ the product topology, giving $\mathcal{F}$ the so-called **point-wise convergence** topology. The reason for this terminology is that a sequence $(f_k)_{k \in \mathbb{N}}$ of functions would converge in Y^X if and only if for every $x \in X$, the sequence $(f_k(x))_{k \in \mathbb{N}}$ converges in Y. However, there are many different ways to define a topology on a function space.

Often a function space $\mathcal{F}$ comes equipped with a norm or metric, as discussed in §2.5. Thus we can define the L^p spaces, for example. More generally, if there is a topology on a set of functions $\mathcal{F}$ such that addition and multiplication by scalars are both continuous operations, then we call $\mathcal{F}$ a **topological vector space**. Unfortunately, these topics take us too far afield, and the interested reader is encouraged to explore the vast field of *functional analysis*.

Exercises

1. Find an explicit homeomorphism $\mathbb{I}^2 \to D_1(\mathbf{0})$ (the closed unit disk centered at the origin).

2. We often use the notation $\mathbb{D}^n$ for the closed unit disk $D_1(\mathbf{0}) \subseteq \mathbb{R}^n$. Prove that $\mathbb{D}^n \approx \mathbb{I}^n$. (*Hint:* Consider a map $f : \mathbb{I}^n \to \mathbb{D}^n$ that scales the distance of points from the origin so that the boundary $\partial\mathbb{I}^n$ gets mapped onto the boundary sphere $\mathbb{S}^{n-1}$ of $\mathbb{D}^n$.)[15]

3. Prove that the projection maps $p_X : X \times Y \to X$ and $p_Y : X \times Y \to Y$ are continuous in the box topology.

4. Describe the product $\mathbb{S}^1 \times \mathbb{S}^1$. It may help to use $\mathbb{S}^1 = \mathbb{I}/\sim$, where $0 \sim 1$.

5. Prove that a product $X = \prod_{k=1}^m X_k$ is connected if and only if each space X_k is connected.

6. Consider spaces X and Y with chosen points $x_0 \in X$ and $y_0 \in Y$. Let $p_X : X \times Y \to X$ and $p_Y : X \times Y \to Y$ be the projection maps, and let $i_X : X \to X \times Y$ and $i_Y : Y \to X \times Y$ be the injection maps defined by $i_X(x) = (x, y_0)$ and $i_Y(y) = (x_0, y)$, respectively. Describe the compositions $p_X \circ i_X$, $p_Y \circ i_Y$, $p_X \circ i_Y$, and $p_Y \circ i_X$.

7. Let p_X, and i_X be as in Exercise 6. Show that $i_X \circ p_X$ is neither surjective nor injective unless $Y = \{y_0\}$.

8. Let $B = B_\epsilon(\mathbf{x}) \subseteq \mathbb{R}^2$. Describe the space $B \times \mathbb{I}$ as a subset of $\mathbb{R}^3$. Note that $B \times \mathbb{I}$ is an example of a **tubular neighborhood** of $\mathbb{I}$ embedded into $\mathbb{R}^3$. Describe a tubular neighborhood of the circle $\mathbb{S}^1$ embedded into $\mathbb{R}^3$.

4.6　*The Infinitude of the Primes

As we may have learned in elementary school, a **prime** number is any whole number that has exactly two distinct divisors, itself and 1. The first few such numbers are $2, 3, 5, 7, 11, 13, 17, \ldots$ (note that 1 is not considered prime because 1 has only *one* divisor). How many primes are there? It's fairly easy to prove there must be infinitely many of them – in fact, you may find six different proofs in *Proofs from the Book* [AZ10].

> **Theorem 4.6.1.** *The set P of prime numbers is infinite.*

---

[15]You might think of this as "inflating" the hypercube into the shape of a hypersphere.

Euclid's Proof

Euclid ($\sim$300 BC), who is regarded as the father of geometry, did not rigorously prove the infinitude of the primes. Euclid *IX:20* simply proves that there are more than three primes, but it is clear what was intended.

Proof. Let $Q = \{p_1, p_2, \ldots, p_r\} \subseteq P$ be any *finite* set of primes. Let n be a whole number such that every p_i is a divisor of n; for instance, n could be the product of all elements of Q. Consider the whole number $n + 1$, and suppose there is a prime $p_i \in Q$ such that p_i is a divisor of $n + 1$. Now since p_i divides both n and $n + 1$, it must also divide 1. This is a contradiction, since the only divisor of 1 is $1 \notin Q$.

By the *Fundamental Theorem of Arithmetic* (Euclid *VIII:31*), $n + 1$ must have at least one prime divisor, but we have just shown that no prime in Q is a divisor of $n + 1$. Thus any finite set of primes fails to contain all of the prime numbers, and so the set of primes P must be infinite. $\square$

Furstenberg's Proof

Hillel (Harry) Furstenberg (b. 1935) found a curious topology on the set of integers $\mathbb{Z}$ that leads to a *topological* proof of the infinitude of the primes. The arguments here are also interesting for their applications in number theory.

As in Example 149, we define certain sets related to multiples and arithmetic sequences. For $a \in \mathbb{Z}$, $b \in \mathbb{N}$, define the set $N_{a,b} = \{a + nb \mid n \in \mathbb{Z}\}$. Define a topology $\mathscr{T}$ on $\mathbb{Z}$ with base $\mathscr{B} = \{N_{a,b} \mid a \in \mathbb{Z}, b \in \mathbb{N}\}$. Before we proceed, it is important to verify that $\mathscr{B}$ does indeed satisfy Definition 4.1.8.

1. For each $a \in \mathbb{Z}$ and for any choice of $b > 0$, we have $a \in N_{a,b}$. Thus $\mathscr{B}$ covers $\mathbb{Z}$.

2. Let $N_{a,b}, N_{c,d} \in \mathscr{B}$ such that $N_{a,b} \cap N_{c,d} \neq \emptyset$, and let $x \in N_{a,b} \cap N_{c,d}$.

   Claim: $x \in N_{x,bd} \subseteq N_{a,b} \cap N_{c,d}$. The first part of the claim is easy to show; since $x = x + 0(bd)$, we have $x \in N_{x,bd}$. The second part of the claim is a set inclusion. Let $y \in N_{x,bd}$ be arbitrary. Then $y = x + k(bd)$ for some $k \in \mathbb{Z}$. Since $x \in N_{a,b}$, we may write $x = a + nb$ for some $n \in \mathbb{Z}$. Similarly, $x = c + md$ for some $m \in \mathbb{Z}$. Now we can rewrite y in two ways.

   $$y = x + k(bd) = a + nb + kbd = a + (n + kd)b \qquad (4.5)$$
   $$y = x + k(bd) = c + md + kbd = c + (m + kb)d \qquad (4.6)$$

   Equation (4.5) demonstrates $y \in N_{a,b}$ (since $n + kd \in \mathbb{Z}$), and (4.6) demonstrates $y \in N_{c,d}$ (since $m + kb \in \mathbb{Z}$). Thus $y \in N_{a,b} \cap N_{c,d}$ which proves the claim $N_{x,bd} \subseteq N_{a,b} \cap N_{c,d}$.

Thus we know that $\mathscr{B}$ generates a topology on $\mathbb{Z}$, but what does this topology *look* like? Intuitively, a nonempty subset U of $\mathbb{Z}$ is considered *open* in $\mathscr{T}$ if every element is part of a complete *arithmetic sequence* contained in U. In fact, every nonempty open set must be infinite, because arithmetic sequences are infinite. Each $N_{a,b}$ is open (being a base set), but what may be surprising is that each $N_{a,b}$ is also closed. You will be asked to prove this in the exercises.

Lemma 4.6.2. *Each set $N_{a,b}$ is closed in $\mathscr{T}$.*

Now let's prove (again) that there are infinitely many prime numbers.

Proof. Let P be the set of all prime numbers. For every $n > 1$, there is at least one prime $p \in P$ such that p is a divisor of n (Fundamental Theorem of Arithmetic). For this choice of p, we have $n \in N_{0,p}$. Note that $-n \in N_{0,p}$ as well. Clearly $0 \in \mathbb{N}_{0,p}$, but $\pm 1 \notin \mathbb{N}_{0,p}$ for any prime p. Thus we have:

$$\mathbb{Z} \setminus \{-1, 1\} = \bigcup_{p \in P} N_{0,p}. \tag{4.7}$$

By DeMorgan's Law for arbitrary unions, (4.7) may be written:

$$\{-1, 1\} = \bigcap_{p \in P} \left(\mathbb{Z} \setminus N_{0,p}\right). \tag{4.8}$$

Each $N_{0,p}$ is closed, by Lemma 4.6.2, so each complement $\mathbb{Z} \setminus N_{0,p}$ is open. Now, if P were finite, then the (finite) intersection in (4.8) would be an open set, implying that $\{-1, 1\}$ is open. This contradicts the fact that every nontrivial open set in the topology $\mathscr{T}$ is infinite. Thus P cannot be finite. □

Exercises

1. Prove that $N_{a,b}$ is closed in Furstenberg's topology $\mathscr{T}$. (*Hint:* Express $N_{a,b}$ as a complement of a union of open sets.)

2. Show that $\{0\}$ is a closed set in the topology $\mathscr{T}$.

3. Show that all points are closed in $\mathscr{T}$. In other words, for every $x \in \mathbb{Z}$, the set $\{x\}$ is closed.

4. Show that every finite subset of $\mathbb{Z}$ is closed in $\mathscr{T}$. (*Hint:* See Exercise 3.)

Supplemental Reading

- Aigner and Ziegler [AZ10], Chapter 1, "Six Proofs of the Infinity of Primes," which inspired §4.6.

- Munkres [Mun00], Chapters 2-3, form the core of elementary point-set topology. Further important topics that fall outside the scope of this text-book are found in Munkres, Chapters 4-8.

- Steen and Seebach [SS95] provide an encyclopedia of topological spaces that display certain properties and not others.

Chapter 5

Surfaces

On the surface of the Earth, there are two degrees of freedom. Simply fix a direction to face; the line forward and backward relative to you is one degree of freedom, while the line left and right through you determines another. Ignoring the scenery, buildings, etc., there is no topological difference between the patch of Earth you are standing on and the one near me; the Earth's surface always looks the same locally (see Figure 5.1). In fact, if we did not realize the Earth is a sphere, we could mistake any particular patch of land or sea as part of a plane,[1] or a cylinder, or even a torus. This property of indistinguishability is key in defining what we mean by a surface. In this chapter we define surfaces; introduce methods for combining, dissecting, and analyzing surfaces; and then prove half of the important theorem that classifies all compact surfaces. Near the end of the chapter we explore vector fields on surfaces other than the familiar plane.

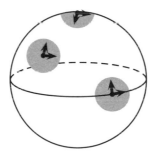

Figure 5.1: There are two degrees of freedom at any point on the surface of the Earth.

---

[1] As early as the sixth century BC, Pythagoras postulated that the Earth is a sphere, though his reasons were more philosophical than empirical. In the fourth century BC, Eratosthenes computed the circumference of the Earth using the observation that shadows were cast at different angles in two distant places at the same time. Eratosthenes's estimate was amazingly accurate.

5.1   Surfaces and Surfaces-with-Boundary

Let's begin with a formal definition of *surface*. We assume that our surfaces are connected and have no boundary edge.

Definition 5.1.1. A **surface** is a connected Hausdorff space S such that for each $x \in S$, there is an open set $U \subseteq S$ with $x \in U$, such that $U \approx B_1(\mathbf{0}) \subseteq \mathbb{R}^2$.

An open set $U \approx B_1(\mathbf{0})$ as in Definition 5.1.1 is called a **disk neighborhood** or **patch**, so we may think of a surface S as a patchwork of overlapping open disks. In this chapter we primarily work with *compact* surfaces. Recall from §4.4 that a compact space is one in which every open cover has a finite subcover. Thus, for a compact surface, it is always possible to fulfill Definition 5.1.1 with a *finite* collection of *patches*. Moreover, surfaces that have *punctures* or open edges, as illustrated in Figure 5.2, are not compact. On the other hand, there is no requirement for "smoothness" – a surface may have corners or sharp edges, as in the surface of a cube.

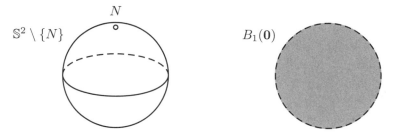

Figure 5.2: The punctured sphere, $\mathbb{S}^2 \setminus \{N\}$, and open disk are noncompact surfaces.

Example 85. Examples of compact surfaces include the sphere $\mathbb{S}^2$, the torus $\mathbb{T}$, and Klein bottle $\mathbb{K}$, among many others. Note that $\mathbb{K}$ cannot be realized in three-dimensional space without *self-intersections* (see, e.g., a typical representation of $\mathbb{K}$ in Figure 1.3). A noncompact surface may extend infinitely, such as the Euclidean plane $\mathbb{R}^2$, or the infinite cylinder, $\mathbb{S}^1 \times \mathbb{R}$, or such a surface may have a puncture and/or open boundary edge, as in the two spaces shown in Figure 5.2.

The following examples are *not* surfaces at all, according to the definition (see Figure 5.3).

- The circle $\mathbb{S}^1$. No open set of $\mathbb{S}^1$ homeomorphic to an open disk. Indeed, we may say $\mathbb{S}^1$ is one dimensional, whereas surfaces are, by definition, two dimensional. The higher-dimensional spheres $\mathbb{S}^3$, $\mathbb{S}^4$, etc., are also not surfaces.

- The closed unit disk $\mathbb{D}^2 = D_1(\mathbf{0}) \subseteq \mathbb{R}^2$. $\mathbb{D}^2$ is not a surface because there are *boundary points* in $\mathbb{D}^2$ that cannot be surrounded by an open set U homeomorphic to an open disk.

- A pair of intersecting planes. Suppose x is any point on the intersection line. Any open set U containing x must contain points that are on both planes, which implies that U is not homeomorphic to an open disk.

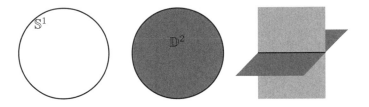

Figure 5.3: The circle, closed disk, and a pair of intersecting planes. None of these spaces is a surface according to Definition 5.1.1.

Consider two related spaces: the open unit disk, $U = B_1(\mathbf{0}) = \{x \in \mathbb{R}^2 \mid d(x, \mathbf{0}) < 1\}$, and the closed unit disk, $\mathbb{D}^2$. According to Definition 5.1.1, U is a noncompact surface, while $\mathbb{D}^2$ is a compact set that is technically *not* a surface. However, *most* points of $\mathbb{D}^2$ do fit the conditions of the definition. Only those points on the boundary edge fail to meet the requirement of belonging to an open set homeomorphic to the open disk; instead, a typical open set containing the boundary point is homeomorphic to the *half-disk* $V = \{(x, y) \in \mathbb{R}^2 \mid d((x, y), 0) < 1, \ y \geq 0\}$, as illustrated in Figure 5.4.

Figure 5.4: An open neighborhood $U \ni z$, for z on the boundary of the closed disk, is homeomorphic to a half-disk V.

> **Definition 5.1.2.** A **surface-with-boundary** is a connected Hausdorff topological space S such that for each $x \in S$, there is an open set $U \subseteq S$ with $x \in U$, such that *either*:
>
> - $U \approx B_1(\mathbf{0}) \subseteq \mathbb{R}^2$ *or*
>
> - $U \approx V = \{(x, y) \in \mathbb{R}^2 \mid d((x, y), 0) < 1, \ y \geq 0\}$.

The terminology *surface-with-boundary* is an unfortunate misnomer, as these spaces are not actually *surfaces* according to the strict definition, unless the boundary is empty. In the literature, the term is usually *surface with boundary* (with no hyphens). While it is important to conform to the established terminology, I compromise in this textbook, erring on the side of clarity, by always writing *surface-with-boundary* (with hyphens).

Example 86. The following are examples of surfaces-with-boundary that are not surfaces.

- The closed cylinder, $\mathbb{S}^1 \times \mathbb{I}$. The boundary consists of two disjoint circles.

- The Möbius strip. The boundary consists of only one circle (see Figure 4.8).

- Suppose S is a surface and U is any open disk neighborhood of a point $x \in S$. Then $S \setminus U$ is a surface-with-boundary.

Throughout the remainder of the text, we use the notation $\mathcal{S}_c$ to denote the collection of all compact connected surfaces and surfaces-with-boundary.

Connected Sum Construction

Surfaces or surfaces-with-boundary may be combined by a process called *connected sum*. Intuitively, the construction is simple: remove an open disk from each of the two spaces, resulting in two surfaces with boundary; then glue the two pieces together along the boundary. A precise definition of this construction is tricky, as is the proof that the construction does not depend on any of the choices made along the way. (*Which open disks? In what way should the boundaries be identified?*) The fact that the connected sum construction is well defined for all compact surfaces is beyond the scope of this text,[2] and we shall take it for granted.

> **Definition 5.1.3.** Suppose $A, B \in \mathcal{S}_c$. Choose $U_1 \subseteq A$ and $U_2 \subseteq B$ open sets, each homeomorphic to an open disk, and let C_1 and C_2 be the boundary circles of U_1 and U_2, respectively. Let $f : C_1 \to C_2$ be a homeomorphism. The **connected sum** of A and B (with respect to U_1, U_2 and f) is the quotient space,
>
> $$A \# B = ((A \setminus U_1) \cup (B \setminus U_2)) / \sim,$$
>
> where $x \sim y$ if and only if $x \in C_1$, $y \in C_2$, and $y = f(x)$.

---

[2]See Proposition 2.6.1 of Bloch [Blo97].

The connected sum operation is *commutative* and *associative*, in the sense that the resulting spaces are homeomorphic. That is:

- $A \# B \approx B \# A, \quad \forall A, B \in \mathcal{S}_c$.

- $(A \# B) \# C \approx A \# (B \# C), \quad \forall A, B, C \in \mathcal{S}_c$.

Example 87. Let's see what $\mathbb{T} \# \mathbb{T}$ looks like. As Figure 5.5 illustrates, the result is a *two-holed torus*, $2\mathbb{T}$.

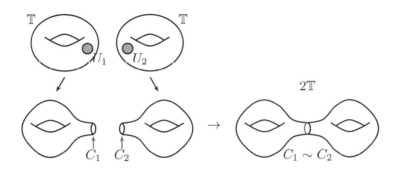

Figure 5.5: $\mathbb{T} \# \mathbb{T} = 2\mathbb{T}$.

Definition 5.1.4. The n-**holed torus**, $n\mathbb{T}$, is the connected sum of n tori. In general, if S is any surface, then nS is the connected sum of n copies of S.

See Figure 5.6 for illustrations of the n-holed torus.

Example 88. Describe each of the following:

(a) $\mathbb{T} \# \mathbb{S}^2$ (b) $4\mathbb{S}^2$ (c) $3\mathbb{T} \# 2\mathbb{T}$

Solution:

(a) $\mathbb{T} \# \mathbb{S}^2 \approx \mathbb{T}$, as illustrated below.

Figure 5.6: *Left*, the two-holed torus, $2\mathbb{T}$. *Right*, the n-holed torus, $n\mathbb{T}$. Images courtesy of Wikimedia Commons (modified).

(b) Just as in example (a), it should be clear that $\mathbb{S}^2\#\mathbb{S}^2 \approx \mathbb{S}^2$. Then $4\mathbb{S}^2 = \left(\left[\mathbb{S}^2\#\mathbb{S}^2\right]\#\mathbb{S}^2\right)\#\mathbb{S}^2 \approx \left(\mathbb{S}^2\#\mathbb{S}^2\right)\#\mathbb{S}^2 \approx \mathbb{S}^2\#\mathbb{S}^2 \approx \mathbb{S}^2$.

(c) $3\mathbb{T}\#2\mathbb{T} = (\mathbb{T}\#\mathbb{T}\#\mathbb{T})\#(\mathbb{T}\#\mathbb{T}) \approx 5\mathbb{T}$.

Example 89. Suppose P is a polyhedron in $\mathbb{R}^3$. Let P' be the union of the edges of P (think of this as a "wireframe"). Now suppose the edges in P' are thickened a bit, and let S be the surface of the "thickened wireframe." (That is, S is the boundary of a *tubular neighborhood* of P'; see §4.5, Exercise 8.) It can be shown that S is always homeomorphic to a connected sum of tori ($n\mathbb{T}$ for some $n \in \mathbb{N}$). Consider the tetrahedron T. Its thickened wireframe can be deformed to $3\mathbb{T}$, as illustrated in Figure 5.7. First flatten the wireframe and then arrange the holes. For what value of n is the surface in Figure 5.8 homeomorphic to $n\mathbb{T}$?

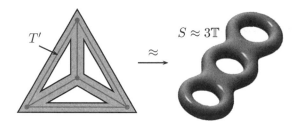

Figure 5.7: The thickened wireframe of a tetrahedron is homeomorphic to $3\mathbb{T}$.

In Example 88, we have shown that $3\mathbb{T}\#2\mathbb{T} \approx 5\mathbb{T}$. It's not too hard to see that $m\mathbb{T}\#n\mathbb{T} \approx (m+n)\mathbb{T}$ in general. It should also be clear that the sphere $\mathbb{S}^2$ acts as an *identity* for the connected sum operation. That is,

$$A\#\mathbb{S}^2 \approx \mathbb{S}^2\#A \approx A, \quad \forall A, B \in \mathcal{S}_c.$$

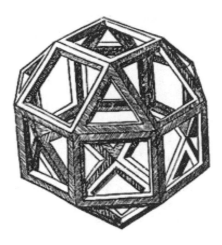

Figure 5.8: The thickened frame of a polyhedron (drawn by Leonardo da Vinci). The surface is homeomorphic to a connected sum of tori. Image courtesy of Wikimedia Commons.

Together with associativity, the existence of an identity makes $\mathcal{S}_c$ into a *monoid* under the connected sum operation. We shall have more to say about this in the next section.

Manifolds

Surfaces are essentially two-dimensional spaces, but Definition 5.1.1 can easily be extended to arbitrary dimensions. We call such spaces *manifolds*.[3]

Definition 5.1.5. Let $n \in \mathbb{N}$. A **manifold** is a connected Hausdorff space M such that for each $x \in M$, there is an open set $U \subseteq M$ with $x \in U$, such that $U \approx B_1(\mathbf{0}) \subseteq \mathbb{R}^n$. The manifold is said to be n-dimensional, or M is called an n-manifold.

There is an analogous definition of **manifold-with-boundary** (try to formulate a reasonable definition yourself, extending Definition 5.1.2).

Example 90. Because manifolds (by our definition) must be connected, the only zero-dimensional manifold is a single point. A one-dimensional manifold must be locally homeomorphic to an open interval. The whole real line $\mathbb{R}$ and the open interval (a, b) are examples of *noncompact* 1-manifolds. A closed interval

---

[3]Perhaps the better term here would be *topological* manifolds. In the literature, much more attention is paid to the more restrictive *smooth* or *differentiable* manifold theory, which falls outside the scope of this text.

$[a, b]$ is compact, but not a manifold; instead, $[a, b]$ is a manifold-with-boundary. The only compact 1-manifold (according to Definition 5.1.5) is the circle $\mathbb{S}^1$.

Example 91. Let's explore a few interesting manifolds and manifolds-with-boundary.

- Recall that $\mathbb{I}^n = \mathbb{I} \times \mathbb{I} \times \ldots \times \mathbb{I} \subseteq \mathbb{R}^n$ is the product of the unit interval with itself n times. This space is not a manifold because it has a boundary whose points cannot be surrounded by an open ball contained in $\mathbb{I}^n$. However, this space, which is called a **hypercube**, is a compact manifold-with-boundary. It is interesting to ponder what a 4-cube ($\mathbb{I}^4$) might look like. This object is called a **tesseract**,[4] and we might visualize it as a cube that has been extended through a fourth dimension (see Figure 5.9). What might $\mathbb{I}^5$ or $\mathbb{I}^6$ look like?

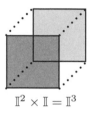
$$\mathbb{I}^2 \times \mathbb{I} = \mathbb{I}^3$$

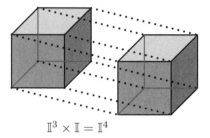
$$\mathbb{I}^3 \times \mathbb{I} = \mathbb{I}^4$$

Figure 5.9: Just as a solid cube may be thought of as a square that has been extended through a third dimension, the tesseract is a cube extended through a fourth dimension.

- The open balls $B_\epsilon(x)$ for $x \in \mathbb{R}^n$ and $\epsilon \in \mathbb{R}^+$ are noncompact manifolds. Their closures $D_\epsilon(x) \approx \mathbb{D}^n$ are compact manifolds-with-boundary (but not manifolds). Note that $\mathbb{I}^n \approx \mathbb{D}^n$ via a deformation that "inflates" the hypercube until all points on the boundary are the same distance from the center (recall Exercise 2 from §4.5).

- The n-sphere $\mathbb{S}^n$, or **hypersphere**, was introduced in Chapter 2 by way of a particular representation of the space, as a *unit* sphere:

$$\mathbb{S}^n = \{x \in \mathbb{R}^{n+1} \mid d(x, \mathbf{0}) = 1\} = \{(x_1, \ldots, x_{n+1}) \mid x_1^2 + \cdots x_{n+1}^2 = 1\}.$$

There are many other useful representations of the n-sphere. It should not be hard to see that for any given $x \in \mathbb{R}^{n+1}$ and $\epsilon > 0$, we have $\partial B_\epsilon(x) = \partial \mathbb{D}^{n+1} \approx \mathbb{S}^n$; that is, the boundary of any open ball or closed

---

[4]The tesseract has fascinated many science fiction authors, making an appearance in Robert Heinlein's short story "And He Built a Crooked House" [Hei41] and Madeleine L'Engle's popular children's novel *A Wrinkle in Time* [L'E62].

disk in $\mathbb{R}^{n+1}$ is an n-sphere. Thus we also get $\mathbb{S}^n \approx \partial \mathbb{I}^{n+1}$ by way of the homeomorphism $\mathbb{I}^n \approx \mathbb{D}^n$.

Now suppose H^+ and $H^- \subseteq \mathbb{S}^n$ are the "northern" and "southern" hemispheres of the unit n-sphere, respectively. That is,

$$
\begin{aligned}
H^+ &= \{(x_1,\ldots,x_{n+1}) \mid x_1^2 + \cdots x_{n+1}^2 = 1 \text{ and } x_{n+1} \geq 0\} \\
H^- &= \{(x_1,\ldots,x_{n+1}) \mid x_1^2 + \cdots x_{n+1}^2 = 1 \text{ and } x_{n+1} \leq 0\}.
\end{aligned}
$$

The quotient space H^+/H^- that collapses the entire southern hemisphere to a single point is also homeomorphic to $\mathbb{S}^n$ (and, by symmetry, $H^-/H^+ \approx \mathbb{S}^n$ as well). In fact, if we let $E = H^+ \cap H^-$, which is the equatorial sphere $\mathbb{S}^{n-1} \subseteq \mathbb{S}^n$, then we have

$$
H^+/E \approx H^-/E \approx \mathbb{S}^n.
$$

Example 22 (in Chapter 2) explains how $\mathbb{S}^n$ can also be regarded as $\mathbb{R}^n \cup \{\infty\}$, where ∞ is the *point at infinity*. Consider now a variation on this idea. Suppose $F : \mathbb{S}^n \setminus \{N\} \to \mathbb{R}^n$ is stereographic projection. The image $F[H^-]$ of the southern hemisphere is a closed disk $\mathbb{D}^n \subseteq \mathbb{R}^n$, as shown in Figure 5.10 (for the two-dimensional case). When all points on the boundary of that disk are identified to a single point, the resulting quotient space is again the n-sphere. That is,

$$
\mathbb{D}^n/\partial \mathbb{D}^n \approx \mathbb{I}^n/\partial \mathbb{I}^n \approx \mathbb{S}^n.
$$

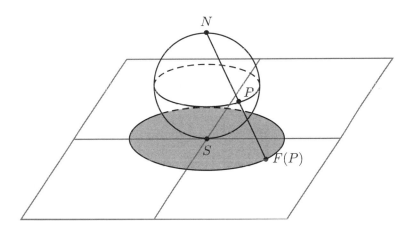

Figure 5.10: Stereographic projection to $\mathbb{R}^2$ sends the southern hemisphere of $\mathbb{S}^2$ to a disk in the plane.

- The n-**torus** $\mathbb{T}^n$ is an n-dimensional manifold defined by $\mathbb{T}^n = \mathbb{I}^n/\sim$, in which the equivalence relation is generated by identifying corresponding points on opposite facets of the hypercube. That is, $(x_1,\ldots,0,\ldots,x_n) \sim$

$(x_1, \ldots, 1, \ldots, x_n)$, with the understanding that the 0 and 1 occur in the same position and all other entries agree. For $n = 1$, this reduces to the usual definition of $\mathbb{S}^1 = \mathbb{I}/\{0 \sim 1\}$, and for $n = 2$, this space is the familiar torus, $\mathbb{T}^2 = \mathbb{T}$.

The "n" in $\mathbb{T}^n$ is a little misleading. The n-torus is not the product of n copies of $\mathbb{T}$. In fact, the n-torus may be expressed better by $(\mathbb{S}^1)^n$, since $\mathbb{T}^n$ is actually the product of n copies of $\mathbb{S}^1$.

- **Projective space** $\mathbb{P}^n$ is defined as a certain quotient space of $\mathbb{R}^{n+1} \setminus \{\mathbf{0}\}$ as in §4.3, Example 76, and Exercise 11). In essence, every point on a line through the origin (except the origin itself) gets identified to a single point. But $\mathbb{P}^n$ can also be defined by identifying all pairs of **antipodal** points on $\mathbb{S}^n$; that is, points on the sphere that differ only by the scalar multiple -1. This means that $\mathbb{P}^n \approx \mathbb{S}^n/(x \sim -x)$, and so representatives for the equivalence classes may be taken just from the upper hemisphere, $H^+ \subseteq \mathbb{S}^n$. There are still antipodal identifications on the equator, leading to another useful way to write projective space:

$$\mathbb{P}^n \approx H^+/(x \sim -x) \approx \mathbb{D}^n/(x \sim -x, \forall x \in \partial \mathbb{D}^n). \qquad (5.1)$$

Looking ahead to Figure 5.24, we can see how these identifications work in $\mathbb{P}^2 = \mathbb{P}$, the *projective plane*. Figure 5.11 shows a representation of $\mathbb{P}^3$.

The connected sum of manifolds may be defined in a manner similar to that of Definition 5.1.3, with $\mathbb{S}^n$ serving as an identity for the connected sums of n-manifolds. However, care must be taken for $n \geq 3$; the *orientation* of each manifold must be considered in order to have a well-defined operation. It would take us too far afield to explore manifolds further in this textbook. The interested reader may consult more advanced texts, such as *Introduction to Topological Manifolds* by John Marshall Lee [Lee00].

Exercises

1. Classify each of the following as surface, surface-with-boundary, or not a surface. If the space is a surface-with-boundary, describe the boundary. In each case, indicate whether the space is compact or not.

   (a) Closed square, $\mathbb{I}^2$.

   (b) Annulus, $\{(x, y) \in \mathbb{R}^2 \mid 1 \leq x^2 + y^2 \leq 2\}$.

   (c) $\mathbb{T} \setminus \{p\}$, where $p \in \mathbb{T}$ is any single point.

   (d) The union of the xy-plane, xz-plane, and yz-plane in $\mathbb{R}^3$.

   (e) Finite open cylinder $\mathbb{S}^1 \times (0, 1)$.

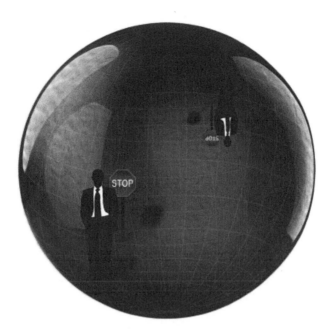

Figure 5.11: Three-dimensional projective space $\mathbb{P}^3$ may be defined as the solid sphere $\mathbb{D}^3$ with antipodal points on the surface identified. If you lived within a small model of $\mathbb{P}^3$, then you could look down the street and see yourself rotated by one half-turn. Image courtesy of Cyrus Rua – used by permission.

2. Consider the capital letters of the alphabet written in block form and thickened to three-dimensional objects (as in Exercise 1 in §1.1). For each letter, identify the value of n such that the letter is homeomorphic to $n\mathbb{T}$.

3. For each of the five Platonic solids (Figure 1.11) P, find the value n such that $P' \approx n\mathbb{T}$, where P' is the *thickened wireframe* of P (see Example 89). What about the object in Figure 5.8?

4. Recall that the Jordan Curve Theorem (Theorem 2.4.8) states that a simple closed curve C separates the plane $\mathbb{R}^2$ into two arc-connected components. Argue, with the help of stereographic projection, that the same happens on the surface of a sphere $\mathbb{S}^2$. Give an example of a simple closed curve on the torus $\mathbb{T}$ that does not separate it.

5. Write a reasonable definition for manifold-with-boundary, analogous to Definition 5.1.2.

6. Show that the connected sum of two connected surfaces is arc-connected (therefore also connected).

7. Show that the connected sum of two compact surfaces is compact. What if one or both of the surfaces are noncompact?

8. Write each connected sum below as $n\mathbb{T}$ for an appropriate value of n. In part (b), assume $k \in \mathbb{N}$.

    (a) $\mathbb{T}\#3\mathbb{S}^2\#2\mathbb{T}\#7\mathbb{S}^2$ (b) $\mathbb{T}\#2\mathbb{T}\#3\mathbb{T}\#\cdots\#k\mathbb{T}$

9. Let $S \in \mathcal{S}_c$ be arbitrary. Describe $S\#\mathbb{D}^2$.

5.2   Plane Models and Words

In this section we develop a powerful combinatorial tool that enables us to work with surfaces in an easier way, eventually allowing us to classify compact surfaces and to understand subtle structures in the surface. Let's revisit the torus, defined as a quotient space as in §4.3.

$$\mathbb{T} = \mathbb{I}^2/\{(0,y) \sim (1,y),\ (x,0) \sim (x,1)\}$$

We represent $\mathbb{T}$ as a square together with the appropriate identifications along the edges as shown in Figure 5.12, with the convention that edges labeled with the same letter should be identified in the direction indicated by the arrowheads.

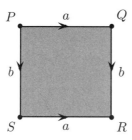

Figure 5.12: Plane model for the torus $\mathbb{T}$ and its realization. After the identification, all four vertices are identified to a single point, $P \sim Q \sim R \sim S$.

This type of diagram – a polygon showing certain pairs of edges identified – is called a **plane model** for the surface or surface-with-boundary, and it exhibits the space as a quotient space of a closed disk (recall $\mathbb{I}^2 \approx \mathbb{D}^2$). Any pair of edges marked with the same letter correspond to a single segment or loop in the space itself. Any segments without a label or with a unique label correspond to edges in the space, and in that case, the space may be a surface-with-boundary, but not a surface.

For each plane model, there is a way to describe the model using a sequence of letters, or a **word**. Simply pick a starting vertex; choose a direction, clockwise or counterclockwise; then list the letters in order, indicating the direction of the edge by placing an "exponent" -1 on the letter if the arrow is backward with respect to the direction you chose. There are many different words for the same plane model. For instance, there are eight equivalent words for the torus in Figure 5.12:

$$aba^{-1}b^{-1}, \qquad ba^{-1}b^{-1}a, \qquad a^{-1}b^{-1}ab, \qquad b^{-1}aba^{-1},$$

$$bab^{-1}a^{-1}, \qquad ab^{-1}a^{-1}b, \qquad b^{-1}a^{-1}ba, \qquad a^{-1}bab^{-1}.$$

Example 92. Describe the topological spaces represented by their plane models below, and find a word for the model. Of the vertices in the original diagram, indicate which ones are identified in the resulting space.

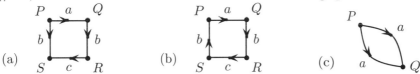

Solution:

(a) Only the left and right sides get glued together. The top and bottom remain as boundaries. This is a surface-with-boundary homeomorphic to the cylinder $\mathbb{S}^1 \times \mathbb{I}$. One word for the model is $abcb^{-1}$. The vertices are identified in pairs: $P \sim Q$ and $R \sim S$.

(b) Again the left and right sides are glued, but this time with a twist. The result is a Möbius strip, as shown in Figure 4.8. One word for the model is $abcb$. The vertices are again identified in pairs, but in the opposite sense: $P \sim R$ and $Q \sim S$.

(c) The opposite sides are identified much like the the two edges of a purse when they get zippered together. This plane model defines the sphere $\mathbb{S}^2$, with corresponding word aa^{-1}. Points P and Q are not identified and may be associated with the north and south pole of the sphere, respectively.

Plane Models and Connected Sum

Suppose we have two plane models representing spaces $A, B \in \mathcal{S}_c$. A plane model for $A \# B$ can be found by the following procedure:

1. Choose a vertex in the plane model for A, and draw a loop at A within the model. Label the loop with a letter not already used in the models for A or B, let's say q. Do the same in the plane model for B, labeling the loop with the same letter, q.

2. Remove the region inside each loop. Then separate each loop at the vertex, thereby creating plane models for $A \setminus U_1$ and $B \setminus U_2$ (where U_1 and U_2 are as in Definition 5.1.3). Now q labels a new edge of each plane model.

3. Join together the two plane models along the edge q. This edge q is now interior to the plane model and so can be erased.

Example 93. Find a plane model and corresponding word for $2\mathbb{T}$.

Solution: The steps are illustrated below. **Caution:** It is important to give distinct labels to the edges of each space.

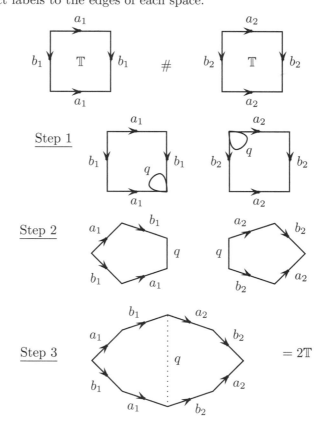

Beginning at the leftmost vertex and going clockwise, a word for $2\mathbb{T}$ is: $a_1 b_1 a_2 b_2 a_2^{-1} b_2^{-1} a_1^{-1} b_1^{-1}$ (there are many other equivalent words).

It is interesting to see how a given representation of a surface relates to the plane model for the surface. For example, how does Figure 5.6 correspond to the octagonal plane model for $2\mathbb{T}$ found in Example 93? The key idea is to find *nonseparating* loops – loops that do not divide the space into disconnected pieces. Some spaces like the sphere have no nonseparating loops, while on the torus $\mathbb{T}$, one can find two loops that can be cut simultaneously and not separate the space (e.g., the loops a and b in Figure 5.12). It turns out that the n-holed torus has a set of $2n$ simultaneously nonseparating loops.

Example 94. By cutting along nonseparating loops, a_1, a_2, b_1, and b_2, Figure 5.13 illustrates how the standard embedding of $2\mathbb{T}$ within $\mathbb{R}^3$ corresponds to its plane model. Keep in mind that the cuts are not actually changing the topology of the surface because we still keep track of the identifications.

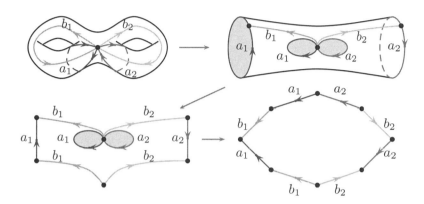

Figure 5.13: Cutting $2\mathbb{T}$ along nonseparating loops to obtain the plane model. First cut along a_1 and a_2 to create a tube with a pair of holes near the middle. Then cut lengthwise along b_1 and b_2 to unwrap the tube, finally opening up the figure into an octagon plane model for $2\mathbb{T}$.

Word Arithmetic

The construction given above for the connected sum of two plane models naturally suggests an operation on words themselves. If $abc\ldots$ is a word for a space A and $zyx\ldots$ is a word for B *having different letters than the word for A*, then the concatenation $abc\ldots zyx\ldots$ is a word for $A\#B$. Furthermore, there are certain rules for rearranging words to make equivalent words (i.e., representing the same space). We will use the notation $\equiv$ to indicate that two words are equivalent.

- For convenience, we often write aa as a^2 and $a^{-1}a^{-1}$ as a^{-2}.

- The letters in words do not **commute** ($abcd \neq bacd$); however, *cyclic permutations* are permissible. A **cyclic permutation** of a word is the result of moving one letter from the start of the word to the end, or vice versa. Thus

$$aba^{-1}b^{-1} \equiv ba^{-1}b^{-1}a \equiv a^{-1}b^{-1}ab \equiv b^{-1}aba^{-1}.$$

This move is equivalent to picking a different starting vertex, or rotating the plane figure.

- If a letter and its inverse appear adjacent to one another, then they *cancel* each other out, *except* in the case of the sphere $\mathbb{S}^2$, whose smallest word has only two letters, aa^{-1}: canceling out these two would result in an *empty word*, which cannot be drawn as a plane model. For example,

$$abc^{-1}dd^{-1}ca^{-1}b^{-1} \equiv abc^{-1}ca^{-1}b^{-1} \equiv aba^{-1}b^{-1}.$$

- If X is a word, the inverse word X^{-1} is obtained by reversing the order and inverting each letter. For example,

$$(abc^{-1}d)^{-1} = d^{-1}cb^{-1}a^{-1}.$$

  This has the effect of choosing to read the labels in the opposite direction, or flipping over the plane model, so the resulting space is not affected: $X^{-1} \equiv X$.

- There is a *split-and-rejoin* operation, which is a useful tool in the classification of surfaces. Choose a location in the word in which to insert a new edge and its inverse, then split the word between the new letter and its inverse to obtain two words (which together still describe a single space). Any equivalence operation mentioned above may then be applied to each subword separately. Join the words again by concatenating at a letter-inverse pair.

$$
\begin{array}{ccc}
XY \equiv Xqq^{-1}Y \equiv & Xq, & q^{-1}Y \\
 & \equiv & \equiv \\
 & Ua, & a^{-1}V \equiv Uaa^{-1}V \equiv UV
\end{array}
$$

  This operation corresponds to a cut in the plane model represented by XY by a curve that separates all edges in X from those in Y.

Example 95. The **Klein bottle** is defined as a quotient space by

$$\mathbb{K} = \mathbb{I}^2/\{(x,0) \sim (x,1),\ (0,y) \sim (1,1-y)\}.$$

Draw a plane model representing $\mathbb{K}$, and then show that the Klein bottle has a plane model whose word has the form x^2y^2.

Solution: According to the definition of $\mathbb{K}$, the top and bottom edges are identified in the same direction; however, the sides are identified with a *twist*. A word corresponding to this plane model is $aba^{-1}b$. Consider the following sequence of moves, along with their depictions in the plane model. In each picture, we have circled the starting vertex to make it easier to identify the corresponding word. The direction of the word is clockwise in each diagram.

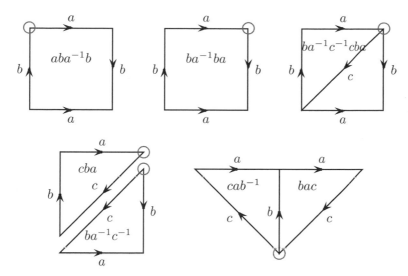

Note that in the fifth diagram above, the triangle $ba^{-1}c^{-1}$ has been flipped to create cab^{-1}, while the word cba has been cyclically permuted to bac, allowing the sides labeled b to be joined together with matching arrow directions. The result is $cab^{-1}bac \equiv caac \equiv aacc - a^2c^2$, having the desired form.

Correspondence between Plane Models and Surfaces

Not every plane model diagram corresponds to an element of $\mathcal{S}_c$. For example, if there are three or more edges labeled by the same letter, then no neighborhood of any point on the common edge would be homeomorphic to a disk. Figure 5.14 illustrates this issue in a space called the *dunce cap*. Any neighborhood of the point x fails to be homeomorphic to a disk; the neighborhood could be described as a "disk with flap."

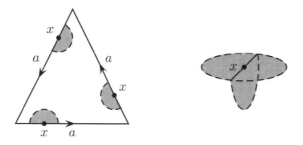

Figure 5.14: The dunce cap.

On the other hand, it can be shown that any plane model diagram in which no letter appears more than twice *does* correspond to a surface or a surface-with-boundary. For simplicity, we will prove the result when it applies to a surface;

the necessary modifications in the case of a surface-with-boundary should be
clear.

Theorem 5.2.1. *Suppose X is a plane model diagram in which each
letter appears exactly twice. Then X represents a compact surface.*

Proof. We must show that each point of the plane model has an open neighborhood U homeomorphic to an open disk. There are three cases to consider. It may
be helpful to refer to Figures 5.15 and 5.16. In what follows, when we say *interior* to X, we mean interior to the plane model diagram before identifications
are made along the boundary.

Case I. Suppose $x \in X$ is in the interior of the plane model. Then an open
neighborhood U of x may be chosen small enough to lie entirely interior to X,
and U is clearly homeomorphic to an open disk.

Case II. Suppose $x \in X$ is on a boundary edge of the plane model, but not
at a vertex. Let the edge have label a. Then, by assumption, there is exactly one
other edge labeled a. Since those two edges are identified in X, a disk neighborhood U surrounding x exists. In terms of the plane model, it is equivalent to
cut and paste a small portion of the model to show how U connects across the
edge a.

Case III. Suppose $x \in X$ is a vertex of the plane model. This is the trickiest
case because x may be identified to many other vertices on the plane model, and
we must build up the disk neighborhood from portions of the model surrounding
all vertices equivalent to x. A priori, we do not know which other vertices belong
to the equivalence class $[x]$, but the procedure below locates them all.

First suppose that the word for X contains no adjacent inverse pair (i.e.,
$\ldots aa^{-1} \ldots$ or $\ldots a^{-1}a \ldots$). If it does, then X can be simplified by zippering
along the inverse pair. For each edge e, let e^- be a chosen point on that edge
near its tail vertex, and let e^+ be on the edge near its head vertex. We shall take a

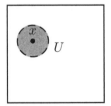
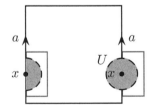

Figure 5.15: Open disk neighborhoods around an interior point and edge point
of a plane model. The diagram on the right shows how a portion of the plane
model can be cut and pasted to form a disk neighborhood around x.

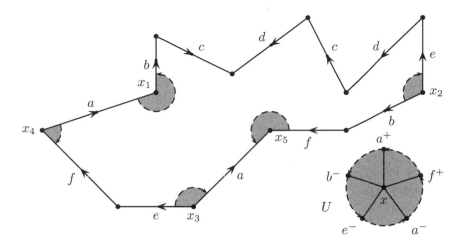

Figure 5.16: Open disk neighborhood U around a vertex x, assembled from regions surrounding all equivalent vertices. This example shows that x is enclosed by a curve containing (in order) $a^+b^-e^-a^-f^+a^+$.

tour of some (or all) of these points that eventually bounds a disk neighborhood of x.

For the purpose of this algorithm, let $x_1 = x$, and let a_1 and a_2 be the two edges meeting at x_1. If x is the tail of a_1, select a_1^-; if x is the head of a_1, select a_1^+. Do the same with respect to a_2. Then draw a small arc in X connecting the selected points. Since edge labels must come in pairs, there must be another edge labeled a_2. Let x_2 be the vertex near to the head (a_2^+) or tail (a_2^-) in the other edge, matching the sign of a_2 that was near x_1. We now know that $x_2 \sim x_1$. Now let a_3 be the label of the other edge meeting x_2. As before, select either a_3^- or a_3^+ depending on how x_2 meets a_3, and draw a small arc in X connecting the selected points. Continue this process until reaching an edge a_n that matches the first edge a_1 *and* meets x in the same way (as head or tail). This completes the tour around vertex x. By cutting and pasting all of these regions together according to the identifications in X, the union of all open regions within the arcs forms an open disk neighborhood of x. □

Unfortunately, what we haven't proven yet is that *every* space $X \in \mathcal{S}_c$ *has* a plane model. This depends on the existence of a *triangulation* for X, which is not a trivial thing to prove in general. We define triangulation in the next section.

Toward Triangulation

Does every surface and surface-with-boundary have a plane model? The answer is *yes*, because every space $X \in \mathcal{S}_c$ has a *triangulation*. Let's first define the term. Let $X \in \mathcal{S}_c$. A **triangulation** of X consists of:

- A finite set $V = \{v_1, v_2, \ldots, v_k\}$ of points $v_i \in X$, called the **vertices** of X.

- A finite set $E = \{e_1, e_2, \ldots, e_\ell\}$, called the **edges** of X, such that each $e_i \subseteq X$ is homeomorphic to $\mathbb{I}$, and the endpoints of each e_i are two distinct points from V. If $i \neq j$, then either $e_i \cap e_j = \emptyset$ or there is a unique $v \in V$ such that $e_i \cap e_j = v$. Each edge $e \in E$ may be described by its endpoints: $e = [v_0, v_1]$.

- A finite set $F = \{f_1, f_2, \ldots, f_m\}$, called the **faces** or **triangles** of X, such that each $f_i \subseteq X$ is homeomorphic to a closed disk, and the boundary of f_i consists of three distinct edges from E. Furthermore, if $f_i \cap f_j \neq \emptyset$ for some $i \neq j$, then $f_i \cap f_j$ is either a unique edge $e \in E$ or a unique vertex $v \in V$. Morever, the triangles *cover* X in the sense that $X = \bigcup F$. Each $f \in F$ may be described by its vertices: $f = [v_0, v_1, v_2]$.

Note that by Definition 5.1.2, every edge can belong to at most two triangles. If X is a surface (rather than surface-with-boundary), then every edge belongs to *exactly* two distinct triangles. Figure 5.17 illustrates a few ways of cutting up the torus that are not triangulations, while Figure 5.18 shows a triangulation of the torus. In fact, many different triangulations exist for a given $X \in \mathcal{S}_c$.

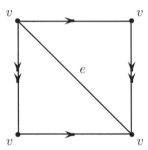 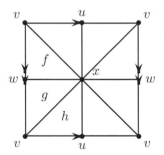

Figure 5.17: Nontriangulations of the torus. *Left*, every edge has both endpoints the same: $e = [v, v]$. *Right*, some pairs of triangles have intersections that are not a single edge or a single vertex. For example, $f \cap g = [w, x] \cup \{v\}$, and $f \cap h = \{x, v\}$.

Theorem 5.2.2. *Every $X \in \mathcal{S}_c$ has a triangulation.*

The proof of Theorem 5.2.2 is not easy. First proved rigorously in 1925 by Radó [Rad25],[5] the result relies on the Jordan Curve Theorem (Theorem 2.4.8), among other things. As a consequence of the theorem, every $S \in \mathcal{S}_c$ has a plane

---

[5]See also Ahlfors and Sario [AS60], Thomassen [Tho92], and Doyle and Moran [DM68].

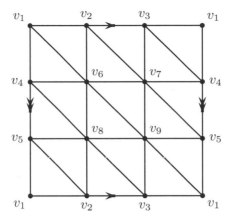

Figure 5.18: A triangulation of the torus using nine vertices. Can you find one that uses fewer vertices?

model. Given a triangulation T for S, with triangles $\{t_1, t_2, \ldots, t_n\}$, build the plane model as follows.

- For each triangle t_k, label its edges and indicate the orientation of each edge so that shared edges receive the same label and orientation.

- Start with $P_1 = t_1$.

- Choose a triangle t_2 that shares an edge with t_1 (which we may do since S is assumed to be a connected surface), and join these triangles at the common edge to form $P_2 - t_1 \cup t_2$. Label the unjoined edges.

- Having constructed $P_k = t_1 \cup t_2 \cup \cdots \cup t_k$, for $k < n$, there must be at least one such triangle t_{k+1} that shares an edge with an existing triangle in P_k. Join t_{k+1} to P_k along that edge, but do not join any other edge of t_{k+1} to P_k. This ensures that P_k remains homeomorphic to a disk.

- Repeat the procedure until all triangles are used. Labels on internal edges should be discarded.

At each step k, the model P_k is a disk with edge labels. Since each edge belongs to at most two triangles, there are no more than two edges in P_k having the same label. Thus P_n is a plane model.

Because of the rather strict rules defining a triangulation, a compact surface can be described unambiguously by a finite set of *combinatorial* data, namely, the list of triangles given in terms of their vertices. In fact, we need only identify each vertex by a number, so a triangle $f = [v_1, v_2, v_3]$ could be represented as $[1, 2, 3]$. This makes it convenient to work with triangulations in computer code, for example.

Exercises

1. Describe the surface or surface-with-boundary represented by each plane model and find a corresponding word, reducing as much as possible. Indicate which vertices are identified in the space.

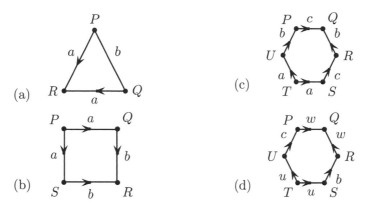

2. Find a plane model and corresponding word representing each of the following spaces.

   (a) $\mathbb{I}^2$ (b) $3\mathbb{T}$ (c) $\mathbb{T}\#\mathbb{K}$ (d) $2\mathbb{K}$

3. Redo Example 93, choosing a different vertex so that the resulting word for $2\mathbb{T}$ is $a_1 b_1 a_1^{-1} b_1^{-1} a_2 b_2 a_2^{-1} b_2^{-1}$.

4. Write a word for the n-holed torus, $n\mathbb{T}$.

5. Use words to show that there is no $Q \in \mathcal{S}_c$, such that $\mathbb{T}\#Q = \mathbb{S}^2$. This demonstrates that $\mathcal{S}_c$ is not a *group* with respect to connected sum.

6. Describe the result when any surface S is joined to the square $\mathbb{I}^2$ by connected sum. How is the word of $S\#\mathbb{I}^2$ related to the word of S?

7. Illustrate the algebraic operations of cyclic permutation and cancellation in a series of plane model pictures.

8. Write a word for the plane model shown in Figure 5.16. Show that the vertices fall into two distinct equivalence classes, $[x]$ and $[y]$. It was shown that x is enclosed by a curve labeled $a^+ b^- e^- a^- f^+ a^+$. Determine a curve bounding a disk neighborhood of y in this figure.

9. Suppose a compact surface has a triangulation with v vertices, e edges, and f faces. Show that $3f = 2e$. Show by example that $3f < 2e$ in a compact surface-with-boundary.

10. Show that any given plane model of a compact surface is equivalent to one in which all of the vertices belong to the same equivalence class. (*Hint:* Suppose there is a vertex y not in the same equivalence class as another vertex x, and suppose y is the head vertex of edge a. Perform the split-and-rejoin operation suggested by the picture below. Why can we assume there is another edge labeled a? How does this procedure eventually eliminate all vertices in the equivalence class of y? What happens if there is only a single vertex in the class of y?)

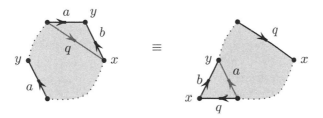

5.3 Orientability

Suppose an ant walks around on the surface of the sphere $\mathbb{S}^2$. We're talking about a three-dimensional ant whose legs touch the outside surface of the sphere as it walks. No matter where the ant walks or for how long it walks, it will never appear on the "inside" of the sphere. To be a little more precise, suppose that the ant is trailing paint as it walks – that way, we can clearly track its path. No matter how long the ant walks, or what kind of meandering path it takes, at the end of the journey we observe that there is paint only on one "side" of the surface (what we call the outside). But now consider an ant walking around on a Möbius strip M. Do this experiment for yourself: Use your pencil or pen to mark your path as you "walk" around on M. If your path is long enough, and stays in roughly the same direction, you will find that you can eventually return to your starting position, but there are now markings on both "sides" of M. This means that M really has only *one* side. The two spaces differ in that $\mathbb{S}^2$ is *orientable* while M is not.

In order to define orientability more precisely, let's first talk about what an orientation is. On the real number line $\mathbb{R}^1$, there are of course only two directions of travel, in order of increasing real numbers (*right*), or decreasing real numbers (*left*). Similarly, there are only two directions of travel on the circle $\mathbb{S}^1$. Let's assume $\mathbb{S}^1$ is represented in the xy-plane as a unit circle. Then at any point of the circle, you may go in order of either increasing angle measure with respect to the x-axis (*counterclockwise*), or decreasing angle measure (*clockwise*). Notice that in each case, we may record the direction using only one bit of information, say, $+1$ for *right* in $\mathbb{R}^1$ and -1 for *left*, or, in the circle, $+1$ for *counterclockwise* and -1 for *clockwise*.

What about an orientation on $\mathbb{R}^2$? This time there are infinitely many distinct directions that a path could take from any given starting point. Suppose you live at the point $(5, 3)$ and you want to tell your friend how to get to your house. Suppose the friend is currently standing at the origin and facing toward the positive x-axis. Then you may tell your friend to go forward five units, turn left, then go forward three units. So long as your friend knows his or her right hand from the left, he or she will end up at your house as expected. On the other hand, another friend who mistakes left for right would have ended up at someone else's house at $(5, -3)$. Even worse, if you and your friend lived on a Möbius strip, then the concepts of left and right could no longer be consistently defined.

An **orientation** at a point on a surface (or surface-with-boundary) is a specification of *left* vs. *right* with respect to *forward* – or, using the familiar directions of *north* (N) and *east* (E) on a compass, we distinguish a *positive* orientation from a *negative* one, as in Figure 5.19. If the smallest angle from E to N is obtained by going counterclockwise, then the orientation is positive. If the smallest angle from E to N is obtained by going clockwise, then negative. But keep in mind that these concepts rely on the observer's choice in how to look "down" onto the surface, which may only make sense in a small neighborhood. We will consider rotations of orientations as equivalent, but reflections are not (see Figure 5.19). Reflecting a positively orientated diagram will result in a negatively orientated diagram and vice versa.

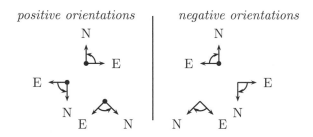

Figure 5.19: Looking down on the page, identify *positive* and *negative* orientations as shown.

We are interested in defining orientations at each point in a way that gives nearby points *consistent* orientations. In other words, close neighbors should be able to agree about what is left and what is right. Sounds pretty straightforward, right? Just take a positively oriented compass throughout the surface and declare each point to have the orientation implied by the compass; however, there are surfaces for which this procedure fails every time.

Consider the Möbius strip M shown in Figure 5.20. Suppose we decide a particular orientation for all points on the vertical segment denoted by L. Now extend the orientation in a consistent manner along the arc indicated by the gray arrows. In other words, take a compass with you and declare the orientation in any small neighborhood along the arc to be the one shown on your compass as you walk around M. By the time you reach L again, every neighborhood of M will have received a consistent orientation *except* any neighborhood of a point on L itself. The key is that there is a loop γ in M such that consis-

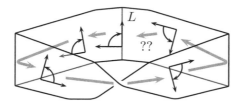

Figure 5.20: There is no way to define the orientation at every point of M such that orientations are consistent throughout M.

tent choices of orientation along γ end up becoming inconsistent (reversed) at $\gamma(1) = \gamma(0)$.

Definition 5.3.1. Let X be a surface or surface-with-boundary.

- A loop γ in X is called an **orientation-preserving loop** if there is a choice of orientations at each point of X so that the orientation remains consistent throughout γ; otherwise, γ is called an **orientation-reversing loop**.

- A space X is called **orientable** if no orientation-reversing loops exist in X; otherwise, X is called **nonorientable**.

Orientability is a topological invariant. Consider a homeomorphism $f : X \to Y$, and any loop γ in X. If γ is orientation preserving, then the choice of orientations at each point of X can be transferred directly to Y via the map f, showing that the loop $f \circ \gamma$ is orientation preserving. Now suppose γ is orientation reversing in X. If $f \circ \gamma$ were orientation *preserving*, then the homeomorphism f^{-1} would imply that $f^{-1} \circ (f \circ \gamma) = \gamma$ is also orientation preserving, contradicting our choice of γ. Thus $f \circ \gamma$ must also be orientation reversing.

Example 96. Show that $\mathbb{R}^2$ is orientable.

Solution: At every point (x, y), choose a positive orientation in which N is in the direction of $(x, y + 1)$, and E is in the direction of $(x + 1, y)$. Every loop in the plane is orientation preserving with respect to these orientations.

Example 97. Determine whether the sphere $\mathbb{S}^2$ is orientable or not.

Solution: It may be intuitively clear that no loop around a sphere could possibly change the orientation, but to prove the statement carefully, we make use of the orientability of the plane $\mathbb{R}^2$.

By removing small open neighborhoods of the north and south poles, the sphere can be regarded as the union of a cylinder and two disks with identifications along the boundaries. If the cylinder is cut (e.g., along segment a in the figure below) and flattened out, then an orientation may be chosen that coincides with a rectangular patch of $\mathbb{R}^2$ (this is how a Mercator projection of the world map works). The disks themselves may be flattened and inherit orientations from $\mathbb{R}^2$. Finally, because of the way that the disks are glued onto the top or bottom of the cylinder to form the sphere, there are only rotations, and never reflections, along any loop. Thus $\mathbb{S}^2$ is orientable.

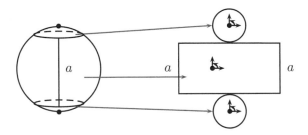

Caution: Just because a space decomposes into orientable pieces does not imply that the space itself is orientable. The Möbius strip can be cut along a single segment so that it becomes $\mathbb{I}^2$, which is orientable. It's in the gluing together, with a twist, that the space becomes nonorientable.

Orientability, Plane Models, and Words

An orientation-reversing loop may be found (or ruled out altogether) by examining a plane model for the space. Figure 5.21 shows a plane model for the Möbius strip M, along with an arc in M. Note that any arc that crosses the edge b must reverse orientation (from our perspective looking down on the square). The orientation at x may begin as positive in the plane model and remain that way as we traverse the arc to the right and arrive at y; however, the orientation flips as we reappear on the left side on the way to z. The points x and z clearly have opposite orientation. Thus it is quite easy to locate an orientation-reversing loop in any plane model that has a pair of edges like those on M.

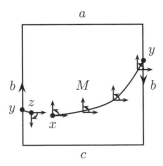

Figure 5.21: The orientation reverses along the arc from x through y to z in M.

Note that from the point of view of a myopic ant traveling along the curve, there is no significant difference between points x and y, so the ant would not experience dramatic change as it passes through point y – with the exception that now the environment around x seems to be inverted by the end of the journey.[6] By appending copies of M to each other along the identified side b, as in Figure 5.22, we may get a better idea how the journey might look to a traveler. Note how the orientation with respect to the traveler never actually changes on the journey from x through y to z. However, because of the identification of b with a twist, it seems that the whole world has been turned upside down.

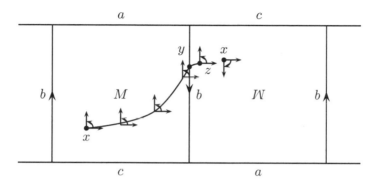

Figure 5.22: Orientation-reversing arc shown on a tiled representation of M.

A word for M is $abcb$. Note how b occurs twice in the same direction on the plane model, and so b occurs twice in the word. Any surface or surface-with-boundary whose word has the form $\ldots x \ldots x \ldots$ or $\ldots x^{-1} \ldots x^{-1} \ldots$ is non-orientable, and a loop that crosses the edge x once is an orientation-reversing loop. Conversely, if every letter of a word occurs only once or only with its inverse, then the space is orientable.

Example 98. The torus $\mathbb{T}$ is orientable because its word, $aba^{-1}b^{-1}$, has only inverse pairs, a and a^{-1}, b and b^{-1}. There is a well-defined orientation on $\mathbb{T}$ such that the orientation is preserved along any loop, as illustrated in Figure 5.23.

The Projective Plane

Let's consider a rather nonintuitive surface called the **projective plane**, $\mathbb{P} = \mathbb{P}^2$. You may recall seeing the projective spaces $\mathbb{P}^n$ in Exercise 11 of §4.3, defined

---

[6]It is interesting to ponder an orientation-reversing loop in our own three-dimensional world. Suppose a one-armed traveler leaves Earth on such a trip. To us here on Earth, when the traveler returns, he is still missing an arm, but now the other one. On the other hand (pun intended), from the traveler's perspective, his original arm has always been there; it's just that now, upon return to Earth, all the signs read backward.

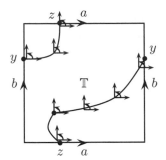

Figure 5.23: The torus is orientable. Orientation is preserved along any loop.

as quotient spaces, and how $\mathbb{P}$ may be regarded as a hemisphere such that antipodal points along the equatorial boundary circle are identified as in (5.1), as illustrated in Figure 5.24. The hemisphere itself may be flattened out into a plane model: then the surface is defined by a simple word, a^2, showing immediately that $\mathbb{P}$ is nonorientable.

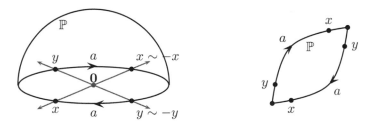

Figure 5.24: Representations of the projective plane. *Left*, $\mathbb{P}$ is the quotient of $\mathbb{R}^3 \setminus \mathbf{0}$, in which every point on a line through the origin is identified. *Right*, by taking representatives on a hemisphere, we find a simple plane model for $\mathbb{P}$ with word a^2.

Suppose $A, B \in \mathcal{S}_c$ are both orientable. Then so is $A \# B$. The easiest way to see this is to consider the words for A and B. If both are orientable, then there are no same-pair letters in A or B; hence $A \# B$ has no same-pair letters in its word (the concatenation of words for A and B). On the other hand, if either one (or both) of A or B is nonorientable, then so is their connected sum. For example, $\mathbb{T} \# \mathbb{P}$ is nonorientable – a word describing its plane model is $aba^{-1}b^{-1}c^2$.

Example 99. Describe $2\mathbb{P}$

Solution: The space $2\mathbb{P} = \mathbb{P} \# \mathbb{P}$ is a nonorientable surface whose word is a^2b^2 (recall that the letters in the words of each space must be distinct), which is a word for the Klein bottle $\mathbb{K}$ (see Example 95). Therefore $2\mathbb{P} \approx \mathbb{K}$.

Exercises

1. Find a word representing $n\mathbb{P}$ for arbitrary $n > 0$.

2. Determine whether the following spaces are orientable or not. If not, show an orientation-reversing loop in a plane model for the space.

   (a) $\mathbb{S}^1 \times \mathbb{I}$

   (b) $3\mathbb{T}$

   (c) $\mathbb{K}$

   (d) $n\mathbb{T}$

   (e) $\mathbb{T}\#\mathbb{P}$

   (f) $3\mathbb{T}\#2\mathbb{P}$

3. Prove $\mathbb{T}\#\mathbb{P} \approx 3\mathbb{P}$.

4. Show that it is possible to classify every finite connected sum of $\mathbb{T}, \mathbb{P}, \mathbb{K}$, and $\mathbb{S}^2$ as one (and only one) of the following.

   - $\mathbb{S}^2$
   - $n\mathbb{T}$
   - $n\mathbb{P}$

   (*Hint:* See Examples 88 and 99, and Exercise 3.) The number $n \geq 1$ in the above is called the **genus** of the surface ($\mathbb{S}^2$ has genus 0). Find the genus of each of the following, and state whether the surface is orientable or nonorientable.

   (a) $3\mathbb{T}\#2\mathbb{P}$

   (b) $m\mathbb{K},\ m > 0$

   (c) $13\mathbb{S}^2\#5\mathbb{T}\#6\mathbb{S}^2$

   (d) $\mathbb{K}\#\mathbb{T}\#\mathbb{P}$

   (e) $a\mathbb{P}\#b\mathbb{T},\ a \geq b > 0$

   (f) $k\mathbb{S}^2,\ k > 0$

5. Show that the connected sum of a nonorientable surface with any other surface is nonorientable.

6. Starting with the plane model for $\mathbb{P}$, divide each edge a into three segments, $a = a_1a_2a_3$. Cut the model at the segments indicated in the figure below to obtain three pieces, P_1, P_2, and P_3. Glue pieces P_1 and P_3 along their common edge a_3a_1, and explain how the resulting figures demonstrate that $\mathbb{P}$ is a Möbius strip whose boundary circle bounds a disk. Can this space be realized within $\mathbb{R}^3$?

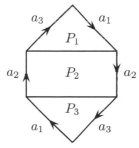

7. An n-twist strip is formed from a strip of paper by twisting one end n times before gluing the ends together. For example, the standard Möbius strip is a one-twist strip, and a two-twist strip is shown at right.

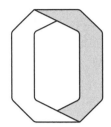

(a) Show that the two-twist strip is orientable.

(b) Determine the orientability of the n-twist strips in general.

5.4   Euler Characteristic

Recall from Chapter 1 that the **Euler characteristic** of a polyhedron P is defined by the formula $\chi(P) = v - e + f$, where v, e, and f are the numbers of vertices, edges, and faces, respectively, of P. Each of the five Platonic solids has Euler characteristic equal to 2. It's no coincidence that the surface of each of these solids deforms to the sphere. In fact, starting with $\mathbb{S}^2$, we may choose any number of vertices on the sphere, connect them with nonintersecting arcs, and by straightening the edges and flattening the faces, we could obtain a new polyhedron P'. Is $\chi(P') = 2$ as well? In other words, is the value of the Euler characteristic somehow *intrinsic* to the sphere itself, not depending on any specific deformation of the sphere to a polyhedron? The answer is yes, but this requires proof. What about the torus or projective plane? In order to address this question, we first need to define explicitly what we mean by *vertices*, *edges*, and *faces* in a topological space.

Cell Decomposition

Recall that a triangulation is a certain way of cutting up a space into more manageable pieces (vertices, edges, and triangular faces). Typically, though, a triangulation must include many faces in order to satisfy the strict requirements about how triangles can meet. A cell decomposition is another way of cutting up a space with fewer restrictions, reducing the amount of data required to accurate "describe" the space. Moreover, this paves the way toward understanding a wider class of spaces than just those in $\mathcal{S}_c$. Our purpose here is not to prove that cell decompositions exist in general (they do not – a simple counterexample is the Hawaiian earring shown in Exercise 1 of §4.3), but if a space has one, then we can use it to calculate the Euler characteristic, among other things.

A *cell* is nothing more than a closed disk $\mathbb{D}^n$ for some dimension n, and a cell complex is a way of building up a space X by attaching cells in a reasonable way. By *reasonable*, we mean that cells should only be glued to each other along their boundaries, not their interiors. For example, edges $\mathbb{D}^1 \approx \mathbb{I}$ may be glued to other edges only at their endpoints, though we do allow multiple edges to share the same endpoints and the endpoints of an edge to be the same point, as shown in Figure 5.25.

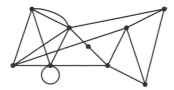

Figure 5.25: Edges meet only at endpoints. The apparent intersections at the interiors of some edges are only an artifact of drawing the figure in two dimensions and are not actual intersections in the space.

Definition 5.4.1. A **cell** in a topological space X is either

- a single point $x \in X$ (called a 0-cell) *or*

- for $n \geq 1$, a closed subspace $C \subseteq X$ (called an n-cell) and a continuous **attaching map** $f : \mathbb{D}^n \to C$ such that f restricts to a homeomorphism on $\text{int}(\mathbb{D}^n)$.

The number n in n-cell is called the **dimension** of the cell.

In any given space X, we typically only care about finitely many cells of a given dimension. Thus, while every single point $x \in X$ qualifies by definition as a 0-cell, we may only be interested in a small number of points $x_1, x_2, \ldots, x_k \in X$. For example, we may consider a polyhedron to consist of some number of vertices (0-cells), edges (1-cells), and faces (2-cells). The entire solid polyhedron is a 3-cell. (What might a 4-cell or 5-cell look like?) Just as a polyhedron is built up from vertices, edges, and faces, many topological spaces may also be built up from cells into what we call *cell complexes*. We must stick to finite-dimensional cell complexes, as the more general case requires more machinery than we have at our disposal.

Definition 5.4.2. A (finite-dimensional) **cell complex** (or simply **complex**) is a topological space X built from cells in the following way:

- There is a sequence of closed subspaces,

$$X^0 \subseteq X^1 \subseteq \cdots \subseteq X^n = X,$$

where each X^k is called the k-**skeleton** of X.

- The 0-skeleton X^0 is the disjoint union of finitely many 0-cells (vertices).

- For $k \geq 1$, X^k is built from attaching finitely many k-cells to X^{k-1}. Each k-cell C is attached via its attaching map $f : \mathbb{D}^n \to C \subseteq X^k$ so that

  1. $f[\partial \mathbb{D}^n] \subseteq X^{k-1}$, and

  2. $f[\mathrm{int}(\mathbb{D}^n)]$ intersects neither X^{k-1} nor the interior of any other k-cell already attached.

At any given stage in the construction, it is possible that no k-cells are added, so that $X^{k-1} = X^k$. The largest n such that X^n is different from X^{n-1} is called the **dimension** of X. A 0-dimensional complex, or 0-complex, is a finite discrete set of points. A 1-complex is simply a finite graph (see §6.1). Surfaces and surfaces-with-boundary are examples of 2-complexes, as a triangulation is a type of cell decomposition.

Example 100. Consider a solid cube C with vertices a, b, c, d, e, f, and g. Consider the cell decomposition on C according to vertices, edges, and faces. Note that C is a three-dimensional cell complex because it includes its solid interior.

- (0-skeleton) $C^0 = \{a, b, c, d, e, f, g, h\}$

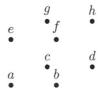

- (1-skeleton) $C^1 = \overline{ab} \cup \overline{ac} \cup \overline{bd} \cup \overline{cd} \cup \overline{ae} \cup \overline{bf} \cup \overline{cg} \cup \overline{dh} \cup \overline{ef} \cup \overline{eg} \cup \overline{fh} \cup \overline{gh}$

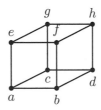

- (2-skeleton) $C^2 = F_1 \cup F_2 \cup F_3 \cup F_4 \cup F_5 \cup F_6$, where each cell F_k is one of the six faces of the cube

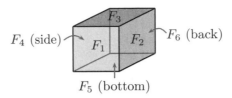

F_4 (side) ~ F_1 F_2 ~F_6 (back)

F_3

F_5 (bottom)

- (3-skeleton) $C^3 = C$, the whole solid cube

The existence of triangulations implies that $\mathcal{S}_c$ is a subset of the collection of all 2-complexes, but not every 2-complex is in $\mathcal{S}_c$. For example, consider the *book* of n pages, defined as follows.

Example 101. The **book of n pages** is the quotient space of n disjoint squares $\mathbb{I}^2$, each identified along one edge as shown in Figure 5.26. It is a two-dimensional cell complex but not a surface or surface-with-boundary unless $n \leq 2$.

Figure 5.26: The book of n pages is a 2-complex but not a surface or surface-with-boundary when $n > 2$.

Example 102. Every sphere $\mathbb{S}^n$ has a cell decomposition. The easiest case, of course, is $\mathbb{S}^0$, which consists of two 0-cells (0-dimensional complex). Suppose now that $n \in \mathbb{N}$, and let $z \in \mathbb{S}^n$ be any fixed point. Since $\mathbb{S}^n \setminus \{z\}$ is homeomorphic to $\mathbb{R}^n$, via stereographic projection, and $\mathbb{R}^n$ is homeomorphic to an open disk, $\mathbb{R}^n \approx \text{int}(\mathbb{D}^2)$, $\mathbb{S}^n$ can be decomposed as a single n-cell attached to a single 0-cell $\{z\}$ by way of an attaching map $f : \mathbb{D}^n \to \mathbb{S}^n$ that identifies the entire boundary of $\mathbb{D}^n$ to z (i.e., $f[\partial \mathbb{D}^n] = \{z\}$). This construction implies that each k-skeleton, for $k = 0, 1, \ldots, n-1$, is identically the solitary point $\{z\}$.

If this seems like "cheating," there is another natural way of building up a cell complex structure on $\mathbb{S}^n$ such that the k-skeletons are all properly k-dimensional. Let $E \subseteq \mathbb{S}^n$ be the equatorial sphere. That is,

$$E = \{x = (x_0, x_1, \ldots, x_{n-1}, 0) \in \mathbb{R}^{n+1} \mid d(x, 0) = 1\} \approx \mathbb{S}^{n-1}.$$

Now $\mathbb{S}^n$ may be built from $\mathbb{S}^{n-1}$ by attaching two n-cells, (i.e., the northern and southern hemispheres H^+ and H^-, as in Example 91; see Figure 5.27). By induction, $\mathbb{S}^n$ has an n-complex structure with exactly two cells in every dimension.

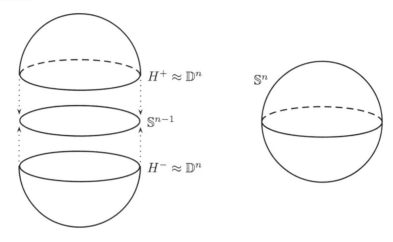

Figure 5.27: The hypersphere $\mathbb{S}^n$ can be built from $\mathbb{S}^{n-1}$ by attaching two n-cells. In this diagram, attaching two disks to a circle $\mathbb{S}^1$ produces the sphere $\mathbb{S}^2$.

Euler Characteristic of a Cell Complex

Every finite-dimensional cell complex has a corresponding Euler characteristic.

> **Definition 5.4.3.** Suppose X is an n-dimensional cell complex with x_k k-cells, for $k = 0, 1, \ldots, n$. The **Euler characteristic** of X (with respect to the given cell decomposition) is defined by
>
> $$\chi(X) = \sum_{k=0}^{n} (-1)^k x_k.$$

If X is a 2-complex, with $x_0 = v$, $x_1 = e$, and $x_2 = f$, then Definition 5.4.3 reduces to the familiar Euler characteristic for polyhedra.

Example 103. Find $\chi(C)$, where C is the solid cube from Example 100.
Solution: C has $x_0 = 8$ (0-cells), $x_1 = 12$ (1-cells), $x_2 = 6$ (2-cells), and $x_3 = 1$ (3-cell), so $\chi(C) = 8 - 12 + 6 - 1 = 1$.

If $S \in S_c$ has a plane diagram, then the diagram itself exhibits a cell decomposition for S with exactly one 2-cell. However, since some of the edges and vertices are identified, we must be careful in our counts of 0- and 1-cells for computing $\chi(S)$.

Example 104. Find $\chi(\mathbb{T})$, $\chi(\mathbb{P})$, and $\chi(\mathbb{K})$ with respect to the cell decompositions shown below.

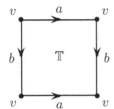 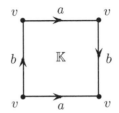

Solution: Based on the plane diagrams, and a careful count of distinct vertices, we find the following.

$$
\begin{aligned}
\chi(\mathbb{T}) &= 1 - 2 + 1 = 0 \\
\chi(\mathbb{P}) &= 1 - 1 + 1 = 1 \\
\chi(\mathbb{K}) &= 1 - 2 + 1 = 0
\end{aligned}
$$

The Euler characteristic is a topological invariant for spaces that have a cell decomposition. However, the proof relies on the theory of *homology* and so falls outside the scope of this textbook.[7] But we can sketch a proof of the famous formula $\chi(P) = 2$ for any polyhedron whose surface is homeomorphic to the sphere (i.e., *simply-connected* polyhedra). In fact, we will prove something a bit stronger. Instead of restricting to *geometric* polyhedra (figures in $\mathbb{R}^3$ made up of flat faces joined at edges and vertices in a certain way), we prove the result for an arbitrary cell decomposition of $\mathbb{S}^2$. See if you can spot a couple places in the proof where we must rely on the Jordan Curve Theorem.

Proposition 5.4.4. $\chi(\mathbb{S}^2) = 2$.

Proof. Suppose $\mathbb{S}^2$ has a cell decomposition (in the sense of Definition 5.4.2). Let F_0 be one of the faces, and let $p \in F_0$. Now since $\mathbb{S}^2 \setminus \{p\} \approx \mathbb{R}^2$ (via

---

[7]We introduce homology in Chapter 8, but a proof that homology itself is a topological invariant requires more machinery than we develop in this text.

stereographic projection), we know that $G = \mathbb{S}^2 \setminus F_0$ admits a cell decomposition for a (bounded) closed disk in $\mathbb{R}^2$. In other words, G is a connected plane figure made up of vertices and noncrossing edges (also known as a *planar graph*; see §6.1), together with the bounded faces determined by the graph. We do not require the edges to be straight segments, nor even to have distinct endpoints (*loops* are allowed). Note that the original cell complex structure on $\mathbb{S}^2$ can be recovered by the inverse stereographic projection and filling in the missing point.

Now let $\mathscr{G}$ denote the set of all cell complexes for closed disks (planar graphs with their bounded regions). For any $G \in \mathscr{G}$, let $|V(G)|$ be the number of vertices, $|E(G)|$ the number of edges, and $|F(G)|$ the number of faces in G. Our goal now is to show that $\chi(G) = 1$ for every $G \in \mathscr{G}$ (recall that G has one less face relative to $\mathbb{S}^2$). Consider the graph G_0 having a single vertex and one loop (edge). The loop, being a simple and closed curve in the plane, bounds a single disk region (face). In this case, we have $|V(G_0)| = 1$, $|E(G_0)| = 1$, and $|F(G_0)| = 1$, so that $\chi(G_0) = 1 - 1 + 1 = 1$. An arbitrary $G \in \mathscr{G}$ may be built up from G_0 by a finite sequence of the following moves.

(a) Connect two existing vertices with an edge.

(b) Place a new vertex on an existing edge.

(c) Deform any edges by stretching, bending, and so forth, but never allowing one edge to cross over another.

For example, suppose we wanted to tack on a new triangle adjacent to edge $\overline{AB}$ in Figure 5.28. First place a new edge that connects A to B, deforming it a bit, and then place a vertex on the new edge.

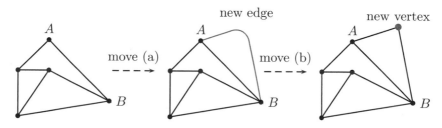

Figure 5.28: Move (a) produces a new edge and face; move (b) produces a new vertex, splitting a single edge into two. Note that move (c) – which is not labeled on the diagram but happened in conjunction with move (b) – is used just to straighten out the edges, but has no effect on the number of cells.

Now suppose G' is obtained from G by move (a). Then $|V(G')| = |V(G)|$, while $|E(G')| = |E(G)| + 1$. The only question is what happens to the number of faces. Let e be the new edge. Either e is on the *outside* of the polygon, in which case a new face has been created, or e is *inside* an existing face, which splits an existing face into two smaller faces. Either way, $|F(G')| = |F(G)| + 1$. Thus

$$
\begin{aligned}
\chi(G') &= |V(G')| - |E(G')| + |F(G')| \\
&= |V(G)| - (|E(G)| + 1) + (|F(G)| + 1) \\
&= |V(G)| - |E(G)| + |F(G)| \\
&= \chi(G).
\end{aligned}
$$

Suppose now that G' is obtained from G by move (b). Then $|V(G')| = |V(G)| + 1$ and $|E(G')| = |E(G)| + 1$, while $|F(G')| = |F(G)|$. Again, the net result is that $\chi(G') = \chi(G)$. Move (c) does not change the number of cells at all.

Thus, for any $G \in \mathscr{G}$, we have $\chi(G) = \cdots = \chi(G_0) = 1$. Finally, adding back in the contribution from the missing face F_0, $\chi(\mathbb{S}^2) = \chi(G) + 1 = 2$. $\qquad\square$

We will take for granted that the Euler characteristic is a topological invariant for spaces on which it is defined. Thus, for example, $\mathbb{S}^2 \not\approx \mathbb{T}$ because $\chi(\mathbb{S}^2) = 2 \neq 0 = \chi(\mathbb{T})$. But χ is not a *complete* invariant for surfaces – Example 104 shows that $\chi(\mathbb{T}) = \chi(\mathbb{K})$ even though $\mathbb{T} \not\approx \mathbb{K}$ (one is orientable; the other is not). On the other hand, it turns out that the combination of Euler characteristic and orientability serves to distinguish all compact surfaces. Note that the Euler characteristic of a connected sum can be computed easily.

Proposition 5.4.5. *Let $A, B \in \mathcal{S}_c$.*

$$
\chi(A\#B) = \chi(A) + \chi(B) - 2
$$

Proof. Suppose $U_1 \subseteq A$ and $U_2 \subseteq B$ are the chosen disks to remove for the connected sum construction. Give A and B triangulations. It can be arranged so that U_1 is the interior of a triangular face and similarly for U_2 in B – possibly by shrinking the open set U_i, so that each is contained *within* a single triangular face, then expanding U_i to fill the triangle. Thus $\chi(A \setminus U_1) = \chi(A) - 1$ and $\chi(B \setminus U_2) = \chi(B) - 1$. When the two spaces are joined at the bounding triangles, three pairs of vertices will be identified, which reduces the overall value of $v - e + f$ by 3, but also three pairs of edges will be identified, which (because of the negative sign on e) yields a compensating increase of 3. Taken together, this shows that

$$
\chi(A\#B) = (\chi(A) - 1) + (\chi(B) - 1) = \chi(A) + \chi(B) - 2.
$$

$\qquad\square$

Example 105.

- $\chi(2\mathbb{T}) = \chi(\mathbb{T}) + \chi(\mathbb{T}) - 2 = -2$

- $\chi(3\mathbb{P}) = \chi(2\mathbb{P}) + \chi(\mathbb{P}) - 2 = (\chi(\mathbb{P}) + \chi(\mathbb{P}) - 2) + \chi(\mathbb{P}) - 2 = -1$
- $\chi(X \# \mathbb{S}^2) = \chi(X) + \chi(\mathbb{S}^2) - 2 = \chi(X)$

Classification of Compact Surfaces

We now take up a theme that showcases how certain topological invariants can be used to completely classify a particular collection of spaces, the compact surfaces. Results of this type are exceedingly rare; for example, there is not a complete classification of compact manifolds in general.[8]

Theorem 5.4.6. *Every compact surface S is homeomorphic to one and only one of the following:*

- $\mathbb{S}^2$,

- $n\mathbb{T}$, *for some $n \in \mathbb{N}$,   or*

- $n\mathbb{P}$, *for some $n \in \mathbb{N}$.*

Proof. The proof is in two main parts. We first show that every space mentioned in the theorem is indeed different. Then we will show how an arbitrary surface can be identified with one of those spaces.

Claim: Each space $\mathbb{S}^2$, $n\mathbb{T}$, and $n\mathbb{P}$ (for various $n \in \mathbb{N}$) is topologically distinct. *Proof of claim:* First, since orientability is a topological invariant, we know that $n\mathbb{P} \not\approx \mathbb{S}^2$ and $n\mathbb{P} \not\approx m\mathbb{T}$ for any choices of $n, m \in \mathbb{N}$. Then the Euler characteristic can be used to distinguish among the orientable spaces and among the nonorientable ones (as Exercise 5 implies).

Now let's tackle the harder part of the theorem. Suppose S is a compact surface. Give S a triangulation (which is possible by Theorem 5.2.2). The triangulation can be used to produce a plane model for the surface along with an associated word, $W = a_1 a_2 \cdots a_k$, in which every letter of the word occurs twice (counting inverses). We may assume that the plane model for S has all vertices identified to a single point of S (see §5.2, Exercise 10). Now let's do a little *word arithmetic*.

1. If $W = aa^{-1}$ or $W = a^{-1}a$, then we are done, and $S \approx \mathbb{S}^2$. Otherwise, move on to step 2.

2. Eliminate all adjacent letter-inverse pairs, including any that arise after a cyclic permutation of the letters. This "clean-up" step should occur after each of the following steps as well.

---

[8]However, see Thurston's Geometrization Conjecture and the Poincaré Conjecture, both proved barely a decade ago by Grigori Perelman. To learn more about the enigmatic Perelman, I suggest *Perfect Rigor* by Masha Gessen [Ges09].

3. Suppose there is a same-letter pair separated by other letters. That is, $W \equiv \cdots aXa \cdots$ or $W \equiv \cdots a^{-1}Xa^{-1}\cdots$, where X is a nonempty string of letters. In what follows, we work with a rather than a^{-1}, but the results are the same. First use a cyclic permutation so that the first occurrence of the letter is at the beginning of the word. Now if $W \equiv aXa$, then, after cyclic permutation, we have $W \equiv a^2X$, and hence may assume that $W \equiv aXaY$, where both X and Y are nonempty strings. Use a split-and-rejoin operation, inverting the first subword and cyclically permuting the second:

$$aXaY \equiv aXq^{-1}\, qaY \equiv qX^{-1}a^{-1}\, aYq \equiv qX^{-1}Yq \equiv q^2X^{-1}Y.$$

Note that after this operation there is one fewer separated same-letter pair. Repeat step 3 until all same-letter pairs are of the form q^2.

4. Now suppose there is an inverse pair within the word. If $W \equiv aXa^{-1}Y$, and if X and Y share no common letters, then W describes the connected sum of two surfaces, one with word X and another with word $a^{-1}Ya \equiv Y$. Thus we may assume X and Y have at least one letter in common. However, since all same-letter pairs are together, we must assume there is another inverse pair $b \cdots b^{-1}$ alternating with $a \cdots a^{-1}$. That is, $W \equiv aXbYa^{-1}Zb^{-1}U$ for some (possibly empty) words X, Y, Z, and U. The following split-and-rejoin steps, together with cyclic permutations and cancellations, eliminate a, a^{-1}, b, and b^{-1} at the expense of including the adjacent pairs $srs^{-1}r^{-1}$.

   - Split-and-rejoin 1:
     $$\begin{aligned} aXbYa^{-1}Zb^{-1}U &\equiv aXq^{-1},\ qbYa^{-1}Zb^{-1}U \\ &\equiv Xq^{-1}a,\ a^{-1}Zb^{-1}UqbY \equiv Xq^{-1}Zb^{-1}UqbY. \end{aligned}$$

   - Split-and-rejoin 2 (after a cyclic permutation):
     $$\begin{aligned} bYXq^{-1}Zb^{-1}Uq &= bYXr^{-1},\ rq^{-1}Zb^{-1}Uq \\ &\equiv YXr^{-1}b,\ b^{-1}Uqrq^{-1}Z \equiv YXr^{-1}Uqrq^{-1}Z. \end{aligned}$$

   - Split-and-rejoin 3 (after a cyclic permutation):
     $$\begin{aligned} q^{-1}ZYXr^{-1}Uqr &\equiv q^{-1}ZYXs,\ s^{-1}r^{-1}Uqr \\ &\equiv ZYXsq^{-1},\ qrs^{-1}r^{-1}U \equiv ZYXsrs^{-1}r^{-1}U. \end{aligned}$$

   Repeat step 4 until all inverse pairs occur as part of a substring like $srs^{-1}r^{-1}$.

5. After completely reducing any remaining adjacent inverse pairs, the word W must either be equivalent to aa^{-1} (the sphere), or must be the concatenation of some number of pairs like $bb = b^2$ and/or substrings like $srs^{-1}r^{-1}$. That is, if S is not a sphere, then S is the connected sum of tori and projective planes. In that case, §5.3, Exercise 4 implies that S is homeomorphic to one of the spaces $n\mathbb{T}$ or $n\mathbb{P}$ for an appropriate $n \in \mathbb{N}$.

$\square$

Exercises

1. Let $n > 1$, and let B be a book of n pages, as defined in Example 101. Find $\chi(B)$.

2. Let T be a solid tetrahedron. Find a natural cell decomposition for T and use it to compute $\chi(T)$. Make a conjecture about $\chi(P)$ for *any* solid convex polyhedron P.

3. Let M be the Möbius strip. Find $\chi(M)$.

4. Referring to Example 102, find $\chi(\mathbb{S}^n)$ for arbitrary $n \in \mathbb{N} \cup \{0\}$,

   (a) using the cell decomposition of $\mathbb{S}^n$ consisting of only two cells total.

   (b) using the cell decomposition of $\mathbb{S}^n$ consisting of two cells in every dimension.

5. Let $n \in \mathbb{N}$.

   (a) Find $\chi(n\mathbb{T})$.　　　　　　　　(b) Find $\chi(n\mathbb{P})$.

6. Let v be the number of vertices, and e be the number of edges in a triangulation of a surface M. Using §5.2, Exercise 9, prove that $e = 3(v - \chi(M))$.

7. Let $n \in \mathbb{N}$. Using the representation of projective space $\mathbb{P}^n$ as a hemisphere of $\mathbb{S}^n$ with antipodal identifications along the boundary (as in Equation 5.1), use induction to build a cell complex structure on $\mathbb{P}^n$ having exactly one cell in each dimension. Use your cell decomposition to compute $\chi(\mathbb{P}^n)$ for arbitrary $n \in \mathbb{N}$.

8. Illustrate the operations outlined in steps 3 and 4 of the proof of Theorem 5.4.6 on plane diagrams.

Supplemental Reading

- Barr [Bar64], Chapters 2–6. Barr introduces the torus, Möbius strip, Klein bottle, and projective plane through hands-on constructions.

- Goodman [Goo05], Chapters 2 and 3.

- Henle [Hen79], Chapter 4.

- Munkres [Mun00], Chapter 12.

- Weeks [Wee02], Part I.

Chapter 6

Applications in Graphs and Knots

Graphs and knots are particular kinds of topological spaces that play important roles in numerous fields of study, including computer science, physics, and biology. While it may be argued that the study of graphs and knots more properly fits into the realm of combinatorics, the language of topology often helps to put the combinatorics into clearer context. In the first two sections, we use cell complexes, embeddings (i.e., injective continuous functions), the Euler characteristic as a topological invariant, and other important topological ideas to study graphs and graph coloring problems. In the remaining sections, we delve into topological knot theory by defining a knot as an embedding $\mathbb{S}^1$ into three-dimensional space, and develop certain invariants to aid in classifying knots. Proofs of many of the results in this chapter are adapted from Goodman [Goo05], but can also be found in many other standard texts on graph theory or knot theory.

6.1 Graphs and Embeddings

Finite graphs are nothing more than one-dimensional cell complexes (recall Definition 5.4.2). Nevertheless, their relatively simple structure conceals a wealth of information. Applications of graphs abound not only in computer science (see Figure 1.4, a graph representing Internet connections), but also in practically every field of scientific study from sociology to psychology, biology, chemistry, and physics. This short section and §6.2 discuss only two small (but important) aspects of graph theory that relate most directly to topology: the problems of embedding graphs in surfaces and associated coloring problems for embedded graphs.[1]

---

[1] For a more comprehensive treatment of graph theory, we refer the reader to one of the many wonderful textbooks on the subject, e.g., Bondy and Murty [BM07] or Harris et al. [HHM08].

Finite Graphs: Definitions

In this section, the term **graph** refers specifically to a (not necessarily connected) 1-complex. That is, a graph G is a cell complex with finitely many vertices V and finitely many edges E such that edges may only intersect at vertices. Our definition allows multiple edges to be attached to the same pair of vertices (*parallel edges*) and edges whose two endpoints are attached to the same vertex (*loops*). Write $V(G)$ (or V) for the set of vertices, and $E(G)$ (or E) for the set of edges of a graph G. The number of vertices and edges are notated $|V(G)|$ (or $|V|$) and $|E(G)|$ (or $|E|$), respectively.

In our graph diagrams, the vertices will be shown as thick dots. While two edges may be drawn on the page in such a way that it seems their interiors might intersect, no intersection is assumed except at the vertices. Figures 5.25 and 6.1 illustrate some example graphs.

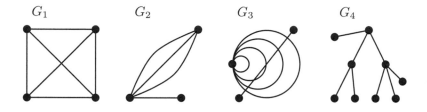

Figure 6.1: $|V(G_1)| = 4$; $|E(G_1)| = 6$ (there is no vertex in the center of the graph, so the edges do not actually meet there). $|V(G_2)| = 3$; $|E(G_2)| = 4$. $|V(G_3)| = 3$; $|E(G_3)| = 5$. $|V(G_4)| = 9$; $|E(G_4)| = 8$. G_3 is not connected, but the other graphs are.

Note that there are many equivalent representations of the same graph; all that matters is that the relationships among vertices and edges are the same. More generally, we say that two graphs G and H are **isomorphic**[2] if there are bijective functions $V(G) \to V(H)$ and $E(G) \to E(H)$ preserving all of the vertex-edge relationships. If such a function exists (called an **isomorphism**) from G to H, then we write $G \cong H$. For example, the three graphs shown in Figure 6.2 are all isomorphic. Isomorphic graphs are always homeomorphic, but two graphs may be homeomorphic as topological spaces without being isomorphic.

Since a graph is a cell complex, it makes sense to talk about its Euler characteristic: $\chi(G) = |V(G)| - |E(G)|$.

Example 106. Find the Euler characteristic of each graph shown in Figure 6.1.

---

[2]In other words, *equivalent* in the realm of graph theory.

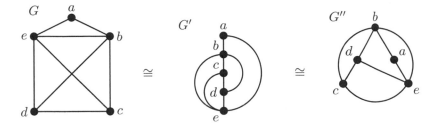

Figure 6.2: Isomorphic graphs: $G \cong G' \cong G''$.

Solution:

$$
\begin{aligned}
\chi(G_1) &= |V(G_1)| - |E(G_1)| = 4 - 6 = -2 \\
\chi(G_2) &= 3 - 4 = -1 \\
\chi(G_3) &= 3 - 5 = -2 \\
\chi(G_4) &= 9 - 8 = 1
\end{aligned}
$$

Note that $\chi(G_1) = \chi(G_3)$ even though the two graphs are not even topologically equivalent (one is connected while the other is not), so χ is not powerful enough to distinguish all graphs.

A **simple graph** is a graph such that each edge is attached to a distinct pair of distinct vertices. In other words, there are no parallel edges or loops. Graphs G_1 and G_4 in Figure 6.1 are simple. A **complete graph** is a simple graph such that every pair of distinct vertices has an edge between them. The complete graph on n vertices is notated K_n. Graph G_1 in Figure 6.1 is a K_4. Additional examples are shown in Figure 6.3.

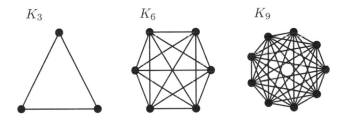

Figure 6.3: Complete graphs K_3, K_6, and K_9.

A graph G is called **bipartite** if there is a grouping of the vertices into sets V_1 and V_2 such that each $V_i \neq \emptyset$, and no edge exists between two points of V_1 or two points of V_2. The **complete bipartite** graph $K_{m,n}$ refers to the simple bipartite graph with $|V_1| = m$, $|V_2| = n$, and *every* possible edge exists connecting points of V_1 to V_2. In the examples shown below, H is bipartite with

$V(H) = V_1 \cup V_2$, where $V_1 = \{a, b, c, d\}$ and $V_2 = \{e, f, g, h, i\}$. The graph on the right is $K_{3,2}$, with vertices partitioned as $V_1 = \{u, v, w\}$ and $V_2 = \{x, y\}$.

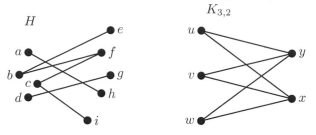

Example 107. Find formulas for $\chi(K_n)$ and $\chi(K_{m,n})$ in general.

Solution: By definition, of K_n, we have $|V(K_n)| = n$. Since there is exactly one edge between distinct pairs of vertices, we have $|E(K_n)| = \binom{n}{2} = \frac{n(n-1)}{2}$. Therefore

$$\chi(K_n) = |V(K_n)| - |E(K_n)| = n - \frac{n(n-1)}{2} = \frac{3n - n^2}{2}.$$

For the complete bipartite graph,

$$\chi(K_{m,n}) = |V(K_{m,n})| - |E(K_{m,n})| = m + n - mn.$$

Suppose v is a vertex of a graph G. The **degree** of v is the number of edges incident to v. For example, every vertex of K_n has degree $n - 1$, since there must be an edge from each vertex to all other vertices. A **cycle graph** is a connected graph such that every vertex has degree 2, which implies that the graph must be homeomorphic to $\mathbb{S}^1$ (see K_3, for example). What is the Euler characteristic for a cycle graph? Suppose C_n is a cycle graph with n vertices. Then $|V(C_n)| = |E(C_n)| = n$, so that $\chi(C_n) = 0$. (By the way, this implies $\chi(\mathbb{S}^1) = 0$, as expected.)

We say that H is a **subgraph** of a graph G if $V(H) \subseteq V(G)$ and $E(H) \subseteq E(G)$ (or, more generally, if there is are injective functions $V(H) \to V(G)$ and $E(H) \to E(G)$ such that all of the structure of H is preserved in the image). A **cycle** in an arbitrary graph G is simply a subgraph of G that is a cycle graph. Similarly, a **path** in G is a subgraph that is isomorphic to the **path graph** P_n, having n edges connected in a row, and we say that path has **length** n. Note that, by this definition, a path of length n in G must visit $n + 1$ distinct vertices.[3] Thus there is a path of length 4 in $K_{3,2}$, but no path of length 5. What is the longest cycle you can find in $K_{3,2}$?

Example 108. Suppose a graph G' is obtained from the graph G by attaching a path graph P_n to G, identifying the endpoints of P_n to distinct vertices in G as shown below. What is the the value of $\chi(G')$ in terms of $\chi(G)$?

---

[3]The graph-theoretic definitions used in this text generally follow those of Harris et al. [HHM08], but with further emphasis on topological properties.

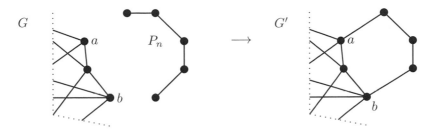

Before the identification, $\chi(P_n) = (n+1) - n = 1$. However, after identifying the two endpoints of P_n to two vertices of G, the contribution to the total Euler characteristic of G' is $(n-1) - n = -1$. Thus $\chi(G') = \chi(G) - 1$.

In what follows, assume G is a graph. Two vertices in G are called **adjacent** if there is an edge in G having those two vertices as endpoints. A **walk** in G is a finite sequence of adjacent vertices, including the edges in between; neither the vertices nor edges have to be distinct. Note that in a simple graph, a walk is completely determined by the ordered list of vertices visited. Topologically, a walk can be described by an arc $\gamma : \mathbb{I} \to G$ such that both $\gamma(0)$ and $\gamma(1)$ are vertices. If the edges of the walk are distinct, then the walk is called a **trail**, and if the vertices are all distinct, then the walk is a path (thus every path is a trail, and every trail is a walk, but not conversely). A graph-theoretic path is equivalent to an **embedding** (i.e., an *injective* continuous function) $\gamma : \mathbb{I} \to G$ such that $\gamma(0), \gamma(1) \in V(G)$. The notation $X \hookrightarrow Y$ is often used for an embedding of a space X into another space Y, so we may write $\gamma : \mathbb{I} \hookrightarrow G$.

A **circuit** in G is a trail that begins and ends at the same vertex (so every cycle in a graph is a circuit, but a circuit may fail to be a cycle). A cycle in G is equivalent to an embedding $\mathbb{S}^1 \hookrightarrow G$ such that $\gamma(0), \gamma(1) \in V(G)$. For example, referring to graph G in Figure 6.2, *abcebedb* is a walk (but not a circuit or trail); *debaccd* is a circuit (but not a cycle); *ecbde* is a cycle, and *decba* is a path.

An **Euler circuit** is a circuit that visits every edge of G. Similarly, an **Euler trail** is a trail that visits every edge of G. In Chapter 1, we discussed the Seven Bridges of Königsberg Problem, which asks whether there is a way to take a walk that crosses each bridge exactly once. This question is equivalent to finding an Euler trail in the graph shown in Figure 1.5, and, as we shall see in the Exercises, the answer is *no*.

Planarity and Surface Embeddings

Some graphs can be drawn on a sheet of paper without edges crossing one another. For example, Figure 6.4 illustrates how K_4 may be drawn without crossings. However, not every graph enjoys this distinction. No matter how you twist and bend the edges and jostle the vertices, the graph K_5 simply cannot be drawn in the plane without crossing edges (*try it!*). Graphs that can be *embedded* in the plane (i.e., drawn with no edges crossing) are called *planar*;

however, it would be equivalent to require that the graph can be embedded on the surface of the sphere. The main reason we want to consider embeddings on $\mathbb{S}^2$ is that we would like to explore embeddings of graphs in other compact surfaces as well.

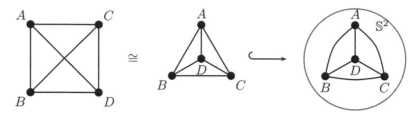

Figure 6.4: K_4 is a planar graph. Every planar graph embeds in the sphere $\mathbb{S}^2$.

Definition 6.1.1. A graph G is called **planar** if G can be embedded into $\mathbb{S}^2$.

Example 109.

- Every complete bipartite graph of the form $K_{2,n}$ is planar. Simply move one of the two points on the left to the right side.

- Of course, $K_{3,2}$ is planar since $K_{3,2} \cong K_{2,3}$, but what about $K_{3,3}$? Suppose there are three houses, A, B, and C, and each one must be connected to three utilities, gas (G), electric (E), and water (W), as in Figure 6.5. If this could be done without any utility lines crossing, then $K_{3,3}$ would be planar. However, no matter how hard we try, no solution to the problem exists. There is no way to avoid crossing lines; that is, $K_{3,3}$ is nonplanar. (It's a good thing we can take advantage of the third dimension to solve this kind of dilemma in real life, e.g., bury the gas line deeper than the water, or put the electric lines on poles above the surface.)

Of course, the fact that we failed to find a planar diagram for a graph does not necessarily mean the graph is nonplanar (maybe we just weren't persistent enough). To see that $K_{3,3}$ is nonplanar requires a bit of combinatorics and the Euler characteristic. First let's talk about embedding graphs on other surfaces besides $\mathbb{S}^2$.

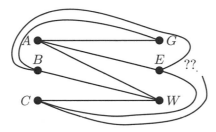

Figure 6.5: Is it possible to get gas, electric, and water lines to each of three houses without crossing lines?

> **Definition 6.1.2.** A graph G is can be **embedded** in a surface S if there is an injective map $G \hookrightarrow S$, and if so, then we say that G is **embeddable** in S.

A graph may be embeddable in may different kinds of surfaces; in fact, a planar graph embeds not only in $\mathbb{S}^2$, but also in any surface S. Every graph, no matter how complicated, embeds into *some* surface (simply attach enough *handles* to the sphere to serve as bridges that allow the segments to avoid internal crossing, as shown in Figure 6.6). On the other hand, for each graph G, there is a *minimal n* such that G embeds into $n\mathbb{T}$; then we call n the **genus** of G. Planar graphs have genus 0.

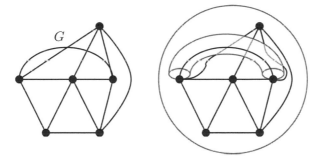

Figure 6.6: The graph G is embeddable into a sphere with one handle attached; that is, a torus. It can be shown that this graph is nonplanar; hence its genus is equal to 1.

Suppose G embeds in a surface S, via a map $f : G \hookrightarrow S$. If each component of $S \setminus f[G]$ is homeomorphic to a disk, then we may regard $f[G]$ as a 1-skeleton in a cell decomposition of S, and in that case we call the embedding **proper**, or that G **properly embeds** in S. A proper embedding provides a ready tool for calculating the Euler characteristic of a surface. Let F be the set of all 2-cells attached to G to form S. Then the formula relating the Euler characteristic of the graph to that of the surface is simply $\chi(S) = \chi(G) + |F|$. We now use this observation to prove that two special graphs K_5 and $K_{3,3}$ (shown in Figure 6.7) are nonplanar.

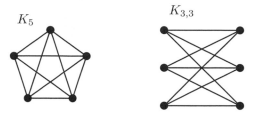

Figure 6.7: K_5 and $K_{3,3}$ are nonplanar graphs.

Theorem 6.1.3. *The graphs K_5 and $K_{3,3}$ are nonplanar.*

Assuming that the Euler characteristic is a well-defined topological invariant, this proof is not very difficult. It relies on a trivial fact about simple graphs embedded into surfaces. Let's count the total number of edges in terms of the number of regions. Since there are no loops or parallel edges, every region must have at least three distinct edges; that is, each region in F corresponds to at least three elements of E. But since each edge occurs in exactly two regions, we get

$$3|F| \leq 2|E|. \tag{6.1}$$

Now let's prove Theorem 6.1.3.

Proof. Suppose that K_5 is planar. Then K_5 embeds into $\mathbb{S}^2$, and let f be the number of regions determined by the embedding. Let $v = V(K_5) = 5$ and $e = E(K_5) = 10$. Then

$$\begin{aligned} \chi(\mathbb{S}^2) &= \chi(K_5) + f \\ 2 &= 5 - 10 + f \\ f &= 7. \end{aligned}$$

But then $3f = 21 \not\leq 2e = 20$, contradicting (6.1). Therefore K_5 is nonplanar.

The proof for showing $K_{3,3}$ is nonplanar follows the same scheme, except that we may assume $4|F| \leq 2|E|$ because $K_{3,3}$ has no cycles of length less than 4. $\qquad\square$

There are, of course, other graphs besides K_5 and $K_{3,3}$ that cannot be embedded in $\mathbb{S}^2$, but it is known[4] that a graph is planar if and only if it does not contain a subgraph *homeomorphic* to K_5 or $K_{3,3}$ (e.g., see Figure 6.8).

---

[4]Kuratowski [Kur30].

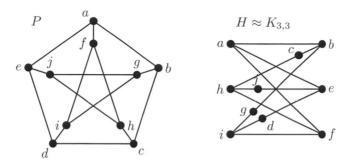

Figure 6.8: The Petersen graph P is nonplanar because there is a subgraph H homeomorphic to $K_{3,3}$. Note that H is not *isomorphic* to $K_{3,3}$ as graphs.

On the other hand, both K_5 and $K_{3,3}$ do embed into $\mathbb{T}$. Figure 6.9 illustrates this for K_5 in the plane diagram for the torus. This embedding cuts up the torus into five regions, each homeomorphic to a disk, and thus yields a consistent result for the Euler characteristic of the torus: $\chi(\mathbb{T}) = 5 - 10 + 5 = 0$. However, we must be careful. An embedding of a graph G need not cut a surface S into disks, and so may not provide a well-defined measure of $\chi(S)$.

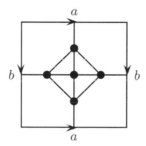

Figure 6.9: K_5 embeds into the torus. This embedding shows four triangles and one 8-gon (the region that "wraps around" the identified edges of $\mathbb{T}$, meeting itself in a number of edges and vertices).

Example 110. Find the smallest $n \in \mathbb{N}$ such that K_n does not embed into $\mathbb{T}$.

Solution: We will find the largest n such that K_n *does* embed into $\mathbb{T}$ (then add 1 to answer the original question). If there is an n such that K_n embeds into a surface, than clearly K_m embeds for every $m \leq n$. Thus it suffices to assume K_n is a proper embedding. Let $v = V(K_n) = n$, and $e = E(K_n) = \frac{n(n-1)}{2}$. Since $K_n \subseteq \mathbb{T}$ is asssumed to be a proper embedding, the number of regions f must be $f = \chi(\mathbb{T}) + e - v = e - v$. Now, since $f \leq \frac{2}{3}e$ in any embedding,

$$e - v = f \leq \frac{2}{3}e \implies \frac{1}{3}e \leq v \implies \frac{n(n-1)}{6} \leq n \implies n(n-7) \leq 0. \quad (6.2)$$

The maximal n satisfying (6.2) is $n = 7$. Thus it is impossible for K_8 to embed into $\mathbb{T}$. All that remains is to demonstrate an actual embedding $K_7 \subseteq \mathbb{T}$; this you will do in Exercise 7.

Note that an embedding, even a proper one, may fail to be a triangulation of a surface.

Example 111. K_3 embeds into $\mathbb{S}^2$, but not as a triangulation. The two triangles formed share all three of their edges. On the other hand, K_4 is a triangulation of $\mathbb{S}^2$ (in fact, it is *minimal*, in the sense that no triangulation with fewer vertices exists for $\mathbb{S}^2$).

Haewood's Number

As you might expect, the more complicated a graph is, the higher genus a surface must be in which to embed the graph. Conversely, the higher genus a surface is, the more complicated a proper embedding must be. In Exercise 7, you will show that K_7 properly embeds into the torus (in fact, as a triangulation), but is there a simpler graph that triangulates $\mathbb{T}$? What about $n\mathbb{T}$ or $n\mathbb{P}$ in general? Haewood's number is a formula for finding the theoretical minimal number of vertices that any triangulation of a surface could have.

> **Theorem 6.1.4.** *Suppose a graph G embeds properly as a triangulation into a surface S. Let $v = |V(G)|$ and $\chi = \chi(S)$. Then*
>
> $$v \geq H_\chi = \left\lfloor \frac{1}{2}\left(7 + \sqrt{49 - 24\chi}\right) \right\rfloor$$
>
> *(where $\lfloor x \rfloor$ is the floor of the number x, the largest integer k such that $k \leq x$). The number H_χ is called **Haewood's number**, and it depends only on the Euler characteristic of the surface.*

Proof. Since G embeds as a triangulation, G is simple. Therefore $e = |E(G)| \leq \binom{v}{2} = \frac{v(v-1)}{2}$. Let $f = |F(G)|$ (number of regions of the triangulation as determined by the embedding). In any triangulation, we have $3f = 2e$. Then, since $f = \frac{2e}{3}$, we have $e = 3(v - \chi(S))$ (see §5.4, Exercise 6).

$$
\begin{aligned}
e = 3(v - \chi(S)) &\leq \frac{v(v-1)}{2} \\
6v - 6\chi(S) &\leq v^2 - v \\
0 &\leq v^2 - 7v + 6\chi(S)
\end{aligned}
$$

Only the positive integer solutions make sense, and using the quadratic formula, we recover Haewood's number:

$$v \geq \frac{7 + \sqrt{49 - 24\chi(S)}}{2} \geq \left\lfloor \frac{7 + \sqrt{49 - 24\chi(S)}}{2} \right\rfloor.$$

☐

The following table shows Haewood's number for some surfaces, both orientable and nonorientable.

S (orientable)	$\mathbb{S}^2$		$\mathbb{T}$		$2\mathbb{T}$		$3\mathbb{T}$		$4\mathbb{T}$
(nonorientable)		$\mathbb{P}$	$2\mathbb{P} = \mathbb{K}$	$3\mathbb{P}$	$4\mathbb{P}$	$5\mathbb{P}$	$6\mathbb{P}$	$7\mathbb{P}$	$8\mathbb{P}$
$\chi(S)$	2	1	0	-1	-2	-3	-4	-5	-6
H_χ	4	6	7	7	8	9	9	10	10

It is important to realize that Haewood's number is just a theoretical lower limit. It may not be possible to realize this minimum in any given surface. Surprisingly, however, the only surface for which the Haewood number is not realizable is the Klein bottle. Exercise 10 asks you to find minimal triangulations of a few surfaces. In order to show that Haewood's number is realizable in general, you would have to find a procedure that builds a minimal triangulation for each space $n\mathbb{T}$, and this falls outside the scope of this elementary discussion.

Exercises

1. The graph G_4 in Figure 6.1 is an example of a **tree**. A tree is a graph that has no cycles. Determine a relationship between $V(T)$ and $E(T)$ for any connected tree T. What is $\chi(T)$ for a connected tree T? What about a tree that consists of k connected components (commonly known as a **forest**)?

2. Show that the graph below is bipartite by finding an appropriate partition of the vertices.

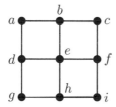

3. A *tripartite* graph is one whose vertex set can be partitioned into three nontrivial subsets, V_1, V_2, V_3, such that no edges exist between vertices within each V_i. If $|V_1| = k_1$, $|V_2| = k_2$, and $|V_3| = k_3$, then find the total number of edges in the *complete* tripartite graph, K_{k_1, k_2, k_3}. What is $\chi(K_{k_1, k_2, k_3})$?

4. Let $c \in \{0, -1, -2, -3, \ldots\}$. Find a bipartite graph B_c such that $\chi(B_c) = c$. (*Hint:* See Exercise 2 and Example 108.)

5. If a graph has an Euler circuit, then the graph is called **Eulerian**.

   (a) The following graphs are Eulerian. Find a specific Euler circuit in each.

      (i) K_5 (ii) $K_{2,4}$

   (b) Prove that a connected but not necessarily simple graph is Eulerian if and only if every vertex has even degree.

   (c) Using the previous result, show that the following graphs are not Eulerian.

      (i) K_6 (iii) G from Figure 6.2
      (ii) $K_{2,3}$ (iv) P from Figure 6.8

   (d) Find a criterion that decides whether K_n is Eulerian. Do the same for $K_{m,n}$.

   (e) Prove that a connected but not necessarily simple graph G has an Euler trail if and only if there are no more than two vertices in G having odd degree. Use this criterion to prove that there is no walk that traverses all seven of the Königsberg bridges exactly once (see Figure 1.5).

6. Finish the proof of Theorem 6.1.3 by filling in the details to show $K_{3,3}$ is nonplanar.

7. Draw an embedding of each of the following graphs into $\mathbb{T}$. (*Hint:* Use the plane diagram of $\mathbb{T}$.)

   (a) K_7 (b) $K_{3,3}$ (c) $K_{3,6}$

8. Draw an embedding of the Petersen graph (P from Figure 6.8) into the projective plane $\mathbb{P}$.

9. Explain why the graph shown embedded into $\mathbb{T}$ below is not a proper embedding.

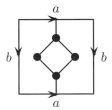

10. According to Theorem 6.1.4, a triangulation of $\mathbb{P}$, $\mathbb{T}$, and $\mathbb{K}$ must have a minimum of 6, 7, and 7 vertices, respectively. There are triangulations of $\mathbb{P}$ and $\mathbb{T}$ that meet the theoretical minimum; find them (using their plane diagrams). The Klein bottle does not have a triangulation with only seven vertices; find one that has eight vertices.

11. The **genus** of a graph G, written $g(G)$, is the smallest number n such that G properly embeds into $n\mathbb{T}$. Using the proof of Theorem 6.1.4 as a model, show that

$$g(K_n) \geq \left\lceil \frac{(n-3)(n-4)}{12} \right\rceil,$$

where $\lceil x \rceil$ is the *ceiling* of the number x, the smallest integer k such that $k \geq x$.

6.2 Graphs, Maps, and Coloring Problems

How many colors does it take to color a map so that adjacent countries receive different colors? It depends – some maps require only two colors, some three, while many others require four. If countries are not connected,[5] then even more colors may be necessary. It also depends on what surface the map is drawn on. We are used to maps drawn on the plane and on the globe, like the map depicted in Figure 6.10, but you might imagine a huge toroidal structure encircling a star, built by an advanced civilization. How many colors would suffice to color an arbitrary map on the torus?

In fact, map coloring and the equivalent problem of graph coloring play a major role in some important applications, including conflict avoidance in scheduling, creating efficient algorithms utilizing limited hardware resources, certain pattern-matching problems, and many others.[6] Before we can answer these questions, we must define precisely what we mean by maps and coloring.

Maps on Surfaces

Bounded maps drawn in the plane are equivalent to those drawn on the surface of $\mathbb{S}^2$. A coloring of a map on $\mathbb{S}^2$ corresponds to a coloring of the same (isomorphic) map on $\mathbb{R}^2$ via stereoscopic projection. Conversely, suppose all bounded maps on $\mathbb{R}^2$ can be colored with N colors; then, for any given map, include one more country consisting of the entire region surrounding the original map (the *ocean*, perhaps). Then the new map with one extra "country" can also be colored with N colors, and by stereographic projection, the extra country becomes a bounded region of $\mathbb{S}^2$. Henceforth we shall restrict to maps that are drawn on compact surfaces.

---

[5]Countries may be enclaves of other countries and/or include exclaves within other countries; see `https://en.wikipedia.org/wiki/Enclave_and_exclave`.

[6]See, for example, `http://mat.gsia.cmu.edu/COLOR/general/ccreview/node2.html`.

Figure 6.10: A coloring of the western United States using four colors. Note that the "four corner" states, Arizona, New Mexico, Colorado, and Utah, only require two colors. Image courtesy of Pixabay (modified).

> **Definition 6.2.1.** A **map** on a surface is a decomposition of S into finitely many 0-cells (vertices), 1-cells (edges), and connected two-dimensional subsets that we will call **regions**. Regions do not have to be homeomorphic to disks. Two regions are considered **adjacent** if they share an edge, but not if they only share a vertex.
>
> A **coloring** of a map is a selection of colors for each region such that adjacent regions have distinct colors.

A map is N-**colorable** if N colors suffice to color the map. Of course, if a map is N-colorable, then it is also $(N+k)$-colorable for any $k \in \mathbb{N}$. Now consider all maps that could possibly be drawn on a surface. It may seem impossible to make any claims at all about the colorability of maps we have never seen (and likely *will* never see), in all of their infinite variation, but this is *precisely* what topology and graph theory can help us to do. It turns out that every surface S has a finite number $\gamma(S)$ such that $N = \gamma(S)$ colors are sufficient to color *any* map drawn on S.

That four colors are sufficient for any map on $\mathbb{S}^2$ was first conjectured by Francis Guthrie in 1852. Many "proofs" followed shortly thereafter, but each had fatal flaws in it. It wasn't until 1976 that Kenneth Appel and Wolfgang Haken announced a valid proof,[7] proving that $\gamma(\mathbb{S}^2) = 4$. In this text, we will only be able show that $4 \leq \gamma(\mathbb{S}^2) \leq 5$. First note that $\gamma(\mathbb{S}^2) \geq 4$ since there are maps that *require* four colors. The simplest such map is shown in Figure 6.11.

---

[7]See https://en.wikipedia.org/wiki/Four_color_theorem for more details on the fascinating history of the four-color problem.

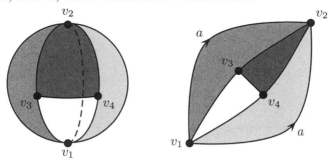

Figure 6.11: *Left*, an embedding of K_4 into $\mathbb{S}^2$. The embedding requires four colors. *Right*, plane model shown. The dashed line going behind the sphere is the identified edge a in the plane model.

Let's begin to address the question of finding $\gamma(S)$ for any given surface S. We must first rule out certain *pathological* maps. If any region borders itself – that is, there is an edge in the map that is internal to a single region – then no amount of colors would suffice to color the map. However, by removing such internal edges, we may alleviate this concern.

Next, an important combinatorial result that leads to the proof that $\gamma(S)$ is finite. In what follows, we let v, e, and f be the numbers of vertices, edges, and regions determined by a graph G embedded into a surface S. The *average number of edges per region* for the embedding is then equal to $\frac{2e}{f}$. We have seen something like this before. For a triangulation, $2e = 3f$, so the average number of edges per region in a triangulation is $\frac{2e}{f} = \frac{2e}{(2/3)e} = 3$, as expected.

Theorem 6.2.2. *Given a surface S, if there exists a number $N > 0$ such that $\frac{2e}{f} < N$ for all maps on S, then $\gamma(S) \leq N$.*

We will prove the result in the case that all regions are homeomorphic to disks, but the result can be generalized to arbitrary maps.

Proof. Suppose $\frac{2e}{f} < N$ for all maps on S. Proceed by induction on the number of regions in an arbitrary map. There are many base cases: If $f \leq N$, then clearly N colors suffice to color the map. Now assume that any map with no more than k (where $k \geq N$) regions is N-colorable, and suppose we have a map with $f = k + 1$ regions. Since the average number of edges per region is strictly less than N, there must be at least one region F_0 having less than N edges. Create a new map by the following procedure:

1. Put a new vertex v_0 in the interior of F_0.

2. Draw edges from v_0 to all vertices of F_0 (nonoverlapping of course).

3. Delete the edges of F_0.

Figure 6.12 illustrates the procedure. Note that the new map obtained this way has one less region. Thus, by the inductive hypothesis, the new map is N-colorable. Simply reverse the process to "inflate" F_0 back into existence, and color F_0 with one of the colors not used in its adjacent regions (there must be an unused color because F_0 borders at most $N-1$ regions). $\square$

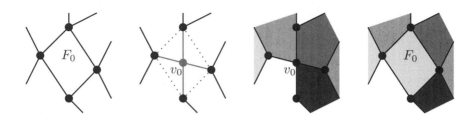

Figure 6.12: Contracting a region. The new map can be N-colored by induction; then the original map can also be N-colored. In this picture, suppose the average edge count per region is $\frac{2e}{f} < 5 = N$.

Theorem 6.2.2 would remain useless if we could not also bound the average number of edges per region universally for all maps on a surface S. Suppose all vertices in the map are at least **trivalent** (i.e., degree 3). This is not a stringent requirement, as we have already ruled out degree 1 vertices (since that would cause an internal edge within a region). A degree 2 (**bivalent**) vertex connects two edges together. Simply removing the vertex and joining the two edges into one continuous edge gets rid of the bivalent vertex. Thus we may assume all vertices in a map are at least trivalent, which leads to the inequality $v \le \frac{2}{3}e \implies v - e \le -\frac{1}{3}e$. Then we have

$$
\begin{aligned}
\chi(S) = v - e + f &\le -\frac{e}{3} + f \\
e &\le 3(f - \chi(S)) \\
\frac{2e}{f} &\le 6\left(1 - \frac{\chi(S)}{f}\right).
\end{aligned}
\tag{6.3}
$$

Now if $\chi(S) \ge 0$ (as for the surfaces $\mathbb{S}^2$, $\mathbb{P}$, $\mathbb{T}$, and $\mathbb{K}$), then we have bounded the average number of edges per region by 6, and hence at most seven colors are required to color any map on a compact surface having a nonnegative Euler characteristic.

Example 112. Determine the theoretical upper bounds on $\gamma(S)$ for $S = \mathbb{S}^2, \mathbb{T}$, and $\mathbb{K}$.

- $\chi(\mathbb{S}^2) = 2$, so $\frac{2e}{f} \le 6\left(1 - \frac{2}{f}\right) < 6$ for any map. That is, the average number of edges per region is strictly less than 6, implying $\gamma(\mathbb{S}^2) \le 6$.

- Since $\chi(\mathbb{T}) = \chi(\mathbb{K}) = 0$, we have $\frac{2e}{f} \leq 6$, which implies that seven colors are sufficient to color maps on either $\mathbb{T}$ or $\mathbb{K}$. The result is sharp for the torus: $\gamma(\mathbb{T}) = 7$, because there is a map on $\mathbb{T}$ that requires seven distinct colors (see Exercise 7). However, it turns out that $\gamma(\mathbb{K}) = 6$. The Klein bottle is simply an anomaly with respect to both graph embeddings and map colorings.

On the other hand, if $\chi(S) < 0$, then (6.3) yields, with the help of the triangle inequality,

$$\frac{2e}{f} \leq \left| 6 - \frac{6\chi(S)}{f} \right| \leq 6 + \frac{6|\chi(S)|}{f} \leq 6 + 6|\chi(S)|, \qquad (6.4)$$

which shows that there is a finite upper bound for $\gamma(S)$ depending only on $\chi(S)$. In fact, we can sharpen this bound substantially, by working with the so-called *dual graph* of a map.

Dual Graphs

In order to analyze maps more efficiently, we construct a related graph called the *dual* of the map. The dual is defined for proper embeddings of a graph into a compact surface, so we must be careful when applying it to a map (which may be regarded as an embedding of a graph, but not necessarily a proper one).

When a graph G embeds into a surface S, then G determines a set of connected regions covering S. A region does not have to be homeomorphic to a disk; however, additional edges and vertices may be added to the embedded graph to make it a proper embedding. For example, if a region is homeomorphic to an annulus, then separate the annulus into two disks as shown in Figure 6.13. The new map can be constructed so that if it is N-colorable, then so must the original map be N-colorable; this is done by making sure one of the disks is large enough to border every region that the original region bordered.

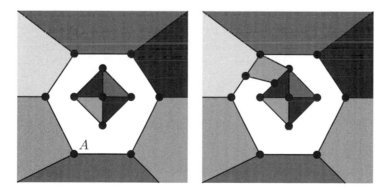

Figure 6.13: Cutting an annular region A to make an equivalent map.

Definition 6.2.3. If a graph G properly embeds in a surface S, then the **dual** graph of G in S is the graph G^* defined by the following.

- There are bijections $f_0 : V(G^*) \to F(G)$, $f_1 : E(G^*) \to E(G)$, and $f_2 : F(G^*) \to V(G)$.

- Two vertices $v_1^*, v_2^* \in V(G^*)$ are adjacent if and only if their corresponding regions in G, $f_0(v_1^*)$ and $f_0(v_2^*)$ are adjacent.

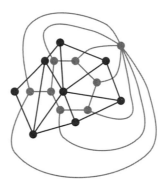

Figure 6.14: A planar graph G in black, and its dual G^* in gray. Note there is also a vertex in G^* corresponding to the *outside* region of G. Both G and G^* are properly embeddable into the sphere $\mathbb{S}^2$.

Implicit in Definition 6.2.3 is the fact that G^* also properly embeds into S. This fact is not too difficult to prove. To construct G^*, begin by putting a new vertex in the interior of each region. Then draw edges connecting vertices in adjacent regions, crossing each edge of G *transversely*.[8] The result is a graph embedded into S (no edge of G^* crosses another edge of G^*). The embedding is proper because the cycle of edges of G^* that surround any given vertex $v \in V(G)$ defines a disk neighborhood of v (*why?*). Thus planar graphs have dual graphs that are planar (see Figure 6.14, for example). Moreover, the dual graph actually *depends* on the choice of embedding. Different embeddings of G into S may result is distinct dual graphs. For this reason, we cannot simply talk about "the dual graph of G" – we must always have a particular embedding in mind first.

Coloring Graphs

Now that we have discussed dual graphs, we return to coloring. Up until now we have only been coloring *maps*, but now we color the graphs themselves. A

---

[8]Transverse, in this case, means that the edges must intersect in a way that forms an angle strictly between $0°$ and $180°$.

graph G is N-**colorable** if colors can be assigned to each vertex in such a way that:

- Any two vertices connected by an edge have distinct colors.

- No more than N colors are used.

> **Theorem 6.2.4.** *A map defined by a graph G properly embedded into a surface is N-colorable if and only if the dual graph G^* is N-colorable.*

The best advantage to working in graph coloring (as opposed to map coloring) is that no particular embedding of the graph G^* is necessary. There is no need to have well-defined *regions*, since it's the *vertices* that are getting colored. Note that we may assume the dual graph G^* is simple. (*Why?* Think about what the map defined by G would look like if G^* has a loop or pair of parallel edges.)

Example 113.

(a) K_n is n-colorable but not $(n-1)$-colorable, since each of the n vertices is adjacent to all of the others.

(b) $K_{3,3}$ is 2-colorable, as shown in Figure 6.15. The corresponding map, embedded in $\mathbb{T}$, has what is known as a *checkerboard coloring* since only two colors are needed, as shown in Figure 6.16. Note that the embedding $K_{3,3} \subseteq \mathbb{T}$ defines three regions (verify this using $\chi(\mathbb{T})$), so the dual $K_{3,3}^*$ has three vertices and six regions.

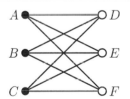

Figure 6.15: A 2-coloring of $K_{3,3}$.

The dual graph can help determine an upper bound on $\gamma(S)$ for surfaces S, which happens to be none other than *Haewood's number*.

> **Theorem 6.2.5.** *Suppose S is a compact surface. If $\chi(S) < 0$, then*
> $$\gamma(S) \leq H_{\chi(S)}.$$

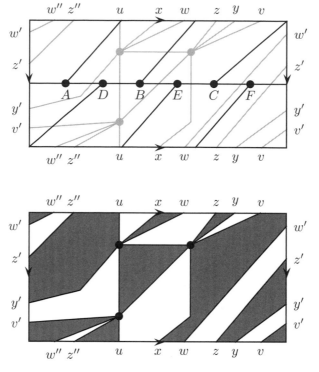

Figure 6.16: *Top,* $K_{3,3}$ embeds into $\mathbb{T}$, with $K_{3,3}^*$ indicated in light gray. *Bottom,* the checkerboard coloring associated to the coloring of $K_{3,3}$ from Figure 6.15.

Proof. By inequality (6.4), there is an upper bound for $\gamma(S)$. Let $N \geq \gamma(S)$ be arbitrary. Clearly any map having $f \leq N$ regions is N-colorable, so assume that $f \geq N + 1$. Then, since $\chi(S) < 0$, we have $\frac{\chi(S)}{f} \geq \frac{\chi(S)}{N+1}$, and together with inequality (6.3), we obtain:

$$\frac{2e}{f} \leq 6 \left(1 - \frac{\chi(S)}{f} \right) \leq 6 \left(1 - \frac{\chi(S)}{N+1} \right).$$

Recall that if the average number of edges per region $\frac{2e}{f}$ in a map is less than $N \in \mathbb{N}$, then the map is N colorable. So let's find the smallest N that satisfies the following inequality.

$$6 \left(1 - \frac{\chi(S)}{N+1} \right) < N \tag{6.5}$$

In what follows, some details are left to the reader in the Exercises. After some algebraic manipulation, we find the positive values of N that satisfy (6.5): $N > \frac{1}{2} \left(5 + \sqrt{49 - 24\chi(S)} \right)$. Now for any real number x, the smallest integer solution to the inequality $N > x$ must be somewhere within the interval $(x, x + 1]$. That is, we can be sure that $N \leq \lfloor x + 1 \rfloor$. This gives

$$N \leq \left\lfloor 1 + \frac{1}{2}\left(5 + \sqrt{49 - 24\chi(S)}\right) \right\rfloor = \left\lfloor \frac{7 + \sqrt{49 - 24\chi(S)}}{2} \right\rfloor = H_{\chi(S)}.$$

In summary, we have found that the average number of edges per region in an arbitrary map on S is strictly less than Haewood's number, and so $\gamma(S) \leq H_{\chi(S)}$. $\qquad\square$

It is truly remarkable that Haewood's number arose in two different computations. This is more than a mere coincidence. In fact, Theorem 6.1.4 and Exercise 11 from §6.1, about certain kinds of graph embeddings, suggest a link via the dual graph. If it could be shown that there is an actual embedding of $K_n \hookrightarrow S$, then the dual graph K_n^* defines a map that must use exactly n colors, so that $\gamma(S) \geq n$. Though it takes some work, such embeddings do exist for $n = H_{\chi(S)}$, except when $S = \mathbb{K}$ (the Klein bottle). Therefore, so long as $\chi(S) < 0$, we have solved the coloring problem: $\gamma(S) = H_{\chi(S)}$. What about compact surfaces S with $\chi(S) \geq 0$? Well it turns out that only $\mathbb{K}$ breaks the mold: $\gamma(\mathbb{K}) = 6$, while $H_{\chi(\mathbb{K})} = 7$. Surprisingly, though, the most difficult case to prove is the sphere, but it is true that $\gamma(\mathbb{S}^2) = H_{\chi(\mathbb{S}^2)} = 4$. In the next section we will only prove that $\gamma(\mathbb{S}^2) \leq 5$.

The "Five-Color" Theorem

Recall that we have already shown that $4 \leq \gamma(\mathbb{S}^2) \leq 6$. The rest of this section is dedicated to showing that $\gamma(\mathbb{S}^2) \leq 5$ (the proof that $\gamma(\mathbb{S}^2) = 4$ is well outside the scope of this text).

Let M be a map on $\mathbb{S}^2$. We first assume that all vertices of M are all trivalent (which would ensure that the dual graph is a triangulation). The general case then follows from a simple transformation of the map. replace every vertex having degree more than 3 with a small region as shown in Figure 6.17, then the theorem would provide a coloring, and finally shrink the region back down to a vertex to obtain a valid coloring of the original map having no more colors than the trivalent map. Furthermore, we may assume that M has at least one region with fewer than six sides (recall Example 112; since $\chi(\mathbb{S}^2) = 2$, we found that the average number of edges per region in any map is strictly less than 6).

The proof is by induction on the number of regions. As a base case, any map with five regions is 5-colorable. Now let $n \geq 5$, and suppose all maps on the sphere with n or fewer regions can be 5-colored. Let M be a map with $n + 1$ regions. If there is a region F with three or four sides, then at most four regions are adjacent to F. Remove any edge of F, thereby joining F to one of the adjacent regions, ignoring any bivalent vertices. Then, by induction, the new map M' (which has n regions) can be 5-colored. Now replace the missing edge, and since at most four colors are used in adjacent regions, F can be colored using the fifth color (see Figure 6.18).

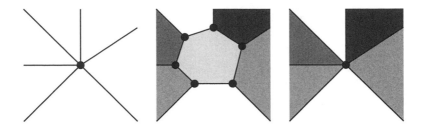

Figure 6.17: Replacing a nontrivalent vertex with a region. After coloring the whole map, the region may be shrunk down to a vertex to give a coloring of the original map.

Figure 6.18: By removing an edge from a region F having three or four sides, the map can be reduced to a five-colorable map. Then the edge may be replaced and F receives a different color than its neighbors.

If M contains no regions having only three or four edges, then there must be at least one region F_0 having five sides. We may label adjacent regions by $F_1, \ldots, F_5$, but keep in mind that the regions need not be distinct. For example, F_1 and F_3 could be the same region, as shown in Figure 6.19. In this case, though, there would be at most four distinct colors in the regions surrounding F_0, and so the same technique works as if F_0 had only four edges (simply delete an edge not belonging to any region that meets F_0 more than once).

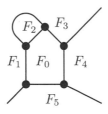

Figure 6.19: A region F_0 having five edges may still be surrounded by only four regions. Here $F_1 = F_3$ represents the same region.

Finally, suppose F_0 is a region surrounded by five distinct regions, $F_1, \ldots F_5$. Now it is possible that some region F_i could adjoin another region F_j; however, there still must be a pair of regions that do not share an edge. Otherwise, if all of $F_1, \ldots, F_5$ share edges, then in the dual graph there would be a K_5 subgraph, but we know that K_5 cannot embed into $\mathbb{S}^2$. Without loss of generality, suppose F_1 and F_3 share no edge. Delete the edges separating F_0 from F_1 and F_3. The resulting map M' has $n-1$ regions, and so can be 5-colored by inductive hypothesis. After replacing the two missing edges, both F_1 and F_3 can receive the same color, leaving the fifth color available for F_0 (see Figure 6.20).

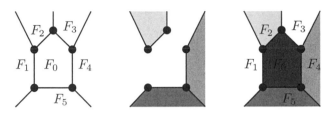

Figure 6.20: Regions F_1 and F_3 do not share an edge.

Exercises

1. Prove that if every vertex of a graph G has at least degree n, then $|V(G)| \leq \frac{2}{n}|E(G)|$.

2. Inequality (6.3) shows that the average edge per region in any map on $\mathbb{S}^2$ is *strictly* less than 6. However, as $f \to \infty$, the average approaches 6. Use this to explain why the sphere cannot be tiled with a finite number of hexagons, but the plane $\mathbb{R}^2$ can be.

3. Prove that the dual of the dual of a graph is isomorphic to the original graph, in the sense that if $G \subseteq S$ is a proper embedding, then $(G^*)^* \cong G$ with respect to the embedding in S.

4. Draw the dual graphs of each of the following (note the surface into which the graph embeds properly).

   (a) G'' from Figure 6.2

   (b) $K_4 \subseteq \mathbb{S}^2$

   (c) $K_{3,2} \subseteq \mathbb{S}^2$

   (d) $K_5 \subseteq \mathbb{T}$

5. A tree T may be embedded properly into $\mathbb{S}^2$, if we consider $\mathbb{S}^2 \setminus T$ as the single region.

   (a) Draw dual graph G_4^*, where G_4 is found in Figure 6.1.

   (b) Describe the dual graph T^* for any tree T.

6. Consider the graphs H_1 and H_2 shown below.

(a) Find the dual graphs of H_1 and H_2 and show that $H_1^* \not\cong H_2^*$, even though $H_1 \cong H_2$.

(b) Find a 3-coloring of H_1. Since $H_2 \cong H_1$, that 3-coloring transfers to H_2. Show that your coloring induces valid colorings of both H_1^* and H_2^*. (Even though we cannot expect the duals of isomorphic graphs to be isomorphic, the associated maps must have the same coloring properties.)

7. Find a map in the torus that requires seven colors. (*Hint:* Find the dual graph of an embedding $K_7 \subseteq \mathbb{T}$.)

8. Determine the theoretical upper bounds on $\gamma(\mathbb{P})$ as provided by inequality (6.3) and Theorem 6.2.2. Show that the bound is sharp by finding a map on $\mathbb{P}$ that requires this many colors. (*Hint:* See Exercise 8 from §6.1.)

9. Determine the least number of colors that are needed to color an arbitrary graph in the following graph families.

   (a) Connected bipartite graphs   (c) Path graphs, P_n

   (b) Connected trees   (d) Cycle graphs, C_n

10. Show that inequality (6.5) is equivalent to the quadratic inequality $N^2 - 5N + (6\chi(S) - 6) > 0$. Then use the quadratic formula to find the roots and solve the inequality.

6.3   Knots and Links

What is a knot? Scouts and sailors learn how to tie together ropes in many different ways (see Figure 6.21), but none of those configurations are what mathematicians would call a knot.[9] When my son asks me to unravel a huge knot that somehow made its way into his shoelaces, I am not practicing *knot theory* – as we shall see, knot theory has more to do with identifying knots rather than untying them. For a mathematician, a *knot* usually means a closed curve in three-dimensional space. Like graphs, knots are essentially one-dimensional objects. Nevertheless, knot theory is a rich field of study full of surprising twists and tangles. This section and the next provides just a brief survey of some important concepts in knot theory.

---

[9]To a mathematician, most everyday "knots" are in fact *tangles*, which are basically knots or links with free ends.

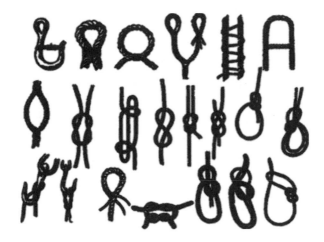

Figure 6.21: Various nautical knots, bends, and splices.

Knots

Why do we care about knots at all? Knots are more than just a fun diversion. They play a fundamental role in our own biology. Long chains of proteins and the strands of our DNA can and do knot in various ways that affect their function. In fact, knots could be the foundation of everything in the universe – some models of string theory incorporate knotted strings, and other areas of theoretical physics use topological quantum field theory (TQFT), which is a certain abstraction of knot theory.[10]

For our purposes, a knot is a smooth, closed curve in three-dimensional space. The term *smooth* comes from differential topology, and it basically means that there are no "sharp corners." More precisely, smooth means that derivatives of all orders exists, and so, in particular, a well-defined tangent line exists at every point of the knot.

> **Definition 6.3.1.** A **knot** K is a smooth embedding of the circle $\mathbb{S}^1$ into $\mathbb{R}^3$.

We typically regard the image of K as the knot itself. To say that $K :$ $\mathbb{S}^1 \hookrightarrow \mathbb{R}^3$ is an embedding implies that $K[\mathbb{S}^1] \approx \mathbb{S}^1$, or, being sloppy, we may state that $K \approx \mathbb{S}^1$ – so every knot is homeomorphic to every other knot and homeomorphic to a circle. Hardly an interesting place to start, topologically speaking. However, what we study in knot theory is the way in which knots can be deformed within the space they inhabit, a concept called **ambient isotopy**.

---

[10]In the late 1800s, Lord Kelvin proposed that the elements themselves were simply distinct knots within the *ether*. Now we know there is no such thing as an ether pervading all of space, and atoms are made of even stranger components – electrons, quarks, gluons, etc.

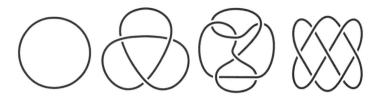

Figure 6.22: Knot diagrams for a few mathematical knots.

Two knots are considered *equivalent* if one can be deformed to the other by ambient isotopy. It turns out that there are infinitely many distinct knots that are *not* equivalent by ambient isotopy, which is much more interesting.

All knots live in three-dimensional space, but they can easily be represented on two-dimensional paper using a **knot diagram** (see Figure 6.22). You may notice "breaks" in the curve where two parts of the knot seem to intersect. These breaks are not part of the knot itself (the curve isn't actually broken up into separate pieces) but indicate how one part crosses over or under another. We'll call these intersections the **crossings** of the knot diagram. Each crossing of a knot diagram must correspond to no more than *two* arcs of the knot going past one another *transversely*.[11]

Figure 6.23: Invalid crossings in a knot diagram.

If a diagram has an isolated invalid crossing, it can be fixed by a small motion, called a **perturbation**, of the knot in space. What valid knot diagrams might result from perturbations of the digrams in Figure 6.23? A knot K will have many different knot diagrams, but there must be a diagram for K having the least number of (valid) crossings.

> **Definition 6.3.2.** The **crossing number** of a knot K is the minimum number of crossings in any knot diagram representing K. The crossing number of K is notated $C(K)$.

Each connected component of the knot diagram is called a **strand**, so a strand starts at an undercrossing and goes to the next undercrossing. Up to three distinct strands are involved at each crossing. Keep in mind that the term "strand" only makes sense for a knot *diagram*, not the knot itself. The simplest

---

[11]Two curves meet transversely if their tangent lines do not coincide at the point of intersection.

kind of knot is called the **unknot**, $\mathbb{U}$, which is characterized by having a knot diagram with only one strand and no crossings, and so $C(\mathbb{U}) = 0$. However, a given knot diagram with many strands and crossings may in fact be equivalent to the unknot.

Example 114. The knot K shown in Figure 6.24 is equivalent to $\mathbb{U}$. Verify this on your own by carefully re-creating the pattern of crossings of K in an actual string (be sure to tie the ends of your string together).

Figure 6.24: An unknot.

What about the knots shown in Figure 6.22? Are any of these equivalent to the others? Convince yourself that they are in fact all distinct knots (though we actually cannot *prove* it until we develop more tools).

Example 115. A **torus knot** is a knot that can be drawn on the surface of $\mathbb{T}$ with no crossings. Torus knots $\mathbb{T}(p, q)$ are characterized by two relatively prime numbers p and q, representing the number of complete turns around the torus in the two independent directions. The requirement that p and q be relatively prime ensures that there is only one component. If $\gcd(p, q) = d$, then $\mathbb{T}(p, q)$ is a **torus link** of d components (see below for more on links). An easy way to construct torus knots (or links) is to start with the plane model for $\mathbb{T}$, and draw p points on the left/right edge and q points on the top/bottom edge. Then, beginning near one of the corners, connect points in a diagonal stripe pattern; if p and q have the same sign, then the segments should have positive slope, while if the sign of p and q differ, then the segments should have negative slope. Figure 6.25 shows the simplest nontrivial torus knot, $\mathbb{T}(3, 2)$.

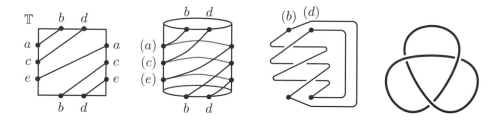

Figure 6.25: The torus knot $\mathbb{T}(3,2)$ is a trefoil. First identify the points a, c, and e by connecting each to its counterpart with a segment that goes "behind" the torus. Then identify b and d with connecting curves. After smoothing out the resulting knot diagram, it becomes clear that this knot is a trefoil.

Oriented Knots

Up to this point all of our knots have been *unoriented*, meaning that there is no preferred direction of travel along the strands of the knot. An **oriented** knot is a knot K together with a specified *orientation*, often denoted by arrowheads on the knot diagram. Two knots oriented K and L are equivalent if there is an ambient isotopy (in $\mathbb{R}^3$) taking K to L while preserving the orientation. Reversing the orientation of a knot may result in an inequivalent knot (as oriented knots), but does not change the equivalence class of the underlying unoriented knot. By convention, the torus knot $\mathbb{T}(p,q)$ has the orientation induced by traversing the segments in plane diagram for $\mathbb{T}$ up and to the right, while $\mathbb{T}(-p,-q)$ has the opposite orientation (see Figure 6.26). However, $\mathbb{T}(p,q)$ and $\mathbb{T}(-p,-q)$ are equivalent as unoriented knots.

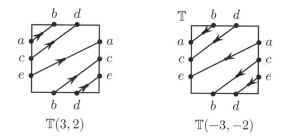

Figure 6.26: Torus knots $\mathbb{T}(p,q)$ and $\mathbb{T}(-p,-q)$ have opposite orientations.

One simple way to describe a knot diagram (oriented or not) is by its **Gauss code**. First choose a starting point and an orientation (if the knot is unoriented). Label the first crossing you encounter as "1," and continue to label new crossings by consecutive integers. When you encounter a crossing for the second time, leave the label as it is. Now, once all of the labels are on the diagram, start the journey over. The Gauss code is a sequence of integers determined in the following manner. For each crossing n you encounter, put n if the strand you are on crosses over the other, or $-n$ if it crosses under. By the time you get back to the starting point, every crossing will have been visited exactly twice, once as an over (n), and once as an under $(-n)$ crossing.

Example 116. Find a Gauss code for the trefoil.

Solution: Based on the starting point and orientation shown in the figure below, the Gauss code is $(1, -2, 3, -1, 2, -3)$.

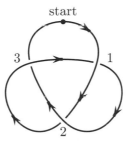

Unfortunately, owing to all of the choices made, there could be many different Gauss codes for any given knot. Moreover, the Gauss code may not determine the knot precisely (see Exercise 9). However, there's an easy fix called the **extended Gauss code**, which takes into account the *sign* of the crossing. In an oriented knot, every crossing is either **positive (right-handed)** or **negative (left-handed)**. Perhaps the easiest way to tell the sign (or *handedness*) of a crossing is to rotate the diagram until the under strand is pointing up. Then the crossing is negative if the over strand goes to the left, and positive if to the right. See Figure 6.27.

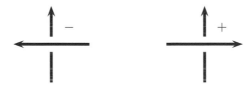

Figure 6.27: *Left to right*, negative and positive crossings.

The extended Gauss code is determined by first labeling the crossings by integers as you would for Gauss code. Now as you traverse the knot, when you encounter crossing n for the first time, put n or $-n$ if it's an over or under crossing, respectively, just as before. However, when you reach crossing n for the second time, put n or $-n$ based on the sign of the crossing. For example, the extended Gauss code of the right trefoil labeled as in Example 116 would be $(1, -2, 3, 1, 2, 3)$, the last three terms being positive because every crossing in the right trefoil is positive.

Example 117. Draw the oriented knot whose extended Gauss code is

$$(1, -2, 3, -4, -2, -1, 5, -6, 4, 3, -6, -5).$$

Solution: There will be six distinct crossings. Start with the four unique crossings 1 *over*, 2 *under*, 3 *over*, 4 *under*.

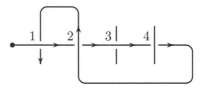

The next number in the code is -2. This means we have to reenter crossing 2 in such a way as to make a negative crossing; in this case, we must come in from the below. The next code number, -1, means that crossing 1 must also be negative.

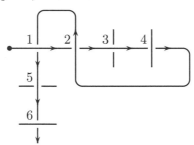

Next we encounter two new labels, 5 and 6. This means we should draw two more crossings at this point, first an over- and then an under-crossing.

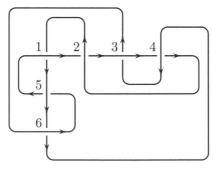

Finally, finish the diagram with crossings 4 and 3 being positive and 6 and 5 being negative.

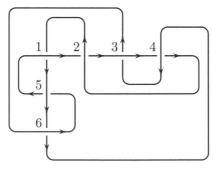

A knot K is called **alternating** if K has a knot diagram such that over-crossings and under-crossings alternate throughout the knot. The Gauss code of an alternating knot diagram alternates between positive and negative labels (but the *extended* Gauss code will not in general). It's important to realize that even though a knot diagram may not be alternating, it still could be equivalent to one that is. Alternating knots are important in knot theory because a *reduced* alternating knot diagram is guaranteed to have the minimal number of crossings,[12] which gives a handy criterion for finding the crossing number $C(K)$. A knot diagram is **reduced** if there are no crossings that separate the diagram into two components. An unreduced knot diagram can easily be reduced by untwisting these kinds of crossings (see Figure 6.28).

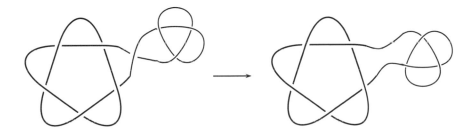

Figure 6.28: Reducing a knot diagram.

On the other hand, there are knots that simply have no alternating diagram at all; those are the **nonalternating** knots. Figure 6.29 shows a nonalternating knot.

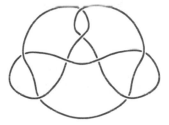

Figure 6.29: A nonalternating knot with eight crossings. Image courtesy of Wikimedia Commons.

---

[12]This was one of three famous conjectures by Tait in the late nineteenth century. Two of the Tait conjectures were proven in a series of papers in the late 1980s by Louis Kauffman [Kau87], K. Murasugi [Mur87a, Mur87b], and Morwen Thistlethwaite [Thi87, Thi88], while the third conjecture was proven in 1991 by W. Menasco and Thistlethwaite [MT91].

Links

A **link** is a particular embedding of one or more copies of $\mathbb{S}^1$ into $\mathbb{R}^3$. All knots are links, but not all links are knots. Links may be *oriented*, which means that an orientation is assigned to each component of the link. An **unlink** of n components is an embedding of n completely separated unknots. Perhaps the simplest link that cannot be separated is the **Hopf link**, consisting of two unknots linked together. Other famous links include the **Whitehead link** and the **Borromean rings** shown in Figure 6.30. Figure 6.31 shows a link in which one component is a trefoil.

Figure 6.30: *Left to right*, the Hopf link, Whitehead link, and Borromean rings. The Borromean rings have the curious property that whenever any single component is removed, the other two components are unlinked.

Figure 6.31: This Celtic design is an example of a two-component link whose components are a trefoil and an unknot. Image courtesy of Pixabay.

Example 118.   Suppose $\gcd(p, q) = d > 1$. Then $\mathbb{T}(p, q)$ is a link of d components. Figure 6.32 shows a $(4, 2)$-torus link.

Chirality

Suppose all of the crossings in a given diagram for a knot (or link) K are reversed – over-crossings become under-crossings, and vice versa. The resulting knot K'

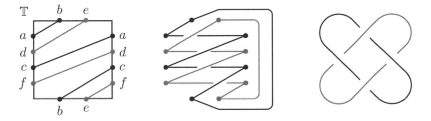

Figure 6.32: The $(4, 2)$-torus link, also known as King Solomon's knot, though it is technically not a knot.

is said to have the opposite **chirality**[13] as K. The mirror image of an oriented knot changes every positive crossing to a negative one, and vice versa.

Often the two knots K and K' are not equivalent to one another, in which case we say that K (or K') is **chiral**. The trefoil is chiral, as the left and right versions are not equivalent (see Figure 6.33). In fact, every nontrivial torus knot $\mathbb{T}(p, q)$ is chiral, its mirror image being $\mathbb{T}(p, -q)$ (see Exercise 5). If K is equivalent to K', then K is called **amphichiral**. The **figure-eight** knot (rightmost knot in Figure 6.33) is amphichiral (see Exercise 6). Note that the right trefoil has only positive crossings (regardless of the orientation), while the left trefoil has only negative crossings, while the figure-eight has two negative and two positive crossings.

Figure 6.33: *Left to right*, left and right trefoil knots, figure-eight knot.

Knot Sums

Given two oriented knots K_1 and K_2, represented by knot diagrams, the **sum** or **composition** of K_1 and K_2 is an oriented knot $K_1 \# K_2$ formed as follows (compare to the *connected sum* of surfaces, Definition 5.1.3):

1. Remove an open arc from the diagram for K_1, without removing any crossings.

---

[13]The word "chirality" comes from the Greek root $\chi\epsilon\iota\rho$ for *hand*. Just as your two hands are mirror images of one another, two knots with opposite chirality are mirror images.

2. Remove an open arc from the diagram for K_2, without removing any crossings.

3. Join the knot diagrams by identifying the endpoints in a way that is consistent with the orientations and does not create any new crossings.

The knot sum is unique in the sense that if H_1 and H_2 are oriented knots equivalent to K_1 and K_2, respectively, then $H_1 \# H_2$ is equivalent to $K_1 \# K_2$. In particular, this means that the construction does not depend on which arcs were removed from either knot. The connected sum of unoriented knots may depend on the choice of orientations, unless at least one of the knots has the property that one orientation is equivalent to the other through ambient isotopy (in which case we call the knot **invertible**). The sum of links is not well defined in general, as there is a choice as to which component of the link from which to remove the arc.

Example 119. The diagram below illustrates a knot sum of oriented knots K_1 and K_2.

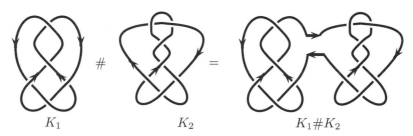

$$K_1 \qquad\qquad K_2 \qquad\qquad K_1 \# K_2$$

The knot sum is *commutative* ($K_1 \# K_2 = K_2 \# K_1, \forall K_1, K_2$) and *associative* (($K_1 \# K_2) \# K_3 = K_1 \# (K_2 \# K_3), \forall K_1, K_2, K_3$). There is also a unit element, the unknot $\mathbb{U}$, which makes the set of oriented knots into a commutative *monoid* with respect to the operation $\#$.

Exercises

1. Prove that 3 is the least crossing number possible for a nontrivial knot by showing the following.

   (a) There are essentially two knot diagrams that have only a single crossing; four if you count mirror images. Draw them and convince yourself that each is equivalent to $\mathbb{U}$.

   (b) How many knot diagrams have exactly two crossings (there may be more than you think). Draw each one and show the moves that prove each is equivalent to $\mathbb{U}$.

Figure 6.34: The Buddha knot.

2. The "Buddha knot," shown in Figure 6.34, is equivalent to U. Verify this either by drawing a series of moves or by creating the knot in a string and manipulating it.

3. Sketch each of the following torus knots and/or links, and state the number of components of each.

   (a) $\mathbb{T}(3,4)$ (b) $\mathbb{T}(2,6)$ (c) $\mathbb{T}(3,6)$ (d) $\mathbb{T}(5,7)$

4. Draw a sequence of moves to show that the link on the left is equivalent to the Borromean rings. Then show the link on the right, in which the bottom pieces have been linked, is equivalent to the unknot and a Hopf link that are not linked to each other. (*Moral of the story:* If you go mountain climbing with Borromean rings, never link two of the rings together.)

5. Verify that the torus knot $\mathbb{T}(3,-2)$ illustrated below represents a left trefoil knot.

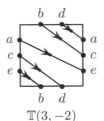

$\mathbb{T}(3,-2)$

6. Draw a sequence of moves that show how the figure-eight knot is amphichiral.

7. Take a long, thin strip of paper and make a Möbius strip by making one half-twist and taping the ends. Then cut down the center of the strip. The result is a single unknotted loop of paper. Now, with a new strip of paper, make three half-twists in the strip. After cutting down the middle, the result is a knot. Which knot is it? Do the same for five half-twists and seven half-twists (you may need a *very* long strip of paper to do this).

8. Find a Gauss code for two knots on the right in Figure 6.22. Then find an extended Gauss code for each of these knots.

9. Label the crossings of a left trefoil in such a way that its Gauss code is the same as the Gauss code for a right trefoil. (This shows that the Gauss code may not distinguish between mirror images.) Next, keeping the labels exactly the same, show that the extended Gauss codes for the left and right trefoil are distinct.

10. Not all sequences of numbers can be realized as the Gauss or extended Gauss code of a knot (or link). Attempt to draw each knot based on its code, and explain why the knot diagram cannot exist.

    (a) *Extended Gauss code:* $(1, -2, 1, -2)$.

    (b) *Gauss code:* $(1, -2, -1, 2)$.

11. Draw a sequence of moves that takes the Whitehead link as pictured in Figure 6.30 to the (equivalent) link shown below.

12. Let T be the right trefoil and T' be its mirror image. Draw the knot sums $T \# T'$ (*square knot*) and $T \# T$ (*granny knot*).

6.4   Knot Classification

The classification of knots began with the work of physicist Peter Guthrie Tait and mathematician Charles Newton Little in the nineteenth century. The *Tait-Little* knot table has been steadily expanding over the years, and errors have been found and corrected. For example, a pair of 10-crossing knot diagrams called 10_{161} and 10_{162} were found to represent the same knot.[14] Much more can be found in standard knot theory texts or online.[15]

Prime Knots

We are familiar with prime numbers. They form the building blocks out of which every natural number can be formed with respect to multiplication. Primes themselves are *indecomposable*, meaning that if a prime $p \in \mathbb{N}$ has a factorization $p = mn$, then either $m = 1$ or $n = 1$ (i.e., one factor must be the *multiplicative unit* element of $\mathbb{N}$). Moreover, there are infinitely many prime numbers (see §4.6), and yet every composite number $n \in \mathbb{N}$ is the product of prime numbers is a *unique* way (up to rearrangement of the factors). The set of knots under the operation of knot sum has much the same kind of structure.

> **Definition 6.4.1.** A knot K is called **prime** if whenever $K = K_1 \# K_2$, then either $K_1 = \mathbb{U}$ or $K_2 = \mathbb{U}$. A knot K that is the sum of two nontrivial knots, K_1 and K_2, is called **composite**, and the knots K_1 and K_2 are **factors** of K.

> **Proposition 6.4.2.**
>
> (a) *Every composite knot K can be written uniquely as a finite knot sum of prime knots (up to rearrangement of the factors).*
>
> (b) *The unknot $\mathbb{U}$ is not composite.*
>
> (c) *There are infinitely many prime knots.*

The proofs of these important results fall outside the scope of this brief survey. Determining whether a given knot diagram represents a composite knot is quite difficult. The key is to find a disk D such that there are nontrivial knotted arcs on both the interior and exterior of D, and such that ∂D contains exactly two points of the the knot diagram. Often the composite knot must be deformed to an equivalent diagram before such a separating disk D can be found. For example, the knot shown in Figure 6.28 is composite, being the knot sum of a five-crossing knot and a trefoil.

Reidemeister Moves and Knot Invariants

So far, we've taken it for granted that you know what kinds of moves on knot or link diagrams correspond to ambient isotopies of $\mathbb{R}^3$. At first it may seem as though there are many ways of moving the strands around. Some moves only

---

[14]This duplication was discovered by Kenneth Perko in 1973, almost 100 years after the knots were first tabulated. The two diagrams are now known as the *Perko pair*.

[15]See the Knot Atlas: http://katlas.org/wiki/Main_Page.

stretch, compress, or move the strands of the diagram without changing the crossings in any way; those kinds of moves are called **planar isotopies**. Other moves change the crossings of the plane diagram without changing the knot or link itself. It turns out that there are only three basic crossing-altering moves. Every ambient isotopy of $\mathbb{R}^3$ that deforms one knot or link into an equivalent one can be described by a finite sequence of planar isotopies and the three basic moves called **Reidemeister moves**.

- Reidemeister move R.I. Twist or untwist a *kink*.

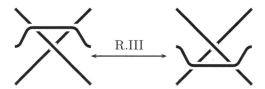

- Reidemeister move R.II. Slide a strand over or under an adjacent parallel strand.

- Reidemeister move R.III. Slide a strand past a crossing. (There are four such moves: the one pictured below, the one in which the middle crossing has the opposite sense, and the mirror images of these.)

While it is easy to see that the three Reidemeister moves do not change the equivalence class of a link, it is quite difficult to prove that only these three moves are sufficient – a fact we will take for granted in this text.[16]

Just as topological invariants can be used to distinguish topological spaces, there are *knot invariants* that serve to distinguish knots (there are also many interesting *link invariants*, but we will discuss only knots in this section).

> **Definition 6.4.3.** Suppose that for every knot there is a measurable quantity or property $\mathscr{Q}$. If the value of $\mathscr{Q}$ is the same on all equivalent knots, then $\mathscr{Q}$ is called a **knot invariant**.

---

[16]For a more thorough discussion and proof of the *Reidemeister Theorem*, see, e.g., Manturov [Man04] or Messer and Straffin [MS06].

Typically a knot invariant is a combinatorial quantity defined on knot diagrams. Thus if two knot diagrams differ with respect to some knot invariant, then they must represent different knots. However, as is generally the case with topological invariants, if two knot diagrams share the same values with respect to a knot invariant, then we still cannot conclude that the knots are the same.

Now suppose we have defined a function $\mathscr{Q}$ on knot diagrams. How can we know whether $\mathscr{Q}$ is a knot invariant? Since the Reidemeister moves are sufficient to change any diagram into an equivalent one, $\mathscr{Q}$ is a knot invariant if and only if it is invariant under each of the three Reidemeister moves and planar isotopy. For example, the total number of crossings in a knot diagram is *not* an invariant. Both R.I and R.II change the number of crossings. (However, the crossing number $C(K)$ *is* a knot invariant because it's defined as the least possible number of crossings over *all* diagrams representing K.)

Definition 6.4.4. A **tricoloring** of a knot diagram is a choice of one of three colors for each strand of the diagram such that,

1. at every crossing, either all three strands have different colors or all are the same color, and

2. all three colors are used.

If a diagram has a tricoloring, then the diagram is called **tricolorable**.

Example 120. A tricoloring of a knot (using the "colors" solid, dashed, and dotted):

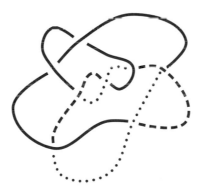

Theorem 6.4.5. *Tricolorability is a knot invariant.*

Proof. We must show that tricolorability is preserved by planar isotopy and the Reidemeister moves. Planar isotopy does not affect the relative position of the crossings, and hence does not affect tricolorability.

R.I. The only valid coloring of a kink uses only a single color, so R.I cannot change the tricolorability of a knot.

R.II. If the two adjacent strands have the same color, then after applying R.II, the new strand created must be colored the same color. If the two adjacent strands have different colors, then after applying R.II, the new strand must be colored the third color.

Conversely, if two parallel adjacent strands overlap as shown above, then the two strands can be consistently colored after pulling them apart. We must only check that no other kind of coloring exists for overlapped strands that would cause inconsistency after being pulled apart. No other combination of colors forms part of a valid tricoloring.

R.III. This is Exercise 4.   □

Example 121. The unknot is obviously not tricolorable, as there is only one strand in the standard knot diagram for $\mathbb{U}$. The trefoil knot is tricolorable (simply color each of the three strands a different color). Since tricolorability is a knot invariant, this proves that the trefoil is not equivalent to the unknot.

Knot Polynomials

Tricolorability is a very limited knot invariant, since its only possible values are *yes* and *no*. Other knot invariants such as the crossing number are difficult to compute because they are defined not in terms of a single given knot diagram, but over the whole class of knot diagrams for a given knot. Over the years, more powerful knot invariants have been developed based only on the combinatorics of the knot diagram.

For example, the sum of signs of the crossings of an oriented knot diagram comes close to being an invariant. But even though this quantity, called the *writhe* of the diagram, is not a true invariant, it will be useful in defining a powerful invariant called the *Jones polynomial* later on in this section.

> **Definition 6.4.6.** Let K be an oriented knot diagram. Let $X(K)$ be the set of all crossings of K and $\sigma(\chi)$ be the sign of a crossing $\chi \in X(K)$. The **writhe** of K is the sum,
> $$w(K) = \sum_{\chi \in X(K)} \sigma(\chi).$$

Based on the definition, it may seem that the writhe of a knot depends on the orientation of the knot diagram. However, reversing orientation of the diagram does not change the sign of any crossing.

Unfortunately, the writhe is *not* a knot invariant because its value changes under R.I. On the other hand, the writhe is invariant under R.II because the signs of the two crossings involved are always opposite and so together contribute 0 to $w(K)$ – the same amount that an uncrossed pair of parallel strands would contribute.

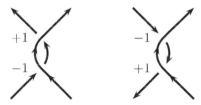

The writhe is also invariant under R.III. Only one diagram is shown below, but you should carefully check that the various choices of orientations and relative positions of the crossings also work.

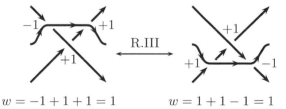

$$w = -1 + 1 + 1 = 1 \qquad w = 1 + 1 - 1 = 1$$

We now turn our attention to *smoothings* of crossings. A crossing χ in the diagram for a link L can be **smoothed** in one of two ways, positively or negatively:

$$\chi \qquad \chi^+ \qquad \chi^-$$

Denote by L_{χ^+} the diagram obtained from L by smoothing χ to χ^+, and similarly for L_{χ^-}. Note that after the smoothings, L_{χ^+} and L_{χ^-} are typically *not* equivalent to L. Here's an easy way to remember the two types of smoothing. If L has an orientation in which χ is a positive crossing, then χ^+ preserves the orientation while χ^- does not, and vice versa: if an orientation on L makes χ a negative crossing, then χ^- will preserve that orientation while χ^+ does not. See the bottom row of Figure 6.35 for the $2^3 = 8$ distinct ways to smooth all three crossings of a trefoil. In what follows, we use the notation $L_1 \sqcup L_2$ to mean that the diagrams L_1 and L_2 are separated (no crossings between L_1 and L_2).

Definition 6.4.7. The **Kauffman bracket** of a knot or link diagram is a Laurent polynomial[17] in the indeterminate A defined recursively as follows:

1. $\langle \mathbb{U} \rangle = 1$.

2. $\langle L \sqcup \mathbb{U} \rangle = (-A^2 - A^{-2})\langle L \rangle$.

3. $\langle L \rangle = A\langle L_{\chi^+} \rangle + A^{-1}\langle L_{\chi^-} \rangle$

Note that rules 1 and 2 of Definition 6.4.7 imply that $\langle n\mathbb{U} \rangle = (-A^2 - A^{-2})^{n-1}$, where $n\mathbb{U}$ is the diagram with n separated copies of the unknot (the unlink of n components). The recursion must terminate since L can have only finitely many crossings and each application of rule 3 (called a **skein relation**) reduces the total number of crossings by 1, and when there are no crossings at all, then rules 1 and 2 can finish the job. However, the recursive definition may lead to an algorithm that runs in *exponential time*. At each step, the number of partially smoothed diagrams doubles.

Example 122. Find $\langle T \rangle$, where T is the right trefoil knot.

Solution: Figure 6.35 illustrates the smoothings of T done recursively. It may help to follow this diagram as we work out the Kauffman bracket below.

---

[17]**Laurent polynomial** is like a polynomial except that negative powers are permitted. In other words, every Laurent polynomial can be written in the form $f(t) = a_m t^m + a_{m-1}t^{m-1} + \cdots + a_1 t + a_0 + a_{-1}t^{-1} + \cdots + a_{-n+1}t^{-n+1} + a_{-n}t^{-n}$, for some nonnegative integers m, n and constants a_k, where $-n \le k \le m$.

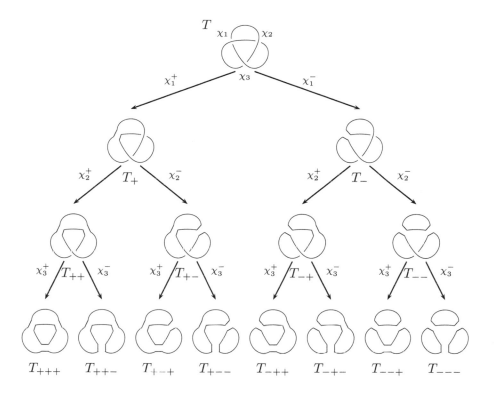

Figure 6.35: Smoothings in the right trefoil knot.

For convenience, we have notated $T_{\chi_k^+}$ by $T_\perp$, etc.

$$
\begin{aligned}
\langle T \rangle &= A\langle T_+ \rangle + A^{-1}\langle T_- \rangle \\
&= A\left(A\langle T_{++} \rangle + A^{-1}\langle T_{+-} \rangle\right) + A^{-1}\left(A\langle T_{-+} \rangle + A^{-1}\langle T_{--} \rangle\right) \\
&= A^2\langle T_{++} \rangle + \langle T_{+-} \rangle + \langle T_{-+} \rangle + A^{-2}\langle T_{--} \rangle \\
&= A^2\left(A\langle T_{+++} \rangle + A^{-1}\langle T_{++-} \rangle\right) + A\langle T_{+-+} \rangle + A^{-1}\langle T_{+--} \rangle \\
&\quad + A\langle T_{-++} \rangle + A^{-1}\langle T_{-+-} \rangle + A^{-2}\left(A\langle T_{--+} \rangle + A^{-1}\langle T_{---} \rangle\right) \\
&= A^3\langle T_{+++} \rangle + A\langle T_{++-} \rangle + A\langle T_{+-+} \rangle + A^{-1}\langle T_{+--} \rangle \\
&\quad + A\langle T_{-++} \rangle + A^{-1}\langle T_{-+-} \rangle + A^{-1}\langle T_{--+} \rangle + A^{-3}\langle T_{---} \rangle \\
&= A^3(-A^2 - A^{-2}) + A(1) + A(1) + A^{-1}(-A^2 - A^{-2}) + A(1) \\
&\quad + A^{-1}(-A^2 - A^{-2}) + A^{-1}(-A^2 - A^{-2}) + A^{-3}(-A^2 - A^{-2})^2 \\
&= -A^5 - A^{-3} + A^{-7}
\end{aligned}
$$

Of course, there are certain shortcuts to finding $\langle L \rangle$ in general. For example, T_{+-} and T_{-+} in Figure 6.35 are equivalent by planar isotopy, and since the Kauffman bracket depends only on crossings, $\langle T_{+-} \rangle = \langle T_{-+} \rangle$. What about different diagrams of the same knot? Unfortunately, the Kauffman bracket does not preserve all three Reidemeister moves, so we cannot conclude that the polynomials are the same. Fortunately, there is an easy fix to this problem: combine the Kauffman bracket with the writhe.

> **Definition 6.4.8.** Let L be a link diagram. The **Jones polynomial** of L is defined as follows. Let
>
> $$[L] = (-A^3)^{-w(L)}\langle L\rangle.$$
>
> Then $J(L)$ is found by replacing A by $t^{-1/4}$ in $[L]$.

While it is not obvious from the definition, the Jones polynomial is indeed a Laurent polynomial, since it can be shown that every term in $[L]$ has degree $4k$ for some $k \in \mathbb{Z}$.

Example 123. Find the Jones polynomial of the right trefoil T.

Solution: We have $\langle T\rangle = -A^5 - A^{-3} + A^{-7}$ from Example 122. The writhe of T is $w(T) = 3$.

$$
\begin{aligned}
J(T) &= [T]_{A=t^{-1/4}} = (-A^3)^{-3}\left(-A^5 - A^{-3} + A^{-7}\right)\Big|_{A=t^{-1/4}} \\
&= \left(A^{-4} + A^{-12} - A^{-16}\right)\Big|_{A=t^{-1/4}} = t + t^3 - t^4
\end{aligned}
$$

> **Theorem 6.4.9.** *The Jones polynomial is a knot invariant. If the orientations in a link have been fixed, then the Jones polynomial is an oriented link invariant.*

Proof. We must show that the value of J is preserved under planar isotopy and the Reidemeister moves. Since J depends only on the crossings, it is certainly invariant under planar isotopy. We should also be careful concerning the orientation of the diagram. If L is a link diagram, then $w(L)$ depends on the orientations chosen for each component; however, if L has only one component, then $w(L)$ is the same regardless of the orientation.

R.I. Suppose L is a diagram having a crossing χ that is a kink. Suppose further that $\sigma(\chi) = 1$ (the negative crossing is handled analogously). Let L' be the diagram resulting from L after R.I has been used to untwist the kink χ. Observe that $L_{\chi^+} = L' \sqcup \mathbb{U}$ and $L_{\chi^-} = L'$. Our goal is to show that $J(L) = J(L')$. It is equivalent to show that $[L] = [L']$. Note that $w(L) = 1 + w(L')$.

$$\begin{aligned}
\langle L \rangle &= A\langle L' \sqcup \mathbb{U} \rangle + A^{-1}\langle L' \rangle \\
&= A\left(-A^2 - A^{-2}\right)\langle L' \rangle + A^{-1}\langle L' \rangle \\
&= -A^3 \langle L' \rangle
\end{aligned}$$

$$\begin{aligned}
\implies [L] = (-A^3)^{-w(L)}\langle L \rangle &= (-A^3)^{-1-w(L')}\left(-A^3\langle L' \rangle\right) \\
&= (-A^3)^{-w(L')}\langle L' \rangle \\
&= [L']
\end{aligned}$$

R.II. Suppose L and L' are related by move R.II, and let χ and ψ be the two crossings that are removed by R.II. We know from above that $w(L) = w(L')$, so it suffices to show that $\langle L \rangle = \langle L' \rangle$. Perform the smoothings of χ and ψ as demonstrated in Figure 6.36. Note that $L_{++} = L_{--}$, $L_{+-} = L'$, and $L_{-+} = L_{++} \sqcup \mathbb{U}$.

$$\begin{aligned}
\langle L \rangle &= A^2\langle L_{++} \rangle + \langle L_{+-} \rangle + \langle L_{-+} \rangle + A^{-2}\langle L_{--} \rangle \\
&= A^2\langle L_{++} \rangle + \langle L' \rangle + \langle L_{++} \sqcup \mathbb{U} \rangle + A^{-2}\langle L_{++} \rangle \\
&= A^2\langle L_{++} \rangle + \langle L' \rangle + (-A^2 - A^{-2})\langle L_{++} \rangle + A^{-2}\langle L_{++} \rangle \\
&= \langle L' \rangle
\end{aligned}$$

Technically we are not done until we have checked the other version of R.II, but the steps are similar.

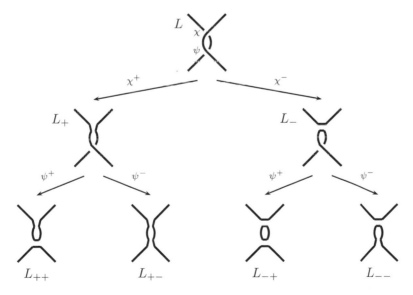

Figure 6.36: Smoothings for the crossed pair of adjacent strands.

R.III. There are many versions of the R.III move, depending on whether the moving strand is above or below the other two, and what type of crossing is in the middle. We will prove the result for one specific R.III move and leave the rest to the diligent reader. Again the writhe is invariant under this Reidemeister move, so we only need to check that $\langle L \rangle = \langle L' \rangle$, where L' is obtained by applying R.III to a link diagram L.

The diagrams involved in the R.III move have three crossings, so it may seem as though we have a lot of work to do. But the great thing about mathematics in general is that prior work can be used to solve later problems. We will use the fact that the Kauffman bracket is invariant under R.II moves. Consider the smoothings of the crossing χ in Figure 6.37. Observe that $L_- = L'_-$ through planar isotopy. Also note that $\langle L_+ \rangle = \langle L'_+ \rangle$ since L'_+ can be obtained from L_+ using two R.II moves. The rest of the argument follows easily.

$$\langle L \rangle = A\langle L_+ \rangle + A^{-1}\langle L_- \rangle = A\langle L'_+ \rangle + A^{-1}\langle L'_- \rangle = \langle L' \rangle$$

$\square$

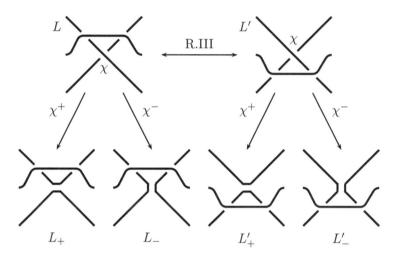

Figure 6.37: Smoothings of crossing χ. The Kauffman brackets are the same before and after the R.III move.

Now that we have shown that the Jones polynomial is a knot/link invariant, it can be used to distinguish knots. For example, it can distinguish between the left and right trefoil, and can prove that the figure-eight knot is not equivalent to the unknot or the trefoil (some things that tricolorability could not accomplish). The Jones polynomial is quite a powerful invariant, but not a *complete* invariant. There are inequivalent knots whose Jones polynomials coincide. However, all prime knots up to nine crossings have distinct Jones polynomials. More powerful invariants are known, including the *HOMFLY polynomial* and Khovanov homology,[18] to name just a couple.

Exercises

1. Suppose L is a link of two oriented components, and let $B(L)$ be the set of all crossings χ that involve strands from both components. Let $\sigma(\chi)$ be the sign of crossing χ. The absolute **linking number** of a two-component link L is defined by

$$\ell(L) = \frac{1}{2} \left| \sum_{\chi \in B(L)} \sigma(\chi) \right|.$$

   (Note that $\ell(L)$ is not necessarily equal to the writhe $w(L)$ since the former only depends on crossings involving both components of the link.)

   (a) Find the linking number of the Hopf link and Whitehead link from Figure 6.30 and the link shown in Figure 6.31. You must specify orientations on each component first. Try this exercise with different choices of orientations.

   (b) Prove that ℓ is a link invariant. Note that move R.I does not need to be checked – why?

   (c) Based only on linking number, which of the links from part (a) are not equivalent to the unlink of two components?

2. Determine whether the following knots are tricolorable or not. If tricolorable, then show a tricoloring.

   (a) Trefoil (left or right)   (c) Knot in Figure 6.29

   (b) Figure-eight knot   (d) $\mathbb{T}(3,4)$

3. Is the knot pictured in Example 114 tricolorable? Explain your reasoning. (*Hint:* The argument should take no more than one line.)

4. Show that tricolorability is preserved by Reidemeister move R.III. Carefully show each case and argue why you have exhausted all the cases.

5. Draw every possible case of an R.III move on an oriented knot diagram. Verify that each one leaves the writhe invariant.

6. Show that the writhe of a link depends on the orientations chosen for each component.

---

[18]The HOMFLY polynomial was named after its co-discoverers, Hoste, Ocneanu, Millett, Freyd, Lickorish, and Yetter [FYH+85]. Khovanov homology is an abstraction of the Jones polynomial into the language of *category theory*. Khovanov homology has connections to theoretical physics by way of *topological quantum field theories (TQFTs)* [Kho00].

7. Determine the writhe of each knot in Figure 6.22 and the figure-eight knot.

8. Find the Kauffman bracket $\langle K \rangle$ and Jones polynomial $J(K)$ for the figure-eight knot.

9. Given a link diagram L with n crossings, each of the 2^n ways of smoothing the crossings is called a **state** of L. Let $S(L)$ be the set of states of L. For each $s \in S(L)$, let s^+ be the number of positive smoothings and s^- the number of negative smoothings required to arrive at state s, and let $|s|$ be the number of components of s. Prove the *state sum* formula for the Kauffman bracket,

$$\langle L \rangle = \sum_{s \in S(L)} A^{s^+ - s^-} (-A^2 - A^{-2})^{|s|-1}.$$

10. Let $K = K[\mathbb{S}^1]$ be a knot embedded in $\mathbb{R}^3$. Prove that $\mathbb{R}^3 \setminus K$ is a 3-manifold[19] (recall Definition 5.1.5).

Supplemental Reading

- Adams [Ada01], *The Knot Book*. An excellent introduction to knot theory.

- Bondy and Murty [BM07].

- Goodman [Goo05], Chapter 4, for maps and graphs, and Chapter 7 for an introduction to knots.

- Harris, Hirst, and Mossinghoff [HHM08]. This text serves as a source text for elementary graph theory.

- Kauffman [Kau87] for the original paper introducing the bracket polynomial and Jones polynomial for knots.

- Manturov [Man04].

- Messer and Straffin [MS06].

- Prasolov [Pra95], Chapter 2.

---

[19]Then the properties of these manifolds can be studied in order to distinguish and classify knots.

Part III

Basic Algebraic Topology

Chapter 7

The Fundamental Group

In §1.1, we discussed two types of equivalence: *homeomorphism* and *homotopy*. Point-set topology is primarily concerned with the former, while *algebraic* topology is built upon the latter. It will be shown that if two spaces are homeomorphic, then they are also homotopic, but the converse is far from true. Indeed, as we'll see shortly, every Euclidean space $\mathbb{R}^n$ is homotopic to a single point. Homotopy is a much rougher measure of *sameness* than even homeomorphism. So what good is homotopy? It turns out that homotopy is easier to work with than homeomorphism in general. This means that homotopy can be used as a practical *topological invariant*, in the sense that if two spaces are *not* homotopic, then they cannot be homeomorphic. In this chapter, we develop the first step toward classifying spaces by their *homotopy type*, the so-called *fundamental group*. The reader who is unfamiliar with group theory should consult §B.1 or a standard text such as Armstrong [Arm88] or Dummit and Foote [DF04].

7.1 Algebra of Loops

In topology, if a difficult problem can be rephrased in terms of well-understood structures, then substantial progress can result. This idea underlies what we call *algebraic topology*. Algebra is the study of mathematical structure arising from operations defined on sets of elements. For example, the set of natural numbers $\mathbb{N}$ has a rich structure with respect to the operations of addition and multiplication.[1] In general, the point-sets of topological spaces do not possess algebraic structures themselves, but there are certain ways to induce a structure from some features of the space. Our first foray into algebraic topology explores the way in which the loops in a space can be regarded as elements in an algebraic structure called a *group*.

---

[1]Technically, $\mathbb{N}$ is a *monoid* with respect to both addition and multiplication – in fact, it is a *commutative semiring*. The integers $\mathbb{Z}$ have something called a *ring* structure with respect to addition and multiplication, while the real numbers $\mathbb{R}$ form an algebraic *field*.

Loops and Homotopy

Recall Definition 2.4.1 of an *arc*, which easily generalizes to any topological space. An **arc** γ in a topological space X is a continuous function, $\gamma : \mathbb{I} \to X$. The **basepoint** of the arc γ is the point $\gamma(0) \in X$. We are most interested in arcs that return to their basepoints; that is, $\gamma(0) = \gamma(1)$. Such an arc is called a *loop* in X and can be regarded as a continuous function $\gamma : \mathbb{I}/\sim \to X$, where $0 \sim 1$. The orientation (direction of travel) for the loop is in the direction of increasing $s \in \mathbb{I}$. Because of the homeomorphism $\mathbb{I}/\sim \approx \mathbb{S}^1$, a loop may be defined as follows.

> **Definition 7.1.1.** A **loop** γ in a topological space X is a continuous function, $\gamma : \mathbb{S}^1 \to X$. The **basepoint** of the loop is the point $x_0 = \gamma(0) \in X$.

Example 124.   Consider the unit circle defined by the **parametric equation**,

$$\mathbb{S}^1 = \left\{ \begin{pmatrix} \cos 2\pi s \\ \sin 2\pi s \end{pmatrix} \mid 0 \leq s \leq 1 \right\}.$$

In fact, this definition of $\mathbb{S}^1$ may be interpreted as a specific loop $\gamma_1 : \mathbb{S}^1 \to \mathbb{S}^1$ such that $\gamma(s) = (\cos 2\pi s, \sin 2\pi s)$. This loop traverses the circle once counterclockwise. A loop that traverses $\mathbb{S}^1$ twice counterclockwise could be defined by

$$\gamma_2 \quad : \quad \mathbb{S}^1 \to \mathbb{S}^1$$
$$\gamma_2(s) \quad = \quad (\cos 2\pi(2s), \sin 2\pi(2s)) ,\ 0 \leq s \leq 1.$$

Similarly, a loop traversing the circle clockwise could be defined by

$$\gamma_{-1} \quad : \quad \mathbb{S}^1 \to \mathbb{S}^1$$
$$\gamma_{-1}(s) \quad = \quad (\cos 2\pi(-s), \sin 2\pi(-s)) ,\ 0 \leq s \leq 1.$$

In this way, one could define loops on $\mathbb{S}^1$ that go around the circle in either direction any number of times,

$$\gamma_n \quad : \quad \mathbb{S}^1 \to \mathbb{S}^1$$
$$\gamma_n(s) \quad = \quad (\cos 2\pi(ns), \sin 2\pi(ns)) ,\ 0 \leq s \leq 1.$$

Note that when $n = 0$, we have $\gamma_0(s) = (\cos 2\pi(0s), \sin 2\pi(0s)) = (1, 0)$ for all $s \in \mathbb{I}$. That is, γ_0 is the **constant loop** at the basepoint $s_0 = (1, 0)$. In practice, the constant loop at a point x_0 is denoted $\mathbf{1}_{x_0}$. When the basepoint is understood, the constant loop may be denoted simply by $\mathbf{1}$.

We say that two loops γ and η in the same space X are **homotopic**, and write $\gamma \simeq \eta$, if the image of γ can be continuously deformed to the image of η. Think of this as a *video* that begins with $\gamma[\mathbb{S}^1]$ and proceeds through time showing how the loop changes, until by the end of the video we see $\eta[\mathbb{S}^1]$. It's important to realize that this morphing of one loop to another must always occur within the same space X. So, for example, if X has a hole, then it may be impossible to deform a loop surrounding that hole into a loop that avoids it. Also, the orientation of a loop, which is determined by direction of increasing t-value, is significant. If a loop encircles a hole in a counterclockwise direction, then that loop may not be homotopic to the same loop traversed in the clockwise direction (see Figure 7.1).

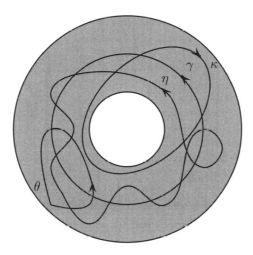

Figure 7.1: Loops in an annulus. Here, $\gamma \simeq \eta$, but $\gamma \not\simeq \theta$ and $\gamma \not\simeq \kappa$.

But how can we describe the *video* showing $\gamma \simeq \eta$ more precisely? What is a video, after all? It's a sequences of images, called *frames*, one for each tiny slice of time.[2] Now imagine that there is an image for every real-number time $t \in \mathbb{I}$. The most important requirement is that if two images are *close* in time (i.e., $t = a$ and $t = a + \epsilon$, where $\epsilon \approx 0$), then the images at those times must also be correspondingly *close*. This is the essence of *continuity*, and suggests how we should define homotopy for loops.

---

[2]In films there are typically at least 24 frames per second so that the illusion of continuous real time can be achieved; however, our movies have infinitely many frames in a finite time period.

Definition 7.1.2. Two loops $\gamma : \mathbb{S}^1 \to X$ and $\eta : \mathbb{S}^1 \to X$ are **homotopic** ($\gamma \simeq \eta$) if there is a continuous function (called a **homotopy**), $h : \mathbb{S}^1 \times \mathbb{I} \to X$, such that

- $h(s, 0) = \gamma(s)$, for all $s \in \mathbb{S}^1$, and

- $h(s, 1) = \eta(s)$, for all $s \in \mathbb{S}^1$.

If in addition the basepoints of γ and η are both $x_0 \in X$, and

- $h(0, t) = x_0$, for all $t \in \mathbb{I}$,

then we say the homotopy is **basepoint preserving**.

A loop homotopy h from γ to η is a continuous map of the cylinder $\mathbb{S}^1 \times \mathbb{I}$ to X, satisfying certain requirements. For each fixed $t \in \mathbb{I}$, $h(s, t)$ is simply a loop, notated by $h_t : \mathbb{S}^1 \to X$ (i.e., $h_t(s) = h(s, t)$). It may help to visualize h as a map taking the unit square into X in such a way that the bottom edge is mapped to the image of the loop $\gamma = h_0$, top edge is mapped to the image of $\eta = h_1$, and if the homotopy is basepoint preserving, then both the left and right edges of the square are mapped into the single point x_0. Each horizontal segment of the square, which corresponds to a constant t-value, is mapped to the loop h_t in X representing a particular frame of the video in which $\gamma[\mathbb{S}^1]$ morphs to $\eta[\mathbb{S}^1]$ (see Figure 7.2).

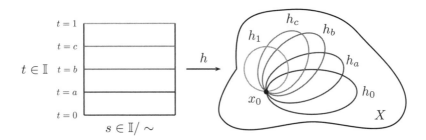

Figure 7.2: A basepoint-preserving homotopy from $\gamma = h_0$ to $\eta = h_1$ in a space X.

Definition 7.1.2 provides a way to consider when two loops are essentially doing the same thing. Since loops are defined by maps $\gamma : \mathbb{S}^1 \to X$, a homotopy is really a way to transform one map into another in a continuous way. This idea can be generalized to any kinds of maps – not just loops.

Definition 7.1.3. Two maps f and g from Z to X are **homotopic** ($f \simeq g$) if there is a continuous function (called a **homotopy**), $h : Z \times \mathbb{I} \to X$, such that

- $h(z, 0) = f(z)$, for all $z \in Z$, and

- $h(z, 1) = g(z)$, for all $z \in Z$.

If in addition there is a subset $Z_0 \subseteq Z$ such that $h(z, t) = f(z) = g(z)$ for all $z \in Z_0$, and $t \in \mathbb{I}$, then h is called a homotopy **relative to** Z_0, or h is said to **preserve** Z_0.

A homotopy $h : Z \times \mathbb{I} \to X$ is like a video showing how the image $f[Z]$ morphs into the image $g[Z]$ within the space X. The case $Z = \mathbb{S}^1$ gives loop homotopy, while the case $Z = \mathbb{I}$ defines *arc* (or *path*) homotopy. Figure 7.3 illustrates an arc homotopy relative to the endpoints of the arcs.

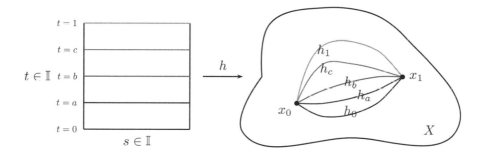

Figure 7.3: An endpoint-preserving homotopy between two arcs h_0 and h_1.

If a map $\gamma : (Z, z_0) \to (X, x_0)$ is homotopic to the constant map (sending all points of Z to the basepoint x_0), then we say γ is **nullhomotopic**. Each Euclidean space $\mathbb{R}^n$, for $n \geq 0$, has the property that *every* map $\gamma : Z \to \mathbb{R}^n$ is nullhomotopic (regardless of the domain space Z).

Definition 7.1.4. A space X is called **contractible** if the identity map of X is nullhomotopic. That is, $\mathrm{id}_X \simeq \mathbf{1}$.

Thus $\mathbb{R}^n$ is contractible. As we shall see, the sphere $\mathbb{S}^n$ is not.

Homotopy Classes

For a given space X, we may consider the set $\mathrm{Loop}(X)$ of all loops in X. Unless X consists of only a finite set of points, $\mathrm{Loop}(X)$ is an incredibly large set – too large to be useful. However, $\mathrm{Loop}(X)$ can be partitioned into useful chunks called **homotopy classes**. First we need to show that homotopy equivalence is an *equivalence relation*. Recall from Definition A.2.2 that an equivalence relation must satisfy *reflexivity* ($\gamma \simeq \gamma$), *symmetry* ($\gamma \simeq \eta \implies \eta \simeq \gamma$), and *associativity* ($\gamma \simeq \eta$ and $\eta \simeq \theta \implies \gamma \simeq \theta$). You will prove these precisely in Exercise 3, but let's gain some intuitive understanding first.

- The statement $\gamma \simeq \gamma$ means that the image $\gamma[\mathbb{S}^1]$ should always be deformable to itself. In terms of a video, this could be accomplished by having every frame be the same image of the loop.

- Given that $\gamma \simeq \eta$, there is a video that begins with the image of γ and ends with the image of η. Simply play the video backward to verify that $\eta \simeq \gamma$.

- Given that $\gamma \simeq \eta$ and $\eta \simeq \theta$, this means there are two videos, Video A showing how γ morphs into η, and Video B showing how η morphs into θ. By simply splicing the two videos together, Video A+B shows how γ transforms into θ. We can even make Video A+B have the same running time as A or B by playing the spliced video at twice the speed (this point becomes important when using the formal definitions of loops as functions $\mathbb{S}^1 = \mathbb{I}/\sim \to X$).

Definition 7.1.5. The **homotopy class** of a loop $\gamma : \mathbb{S}^1 \to X$ is the set of all loops homotopic to γ,

$$[\gamma] = \{\eta : \mathbb{S}^1 \to X \mid \gamma \simeq \eta\}.$$

Example 125. Let $n \in \mathbb{N}$. Prove that in $\mathbb{R}^n$, there is only one homotopy class.

Solution: Let $[\mathbf{0}]$ be the equivalence class of the constant loop at the origin of $\mathbb{R}^n$, and let $\gamma : \mathbb{S}^1 \to \mathbb{R}^n$ be an aribtrary loop. We will show that $\gamma \in [\mathbf{0}]$. Define a map h (candidate for our homotopy from γ to $\mathbf{0}$) by:

$$
\begin{aligned}
h &: \quad \mathbb{S}^1 \times \mathbb{I} \to \mathbb{R}^n \\
h(s,t) &= (1-t)\gamma(s).
\end{aligned}
$$

First note that h is well defined since $\gamma(s) \in \mathbb{R}^n$ for all $s \in \mathbb{S}^1$, and $(1-t)\gamma(s)$ simply scales the vector $\gamma(s)$ by a real number. The function h is continuous since γ is continuous and γ is being scaled by a continous function, $1 - t$.

- $h(s,0) = (1-0)\gamma(s) = \gamma(s)$, $\forall s \in \mathbb{S}^1$

- $h(s,1) = (1-1)\gamma(s) = \mathbf{0}$, $\forall s \in \mathbb{S}^1$

Thus a homotopy exists, proving $\gamma \simeq \mathbf{0}$ and hence $\gamma \in [\mathbf{0}]$.

This type of homotopy is called a *straight-line homotopy* because for any fixed s, $h(s,t)$ follows a straight line as t ranges from 0 to 1. Each loop h_t is a scaled copy of γ.

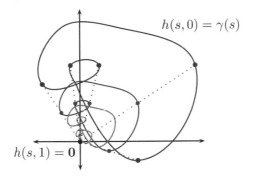

$$h(s,0) = \gamma(s)$$

$$h(s,1) = \mathbf{0}$$

Operations on Loops

Now let's suppose a basepoint $x_0 \in X$ has been fixed. Consider the set of homotopy classes of loops in X based at x_0. Two loops $\gamma, \eta \in \text{Loop}(X, x_0)$ may be **composed**, forming a new loop, called the **product** loop $\gamma \cdot \eta$. Intuitively, we may think of a loop as a journey from x_0 back to x_0 in finite time. When two loops γ and η are composed, this means that first we venture through X via the loop γ, and then after returning to the starting point, set back out on another journey via η. But if every loop must always take the same amount of time (being parametrized by the same space $\mathbb{S}^1$), then we simply require the speed of the journey to be twice as great in order to traverse both γ and η in the required time. This leads to a formal definition for the product of two loops.

> **Definition 7.1.6.** Let $\gamma, \eta \in \text{Loop}(X, x_0)$. The **product loop** is a loop $\gamma \cdot \eta : \mathbb{S}^1 \to X$ defined as follows:
>
> $$(\gamma \cdot \eta)(s) = \begin{cases} \gamma(2s), & s \in [0, 1/2]; \\ \eta(2s - 1), & s \in (1/2, 1]. \end{cases}$$

First note that the product loop is always well defined. To verify this, we must show that $\gamma \cdot \eta$ is a *continuous* map $\mathbb{S}^1 \to X$. Both γ and η are continuous; the key is what happens at $s = 1/2$.

$$\lim_{s \to \frac{1}{2}^-} (\gamma \cdot \eta)(s) = \lim_{s \to \frac{1}{2}^-} \gamma(2s) = \gamma(1) = x_0$$

$$\lim_{s \to \frac{1}{2}^+} (\gamma \cdot \eta)(s) = \lim_{s \to \frac{1}{2}^+} \eta(2s - 1) = \eta(0) = x_0$$

Thus $\lim_{s \to 1/2}(\gamma . \eta)(s) = x_0$, which agrees with the value $(\gamma . \eta)(1/2) = \gamma(1) = x_0$, proving continuity of $\gamma . \eta$. Moreover, $\gamma . \eta$ is a loop based at x_0 since $(\gamma . \eta)(0) = \gamma(0) = x_0$ and $(\gamma . \eta)(1) = \eta(1) = x_0$.

Consider the constant loop at x_0, which we denote by $\mathbf{1}$:

$$\mathbf{1}(s) = x_0, \quad \forall s \in \mathbb{S}^1.$$

Suppose $\gamma \in \mathrm{Loop}(X, x_0)$. What is $\gamma . \mathbf{1}$? According to Definition 7.1.6, $\gamma . \mathbf{1}$ is the loop that traverses $\gamma[\mathbb{S}^1]$ *twice* as fast, then stops and remains at the base-point for the second half of the time interval. It seems reasonable to guess that $\gamma . \mathbf{1} \simeq \gamma$, but in order to prove this claim, we must find a specific homotopy.

The analogy of homotopy as a "video" seems to break down in this example, because the image $(\gamma . \mathbf{1})[\mathbb{S}^1]$ looks just like the image $\gamma[\mathbb{S}^1]$. This is because we see the images are static. But, in fact, it's better to think of each frame as a little video in and of itself. There are really *two* independent time dimensions, if you will – one parametrized by $t \in \mathbb{I}$ and the other by $s \in \mathbb{S}^1 \approx \mathbb{I}/\sim$. So if we start the video and immediately hit the *pause* button, fixing $t = 0$ constant, then we could let s run from 0 to 1 to see how a point would progress along the loop γ. At any point $t \in \mathbb{I}$, if we hit pause, the frame shows another loop in which we could let s run from 0 to 1 to track how a point progresses along the loop, as Figure 7.4 illustrates.

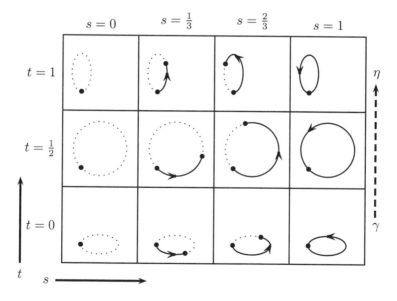

Figure 7.4: A homotopy $h(s, t)$ visualized as an array of frames with respect to two parameters; $t \in \mathbb{I}$ parametrizes the sequence of loops transitioning from γ to η (*bottom to top*); and when t is fixed, $s \in \mathbb{S}^1$ parametrizes the motion of a point along a particular loop in the sequence (*left to right*).

Thinking of a possible video that could illustrate $\gamma \cdot \mathbf{1} \simeq \gamma$, all of the frames would seem to show the static image of $\gamma[\mathbb{S}^1]$; however, at any fixed time $t \in \mathbb{I}$, the motion of the point along the loop will be different. When $t = 0$, we need the point to travel along γ in double-time then rest at x_0 (this is $\gamma \cdot \mathbf{1}$). By the end of the video ($t = 1$), the motion of the point is normal speed along γ. Therefore we expect that at intermediate times $0 < t < 1$, the loop $h_t : \mathbb{S}^1 \to X$ should be defined in such a way that a point travels along γ at some speed between normal and twice as fast, and then rests at x_0 when it is done traversing γ. How can we devise a homotopy h that accomplishes this task? Let's begin by drawing a square as in Figure 7.2. The bottom represents $\gamma \cdot \mathbf{1}$, while the top represents γ. From bottom to top (in order of increasing t), we need to "get rid" of the $\mathbf{1}$ contribution, and the easiest way to do this is by drawing a straight line as shown in Figure 7.5.

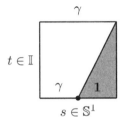

Figure 7.5: Homotopy $\gamma \cdot \mathbf{1} \simeq \gamma$.

Now to build the function $h(s, t)$. Let's assume that at any fixed time t, there are parameters m and k such that:

$$h(s, t) = h_t(s) = \begin{cases} \gamma(ms), & s \in [0, k]; \\ x_0, & s \in (k, 1]. \end{cases}$$

Assuming that k varies linearly in t (based on our choice to draw a line, rather than a parabola or circular arc, e.g., in Figure 7.5), we find an expression for k in terms of t using simple algebra. Since k must start at $k = 1/2$ when $t = 0$ and end at $k = 1$ when $t = 1$, we have $k = \frac{t+1}{2}$. Then we use this information to determine m. Assuming t is fixed, the loop h_t should return to $\gamma(1)$ when $s = \frac{t+1}{2}$; that is, $ms = 1$. This implies that $m = \frac{2}{t+1}$ and leads to our proposed homotopy, which is defined for all $s \in \mathbb{S}^1$ and $t \in \mathbb{I}$.

$$h(s, t) = \begin{cases} \gamma\left(\frac{2s}{t+1}\right), & s \in \left[0, \frac{t+1}{2}\right] \\ x_0, & s \in \left(\frac{t+1}{2}, 1\right] \end{cases} \tag{7.1}$$

We still must verify that (7.1) satisfies the conditions required to be a homotopy from $\gamma \cdot \mathbf{1}$ to γ.

- First verify that $h(s,0) = [\gamma \cdot \mathbf{1}](s)$, $\forall s \in \mathbb{S}^1$. When $t = 0$, (7.1) reduces to

$$h(s,0) = \begin{cases} \gamma(2s), & s \in \left[0, \frac{1}{2}\right]; \\ x_0, & s \in \left(\frac{1}{2}, 1\right]. \end{cases} \tag{7.2}$$

Since $\mathbf{1}(s) = x_0$ for all s, we see that (7.2) is exactly $\gamma \cdot \mathbf{1}$.

- Next verify that $h(s,1) = \gamma(s)$, $\forall s \in \mathbb{S}^1$. Simply observe what happens to (7.1) when $t = 1$ is substituted: $h(s,1) = \gamma\left(\frac{2s}{1+1}\right) = \gamma(s)$ for all $s \in \left[0, \frac{1+1}{2}\right] = [0,1]$.

- Verify the continuity of h. Since γ and $\mathbf{1}$ are both continuous, it is sufficient to show that the limits agree as s approaches $c = \frac{t+1}{2}$ from the left and right.

$$\lim_{s \to c^-} h(s,t) = \lim_{s \to c^-} \gamma\left(\frac{2s}{t+1}\right) = \gamma\left(\frac{2\left(\frac{t+1}{2}\right)}{t+1}\right) = \gamma(1) = x_0 \tag{7.3}$$

$$\lim_{s \to c^+} h(s,t) = \lim_{s \to c^+} x_0 = x_0 \tag{7.4}$$

Equations (7.3–7.4) prove continuity of h.

In fact, the homotopy h is even *basepoint preserving*. Let's verify this now.

$$h(0,t) = \begin{cases} \gamma(0), & s \in \left[0, \frac{t+1}{2}\right] \\ x_0, & s \in \left(\frac{t+1}{2}, 1\right] \end{cases} \tag{7.5}$$

Now since $\gamma(0) = x_0$, we see that (7.5) is the constant function $\mathbf{1}(s) = x_0$.

Through a similar though much more tedious computation, we can prove that $(\gamma \cdot \eta) \cdot \theta \simeq \gamma \cdot (\eta \cdot \theta)$ for arbitrary $\gamma, \eta, \theta \in \mathrm{Loop}(X, x_0)$. Note that in the product $(\gamma \cdot \eta) \cdot \theta$, loops γ and η are traversed in a quarter of the original time, while θ is traversed in half the time; in the product $\gamma \cdot (\eta \cdot \theta)$, γ is traversed in half the original time, while both η and θ are traversed in a quarter of the time. The corresponding picture and homotopy are shown in Figure 7.6. We have yet to show that h is a valid homotopy, but this is straightforward (see Exercise 5).

Now consider traversing a loop defined by γ, but in the opposite sense. The result is called the *inverse* loop for γ.

Definition 7.1.7. Let $\gamma \in \mathrm{Loop}(X, x_0)$. The **inverse** of γ is a loop $\gamma^{-1} : \mathbb{S}^1 \to X$ defined by $(\gamma^{-1})(s) = \gamma(1 - s)$.

In this text, we have already encountered a notation like this, and it is very important not to confuse the notations (especially since they look exactly the same). For example, when talking about functions, we use f^{-1} to denote an *inverse image* or *inverse function* for f. Neither of these two interpretations is

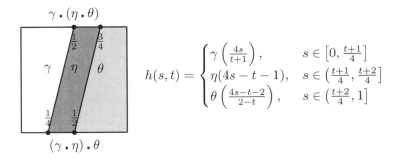

$$h(s,t) = \begin{cases} \gamma\left(\frac{4s}{t+1}\right), & s \in \left[0, \frac{t+1}{4}\right] \\ \eta(4s - t - 1), & s \in \left(\frac{t+1}{4}, \frac{t+2}{4}\right] \\ \theta\left(\frac{4s-t-2}{2-t}\right), & s \in \left(\frac{t+2}{4}, 1\right] \end{cases}$$

Figure 7.6: Here $(\gamma \cdot \eta) \cdot \theta \simeq \gamma \cdot (\eta \cdot \theta)$ via the homotopy h shown above.

valid for γ^{-1}. Here the inverse loop γ^{-1} is *not* a function from X back to $\mathbb{S}^1$. Instead, γ^{-1}, just like γ, is also a function $\mathbb{S}^1 \to X$, so when we form the *loop product* $\gamma \cdot \gamma^{-1}$ or $\gamma^{-1} \cdot \gamma$, we do not mean *function composition* as in $\gamma \circ \gamma^{-1}$ or $\gamma \circ \gamma^{-1}$.

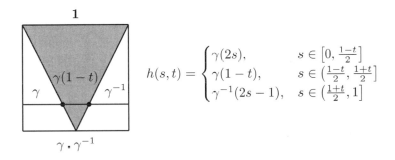

$$h(s,t) = \begin{cases} \gamma(2s), & s \in \left[0, \frac{1-t}{2}\right] \\ \gamma(1-t), & s \in \left(\frac{1-t}{2}, \frac{1+t}{2}\right] \\ \gamma^{-1}(2s-1), & s \in \left(\frac{1+t}{2}, 1\right] \end{cases}$$

Figure 7.7: Here $\gamma \cdot \gamma^{-1} \simeq \mathbf{1}$ via the homotopy h shown above.

Now it should come as no surprise that $\gamma \cdot \gamma^{-1} \simeq \mathbf{1}$ and $\gamma^{-1} \cdot \gamma \simeq \mathbf{1}$, but as always, these kinds of statements require proof. We will prove the first one and leave the second one to the reader. The idea for our homotopy is simple: How does one transition from a journey along γ and back again to a "journey" that goes nowhere? By only taking the journey along γ partway. For example, $h_{1/2}$ should be the loop defined by traversing γ only halfway, then turning around and returning home, thereby traversing the last half of γ^{-1} as well. Note that in order to take each round-trip in the same time period, the traveler could simply wait around a bit in the middle. For example, when $t = 1/2$, traverse the first half the loop γ, wait at the point $\gamma(1/2)$ for some time, and then traverse the last half of the loop γ^{-1} (which is just the first half of γ backward). By the time the last frame of the video is reached ($t = 1$), the point travels *no* distance along γ and simply waits at $\gamma(0) = x_0$ for the entire time period. See Figure 7.7 for the explicit homotopy and associated picture. You should verify on your own that h is indeed a valid basepoint-preserving homotopy, proving $\gamma \cdot \gamma^{-1} \simeq 1$ (see Excercise 6).

Exercises

1. Modify the argument from Example 125 to show that $\mathbb{R}^n$ is contractible. (*Hint:* Consider an arbitrary map $Z \to \mathbb{R}^n$.)

2. Let $r \geq 1$. Prove that $\mathbb{D}^r$ is contractible.

3. Using Definition 7.1.2, prove that homotopy is an equivalence relation on the set $\mathrm{Loop}(X)$. Then show that basepoint-preserving homotopy is an equivalence relation on the set $\mathrm{Loop}(X, x_0)$.

4. Prove that $1 \cdot \gamma \simeq \gamma$, where $1 \in \mathrm{Loop}(X, x_0)$ is the identity loop and $\gamma \in \mathrm{Loop}(X, x_0)$ is arbitrary.

5. Verify that the function h in Figure 7.6 is a basepoint-preserving homotopy from $(\gamma \cdot \eta) \cdot \theta$ to $\gamma \cdot (\eta \cdot \theta)$ by following the steps outlined below.

   (a) Show that $h(s, 0) = [(\gamma \cdot \eta) \cdot \theta](s)$, $\forall s \in \mathbb{S}^1$.

   (b) Show that $h(s, 1) = [\gamma \cdot (\eta \cdot \theta)](s)$, $\forall s \in \mathbb{S}^1$.

   (c) Verify the continuity of h. Since each component function γ, η, and θ are assumed continuous, you just need to show that the limits agree as s approaches from the left and right at each point $s = \frac{t+1}{4}$ and $s = \frac{t+2}{4}$.

   (d) Show that $h(0, t) = x_0, \forall t \in \mathbb{I}$.

6. Verify that the function h in Figure 7.7 is a basepoint-preserving homotopy from $\gamma \cdot \gamma^{-1}$ to the constant map 1 (follow similar steps as those in Exercise 5).

7. The product defined in Definition 7.1.6 may be extended to arcs, so long as the endpoint of the first arc coincides with the initial point of the second arc. That is, if $\alpha : \mathbb{I} \to X$ and $\beta : \mathbb{I} \to X$ such that $\alpha(1) = \beta(0)$, then the product

$$(\alpha \cdot \beta)(s) = \begin{cases} \alpha(2s), & s \in [0, 1/2] \\ \beta(2s - 1), & s \in (1/2, 1] \end{cases} \tag{7.6}$$

is an arc in X from $\alpha(0)$ to $\beta(1)$. We say that arcs α and β are homotopic if $\alpha(0) = \beta(0)$, $\alpha(1) = \beta(1)$, and there is an endpoint-preserving homotopy from α to β (recall Figure 7.3).

   (a) Show that the arc product is associative up to homotopy. That is, if α, β, γ are three arcs such that $\alpha(1) = \beta(0)$ and $\beta(1) = \gamma(0)$, then $(\alpha \cdot \beta) \cdot \gamma \simeq \alpha \cdot (\beta \cdot \gamma)$.

(b) Give a precise description of an **inverse** arc for a given arc $\alpha : \mathbb{I} \to X$. Note that an inverse arc α^{-1} in this context should satisfy: $\alpha \cdot \alpha^{-1} \simeq \mathbf{1}_{\alpha(0)}$ and $\alpha^{-1} \cdot \alpha \simeq \mathbf{1}_{\alpha(1)}$.

7.2 Fundamental Group as Topological Invariant

In §7.1 we have shown that the set $\text{Loop}(X, x_0)$ has a well-defined product, a special element called $\mathbf{1}$ so that $\gamma \cdot \mathbf{1}$ is homotopic to γ, a type of associativity property for the product in which the two ways of associating three loops in a product may not yield *equal* loops but give *homotopic* loops, and the existence of loops that act as inverses, at least up to homotopy. The operations seem to indicate that individual loops are not as important as their *homotopy classes*. In this section we define the *fundamental group*, which is a group structure on the set of homotopy classes of loops in a space X. The fundamental group turns out to be a particularly powerful topological invariant.

The Fundamental Group

We now introduce notation for the set of all homotopy classes of based loops in a space.

> **Definition 7.2.1.** Let X be a space and $x_0 \in X$ a fixed basepoint. The **fundamental group** of X, based at x_0, is defined as a set by
>
> $$\pi_1(X, x_0) = \text{Loop}(X, x_0)/ \simeq,$$
>
> where $\simeq$ is homotopy equivalence. In other words, each element $[\gamma] \in \pi_1(X, x_0)$ is the homotopy class of the loop γ.

Note that Definition 7.2.1 only defines $\pi_1(X, x_0)$ as a *set*. The following theorem specifies a product structure that makes this set into a group. The key is that the loop product given in Definition 7.1.6 remains well defined when applied to homotopy classes of loops, and the *induced* product satisfies the three axioms required for a group structure (see §B.1 for details).

> **Theorem 7.2.2.** *The set $\pi_1(X, x_0)$ is a group whose product is induced by the product of loops in $\text{Loop}(X, x_0)$.*
>
> $$[\gamma] \cdot [\eta] = [\gamma \cdot \eta]$$

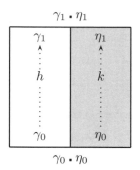

Figure 7.8: Homotopies h and k may be combined to produce a piecewise homotopy H, showing $\gamma_0 \cdot \eta_0 \simeq \gamma_1 \cdot \eta_1$.

Proof. First we must verify that the product is well defined on equivalence classes. Let $\gamma_0, \gamma_1 \in [\gamma]$, and $\eta_0, \eta_1 \in [\eta]$. That is, $\gamma_0 \simeq \gamma_1$, and $\eta_0 \simeq \eta_1$. Suppose $h : \mathbb{S}^1 \times \mathbb{I} \to X$ is a homotopy from γ_0 to γ_1, and $k : \mathbb{S}^1 \times \mathbb{I} \to X$ is a homotopy from η_0 to η_1. Define a new homotopy $H : \mathbb{S}^1 \times \mathbb{I} \to X$ by

$$H(s,t) = \begin{cases} h(2s,t), & s \in [0,1/2]; \\ k(2s-1,t), & s \in (1/2,1]. \end{cases}$$

Note that $H(s,0) = \gamma_0 \cdot \eta_0$ and $H(s,1) = \gamma_1 \cdot \eta_1$. It is easy to verify that H is a continuous (indeed, $H(1/2,t) = h(1,t) = k(0,t) = x_0$ is constant), basepoint-preserving map, a homotopy from $\gamma_0 \cdot \eta_0$ to $\gamma_1 \cdot \eta_1$ (see Figure 7.8). This shows that $[\gamma_0 \cdot \eta_0] = [\gamma_1 \cdot \eta_1]$; in other words, $[\gamma] \cdot [\eta] = [\gamma \cdot \eta]$ is a well-defined product on $\pi_1(X, x_0)$.

The three group axioms follow from our work in §7.1:

I. *(Associativity)* For all $\gamma, \eta, \theta \in \mathrm{Loop}(X, x_0)$, $(\gamma \cdot \eta) \cdot \theta \simeq \gamma \cdot (\eta \cdot \theta)$. Therefore $([\gamma] \cdot [\theta]) \cdot [\theta] = [\gamma] \cdot ([\eta] \cdot [\theta])$ for all $[\gamma], [\eta], [\theta] \in \pi_1(X, x_0)$.

II. *(Existence of identity)* Let $\mathbf{1}$ be the constant loop at x_0. Then $[\mathbf{1}] \in \pi_1(X, x_0)$ is the identity element, since for all $\gamma \in \mathrm{Loop}(X, x_0)$, $\gamma \cdot \mathbf{1} \simeq \mathbf{1} \cdot \gamma \simeq \gamma$ implies $[\gamma] \cdot [\mathbf{1}] = [\mathbf{1}] \cdot [\gamma] = [\gamma]$ for all $[\gamma] \in \pi_1(X, x_0)$.

III. *(Existence of inverses)* For any $[\gamma] \in \pi_1(X, x_0)$, the homotopy class $[\gamma^{-1}]$ is an inverse. That is, $[\gamma]^{-1} = [\gamma^{-1}]$. This follows from the fact that $\gamma \cdot \gamma^{-1} \simeq \mathbf{1} \simeq \gamma^{-1} \cdot \gamma$ for all $\gamma \in \mathrm{Loop}(X, x_0)$; hence $[\gamma] \cdot [\gamma^{-1}] = [\mathbf{1}] = [\gamma^{-1}] \cdot [\gamma]$ for all $[\gamma] \in \pi_1(X, x_0)$.

$\square$

In general the product of homotopy classes is *noncommutative*; in other words, typically we have $[\gamma] \cdot [\eta] \neq [\eta] \cdot [\gamma]$. This should be expected since the order in which loops are traversed actually matters, and more to the point, we cannot generally produce a homotopy from $\gamma \cdot \eta$ to $\eta \cdot \gamma$. Consequently, the group $\pi_1(X, x_0)$ is usually not *abelian*.[3]

We note without proof that the product structure defined in Theorem 7.2.2 is also well defined with respect to *arc products* (as defined in §7.1, Exercise 7). However, this time the product $\alpha \cdot \beta$ only makes sense when $\alpha(1) = \beta(0)$. When the arcs are compatible, then the product of homotopy classes $[\alpha] \cdot [\beta] = [\alpha \cdot \beta]$ satisfies associativity. Moreover, for any loop α, there is a kind of "inverse" $[\alpha]^{-1} = [\alpha^{-1}]$, but there is no *unique* identity.[4]

At this point, we have defined $\pi_1(X, x_0)$ for a topological space X with chosen basepoint x_0. What happens if we decided to use a different basepoint $x_1 \in X$? So long as X is arc-connected, the choice of basepoint is immaterial in the sense that the corresponding fundamental groups are isomorphic.

Theorem 7.2.3. *If X is arc-connected, then $\pi_1(X, x_0) \cong \pi_1(X, x_1)$ for any two basepoints $x_0, x_1 \in X$.*

The idea behind the proof is to choose an arc connecting x_0 and x_1, and use that arc to set up a bijection between loops based at x_0 with those based at x_1. See Figure 7.9.

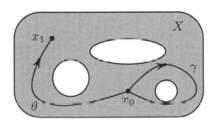

Figure 7.9: The loop γ based at x_0 can be extended to a loop $\theta^{-1} \cdot \gamma \cdot \theta$ based at x_1 which first traverses the arc θ^{-1} from x_1 to x_0, then traverses γ, and finally θ to arrive back at x_1.

Proof. Let $\theta : \mathbb{I} \to X$ be an arc in X such that $\theta(0) = x_0$ and $\theta(1) = x_1$. Define a function on fundamental groups:

$$
\begin{aligned}
f &: \quad \pi_1(X, x_0) \to \pi_1(X, x_1) \\
f([\gamma]) &= \quad [\theta^{-1} \cdot \gamma \cdot \theta].
\end{aligned}
$$

In other words, $f([\gamma])$ is represented by a loop beginning at x_1, traversing θ (in the reverse sense) to get to x_0, then going around the loop γ, and finally

---

[3]Recall that an **abelian** group is a group whose product is commutative.

[4]This kind of structure is called a **groupoid**. For more about how groupoids may be useful in homotopy theory, see Ronald Brown's *Topology and Groupoids* [Bro06].

returning to x_1 via θ. First verify that f is a group homomorphism:

$$
\begin{aligned}
f([\gamma]) \cdot f([\eta]) &= [\theta^{-1} \cdot \gamma \cdot \theta] \cdot [\theta^{-1} \cdot \eta \cdot \theta] \\
&= [\theta]^{-1} \cdot [\gamma] \cdot [\theta] \cdot [\theta]^{-1} \cdot [\eta] \cdot [\theta] \\
&= [\theta^{-1}] \cdot [\gamma] \cdot [\eta] \cdot [\theta] \\
&= [\theta^{-1} \cdot (\gamma \cdot \eta) \cdot \theta] \\
&= f([\gamma \cdot \eta]).
\end{aligned}
$$

The fact that f is an isomorphism follows from the existence of an inverse homomorphism,

$$
\begin{aligned}
g &: \quad \pi_1(X, x_1) \to \pi_1(X, x_0) \\
g([\gamma]) &= [\theta \cdot \gamma \cdot \theta^{-1}].
\end{aligned}
$$

You will verify in Exercise 1 that g is indeed the inverse for f. $\square$

Theorem 7.2.3 shows that the choice of basepoint does not matter when X is arc-connected. In this chapter, we only work with arc-connected spaces, and so it makes sense to use the less cumbersome notation $\pi_1(X)$ for the fundamental group of X. However, there is no *natural* way to make the choice of connecting arc between different basepoints. Even though $\pi_1(X, x_0) \cong \pi_1(X, x_1)$, there is no canonical choice for the isomorphism $f : \pi_1(X, x_0) \to \pi_1(X, x_1)$. While this may seem like a minor annoyance, it can actually cause serious trouble, and so in this text we assume that whenever a space X is considered, we have in mind a specific *fixed* basepoint $x_0 \in X$.

Induced Homomorphisms

Now let's get back to the fundamental group. Suppose X and Y are topological spaces and there is a continuous map $f : X \to Y$. Let $\gamma \in \text{Loop}(X)$; so $\gamma : \mathbb{S}^1 \to X$. Simply by composing with f, we obtain a continuous function $f \circ \gamma : \mathbb{S}^1 \to Y$; in other words, $f \circ \gamma \in \text{Loop}(Y)$. The situation may be expressed succinctly using the following diagram.

$$
\begin{array}{ccc}
\mathbb{S}^1 & \xrightarrow{\ \gamma\ } & X \\
 & {\scriptstyle f \circ \gamma}\searrow & \downarrow {\scriptstyle f} \\
 & & Y
\end{array}
\tag{7.7}
$$

Diagram (7.7) is an example of a **commutative diagram**. In any given diagram, the **objects** (in this case, $\mathbb{S}^1$, X, and Y) can be sets, groups, spaces, etc., and the arrows, called **maps** or **morphisms**, indicate specific relationships among the objects, often by way of functions, homomorphisms, continuous maps, etc. What makes the diagram *commutative* is that if there is more than one way to get from one object to another, then the maps or compositions of maps must be equal.[5] Diagram (7.7) simply expresses the fact that the composition of f and γ is a map from $\mathbb{S}^1$ into Y. If, in addition, f takes the basepoint $x_0 \in X$ to the basepoint $y_0 \in Y$, then composition with f defines a function,

$$f_* \; : \; \operatorname{Loop}(X, x_0) \to \operatorname{Loop}(Y, y_0)$$
$$f_*(\gamma) \; = \; f \circ \gamma.$$

Now suppose $\gamma_0, \gamma_1 \in [\gamma]$, where $[\gamma] \in \pi_1(X)$ (with basepoint $x_0 \in X$ understood). That is, γ_0 and γ_1 are homotopic in X via some homotopy $h : \mathbb{S}^1 \times \mathbb{I} \to X$. Will $f \circ \gamma_0$ and $f \circ \gamma_1$ also be homotopic in Y? Consider the composition $f \circ h$.

$$\begin{array}{ccc} \mathbb{S}^1 \times \mathbb{I} & \xrightarrow{\ h\ } & X \\ & {\scriptstyle f \circ h} \searrow & \downarrow {\scriptstyle f} \\ & & Y \end{array} \tag{7.8}$$

Since f and h are continuous, so is their composition, and that implies that $f \circ h$ is a homotopy between some pair of loops, $(f \circ h)(s, 0)$ and $(f \circ h)(s, 1)$. But

$$(f \circ h)(s, 0) = f(h(s, 0)) = f(\gamma_0(s)) = (f \circ \gamma_0)(s), \text{ and}$$

$$(f \circ h)(s, 1) = f(h(s, 1)) = f(\gamma_1(s)) = (f \circ \gamma_1)(s).$$

Thus $f \circ \gamma_0 \simeq f \circ \gamma_1$ via the homotopy $f \circ h$. In particular, this proves that every loop in the homotopy class $[\gamma] \in \pi_1(X)$ gets mapped via composition with f to a member of the homotopy class $[f \circ \gamma] \in \pi_1(X)$. We now have a well-defined function from one fundamental group to another *induced* by a given continuous function between the spaces. In fact, we've hit the jackpot: this function turns out to be a *group homomorphism*. In what follows, the notation $f : (X, x_0) \to (Y, y_0)$ means that $f : X \to Y$ is a continuous map of topological spaces such that $f(x_0) = y_0$.

Definition 7.2.4. Suppose $f : (X, x_0) \to (Y, y_0)$. The **induced homomorphism** $f_* : \pi_1(X) \to \pi_1(Y)$ is defined for all $[\gamma] \in \pi_1(X)$ by:

$$f_*([\gamma]) = [f \circ \gamma] \in \pi_1(Y).$$

Exercise 2 asks you to prove that f_* is indeed a group homomorphism. The following properties are quite useful.

---

[5]Commutative diagrams are the bread and butter of *category theory*. The language of category theory is incredibly useful in algebraic topology and related areas of mathematics, even though when introduced in the first half of the twentieth century, some mathematicians referred to it as the theory of *general abstract nonsense* [EML45, ML71].

Theorem 7.2.5.

- *Suppose $f : (X, x_0) \to (Y, y_0)$, and $g : (Y, y_0) \to (Z, z_0)$. Then $(g \circ f)_* = g_* \circ f_*$.*

- *If $\mathrm{id}_X : (X, x_0) \to (X, x_0)$ is the identity map, then $(\mathrm{id}_X)_* = \mathrm{id}_{\pi_1(X)}$.*

In words, Theorem 7.2.5 states that the induced map of a composition is the composition of the individual induced homomorphisms, and the induced map of an identity map is the identity homomorphism.[6] What properties does f_* have if $f : X \to Y$ is a homeomorphism (i.e., bijective and bicontinuous)? There would be a continuous map $g : Y \to X$ such that $g \circ f$ and $f \circ g$ are identities on X and Y, respectively. After inducing to maps of fundamental groups, the resulting homomorphisms must also be inverse to one another (see also Figure 7.10).

$$g_* \circ f_* = (g \circ f)_* = (\mathrm{id}_X)_* = \mathrm{id}_{\pi_1(X)}$$
$$f_* \circ g_* = (f \circ g)_* = (\mathrm{id}_Y)_* = \mathrm{id}_{\pi_1(Y)}$$

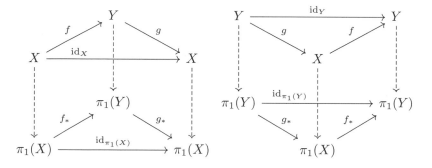

Figure 7.10: The diagrams shown should be read from the top down. Dashed arrows indicate the operation of sending each space to its fundamental group. Each of the four triangles is a separate commutative diagram. The top left triangle indicates $g \circ f = \mathrm{id}_X$, which gets induced to the bottom left triangle, $g_* \circ f_* = \mathrm{id}_{\pi_1(X)}$, and similarly the top right triangle, $f \circ g = \mathrm{id}_Y$, gets induced to the bottom right triangle, $f_* \circ g_* = \mathrm{id}_{\pi_1(Y)}$.

---

[6]In the language of category theory, Theorem 7.2.5 implies that the fundamental group is a **functor** from the category of based topological spaces to the category of groups, which takes each based space (X, x_0) to the group $\pi_1(X)$, and each based map $f : (X, x_0) \to (Y, y_0)$ to the induced map $f_* : \pi_1(X) \to \pi_1(Y)$.

> **Theorem 7.2.6.** *The fundamental group is a topological invariant.*

Proof. If $f : X \to Y$ is a homeomorphism, then f_* is invertible; hence $f_* : \pi_1(X) \to \pi_1(Y)$ is a group isomorphism. □

The fact that the fundamental group is a topological invariant means that we can use it to distinguish spaces. If $\pi_1(X) \not\cong \pi_1(Y)$, then $X \not\approx Y$. However, the fundamental group is not a complete invariant, in the sense that there are spaces $X \not\approx Y$ but with $\pi_1(X) \cong \pi_1(Y)$. In fact, π_1 cannot distinguish the Euclidean spaces $\mathbb{R}^n$ for various $n \in \mathbb{N}$. Example 125 shows that $\pi_1(\mathbb{R}^n) = \{[\mathbf{0}]\}$ for any $n \in \mathbb{N}$.

If $\pi_1(X)$ consists of only one element (the homotopy class of the constant loop), then we say $\pi_1(X)$ is **trivial**, which is typically denoted by 1 (or by 0 if all groups under consideration are abelian), and we call the space X **simply connected**. For $n \geq 2$, $\pi_1(\mathbb{S}^n)$ is trivial, so every sphere of dimension at least 2 is simply connected. On the other hand, $\pi_1(\mathbb{S}^1)$ is infinite, as we shall see in §7.3.

Exercises

1. Suppose f and g are defined as in the proof of Theorem 7.2.3. Verify that $g \circ f$ is the identity on $\pi_1(X, x_0)$, and $f \circ g$ is the identity on $\pi_1(X, x_1)$.

2. Let $f : X \to Y$ be a continuous map. Prove that the induced map $f_* : \pi_1(X) \to \pi_1(Y)$ is a group homomorphism.

3. Prove Theorem 7.2.5.

4. Let $r \geq 1$. Prove that $\pi_1(\mathbb{D}^r) = 1$.

5. Suppose X is a radially convex subspace of $\mathbb{R}^n$ (recall §2.4). Prove that X is simply connected.

7.3 Covering Spaces and the Circle

How many ways are there to draw a loop in the circle $\mathbb{S}^1$? This is actually not a trivial question. Remember that a mathematical loop $\gamma : \mathbb{S}^1 \to X$ can wind around the space X in any crazy way. However, we may intuitively understand that the loops in $\mathbb{S}^1$ should be classified by how many complete times they go around the circle, ignoring any backtracking, so perhaps $\pi_1(\mathbb{S}^1)$ is in one-to-one correspondence with the set of integers $\mathbb{Z} = \{\dots, -2, -1, 0, 1, 2, 3, \dots\}$. The product of a loop going m times around $\mathbb{S}^1$ and a loop going n times (in the same direction) around $\mathbb{S}^1$ is clearly a loop going $m + n$ times around the circle, so there seems to be a correspondence with the group structure of the integers: $\mathbb{Z}$ is a group whose "product" operation is simply addition. However, this all takes a very careful argument. We will require more mathematical machinery, including the idea of a *covering space*, which also becomes useful for other results in algebraic topology.

Covering Spaces

Consider the definition of the torus $\mathbb{T}$ as a square with identifications along the boundary. An equivalent way to define $\mathbb{T}$ would be as a quotient space of the plane, $\mathbb{T} \approx \mathbb{R}^2/\sim$, where for any $(x, y) \in \mathbb{R}^2$, we identify $(x, y) \sim (x+m, y+n)$ for arbitrary $m, n \in \mathbb{Z}$. Think of the plane as divided into infinitely many copies of squares joined at their boundaries. When an arc starts in one square and exits over the right boundary line, it enters the left side of the neighboring square. On the other hand, since all of these squares are identified, the arc is really reentering the *same* square from the left, as we would expect[7] in the usual plane diagram for $\mathbb{T}$. What we have defined is called a *covering map*, $p : \mathbb{R}^2 \to \mathbb{T}$, which takes $(x, y) \in \mathbb{R}^2$ to the equivalence class of (x, y) in $\mathbb{R}^2/\sim$. See Figure 7.11 for an illustration of the map p.

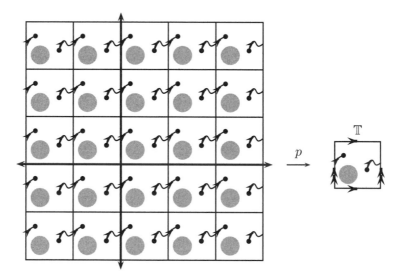

Figure 7.11: The preimage of an arc in $\mathbb{T}$ with respect to the covering map $p : \mathbb{R}^2 \to \mathbb{T}$. Also shown: $p^{-1}[U]$, where U is a small, open disk neighborhood in the torus. Note that $p^{-1}[U]$ (in gray) consists of infinitely many disjoint copies of the open disk.

---

[7]Note that this is exactly the same idea as in Figure 1.19 from §1.2.

Definition 7.3.1. A map of topological spaces, $p : E \to X$, is a **covering** map if for each $x \in X$, there is an open neighborhood $U \subseteq X$ containing x such that $p^{-1}[U] \subseteq E$ is the union of disjoint open sets,

$$p^{-1}[U] = \bigcup_{\alpha} V_{\alpha},$$

and such that p restricts to a homeomorphism $V_{\alpha} \stackrel{\approx}{\to} U$ for each α.

While Definition 7.3.1 may seem esoteric at first, it can be summed up in an intuitive way using the example $p : \mathbb{R}^2 \to \mathbb{T}$ from above. A covering $p : E \to X$ looks *locally* like a stack of identical patches of E mapped onto a single patch of X, just as the infinitely many open disks V_{α} in Figure 7.11 all get mapped by p to a disk neighborhood U in the torus. The main reason to introduce covering maps is that they serve as a way to talk about homotopy in a more familiar context. For example, it may be very difficult to understand when two loops are homotopic within a torus, but arcs in $\mathbb{R}^2$ whose images are loops in $\mathbb{T}$ may be analyzed quite easily. The key is to know how loop homotopies in the **base** space X correspond to arc homotopies in the covering space E when $p : E \to X$ is a covering map.

Suppose $p : E \to X$ is a covering map, and $f : Z \to X$ is any map. We say that $\widetilde{f} : Z \to E$ is a **lift** of f if $p \circ \widetilde{f} = f$. The relationship can be expressed by the commutative diagram shown below.

$$\begin{array}{ccc} & & E \\ & \nearrow^{\widetilde{f}} & \downarrow^{p} \\ Z & \xrightarrow{f} & X \end{array} \tag{7.9}$$

The main idea of the following *Homotopy Lifting Property* is that homotopies defined on a space X can always be lifted to related homotopies on a covering space for X.

Theorem 7.3.2 (Homotopy Lifting Property). *Suppose $p : E \to X$ is a covering map. Given any map $h : Z \times \mathbb{I} \to X$ and a lift $\widetilde{h_0} : Z \times \{0\} \to E$ of $h_0 : Z \times \{0\} \to X$, there is a unique map $\widetilde{h} : Z \times \mathbb{I} \to E$ lifting h, and restricting to $\widetilde{h_0}$ on $Z \times \{0\}$.*

Equivalently, the commutative diagram on the left can always be augmented by the dashed arrow to form the commutative diagram on the right.

$$\begin{array}{ccc} Z \times \{0\} & \xrightarrow{\widetilde{h_0}} & E \\ \downarrow{\scriptstyle \subseteq} & \searrow^{h_0} & \downarrow^{p} \\ Z \times \mathbb{I} & \xrightarrow{h} & X \end{array} \implies \begin{array}{ccc} Z \times \{0\} & \xrightarrow{\widetilde{h_0}} & E \\ \downarrow{\scriptstyle \subseteq} & \nearrow^{\widetilde{h}} & \downarrow^{p} \\ Z \times \mathbb{I} & \xrightarrow{h} & X \end{array} \tag{7.10}$$

The proof is a bit technical; however, it uses nothing more the definition of a covering space and compactness of $\mathbb{I}$ to build up the required map. It may help to refer to Figure 7.12 as you read through the following argument.

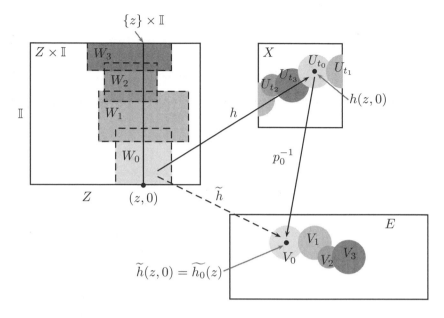

Figure 7.12: Building $\widetilde{h} : Z \times \mathbb{I} \to E$ one neighborhood at a time. This diagram shows four "box" neighborhoods covering $\{z\} \times \mathbb{I}$, but of course there could be many more in general. As suggested by this figure, the covering map p could be understood as "wrapping" the space E around X, and so p^{-1} "unwraps" X locally.

Proof. Let $z \in Z$ be arbitrary. Observe that for each $t \in \mathbb{I}$, we have $h(z,t) \in X$, and since $p : E \to X$ is a covering map, there is an open neighborhood $U_t \subseteq X$ with $h(z,t) \in U_t$ such that $p^{-1}[U_t]$ is the disjoint union of subsets of E each homeomorphic to U_t. Now $h^{-1}[U_t] \subseteq Z \times \mathbb{I}$ is open (by continuity of h), so there is are open neighborhoods $N_t \subseteq Z$ and $I_t \subseteq \mathbb{I}$ such that $(z,t) \in N_t \times I_t \subseteq h^{-1}[U_t]$ (recall that the products of open sets form a base for the topology on a product of two spaces). Furthermore, we may assume there is a number $\epsilon_t \in \mathbb{R}^+$ such that $I_t = (t - \epsilon_t, t + \epsilon_t) \cap \mathbb{I}$.

At this point we have constructed an open set $N_t \times I_t$ for every point $t \in \mathbb{I}$, and the collection of these sets covers the compact subset $\{z\} \times \mathbb{I} \subseteq Z \times \mathbb{I}$ (*why is this subset compact?*). By compactness, finitely many of the sets $N_t \times I_t$ cover $\{z\} \times \mathbb{I}$, say, $\{W_k = N_{t_k} \times I_{t_k}\}_{k=0,1,\ldots,r}$, where $0 \leq t_0 < t_1 < t_2 < \cdots < t_r \leq 1$. We may assume that $I_{t_{k'}} \not\subseteq I_{t_k}$ for any $k' \neq k$; otherwise, $W_{k'} = N_{t_{k'}} \times I_{t_{k'}}$ can be omitted from the cover. As a consequence, $I_{t_k} \cap I_{t_{k+1}} \neq \emptyset$ for each $k = 0, 1, \ldots, r - 1$ (*why?*).

The construction of $\widetilde{h} : Z \times \mathbb{I} \to E$ is done *inductively*. An inductive construction is much like a proof by induction. There is a *base step*, which defines an initial part of the construction. Then we may assume the existence of a *partial* construction satisfying whatever conditions are required. If it can be shown that it is possible to extend this partial construction by one more step, and so long as the number of such steps is finite, then the construction can be carried out to completion.

As a base step, let $V_0 \subseteq p^{-1}[U_{t_0}] \subseteq E$ be the open set homeomorphic to U_{t_0} such that $\widetilde{h_0}(z) \in V_0$. There is only one choice for V_0, since if another such set V_0' existed, then $\widetilde{h_0}(z) \in V_0 \cap V_0'$, contradicting the fact that the open sets comprising $p^{-1}[U_{t_0}]$ must be disjoint. The homeomorphism $V_0 \to U_{t_0}$ is the restriction $p|_{V_0}$ of the covering map p to the set V_0, which we will call p_0. Define $\widetilde{h}$ on W_0 as the composition: $p_0^{-1} \circ h$. This composition is well defined because $h[W_0] \subseteq U_{t_0}$, and p_0^{-1} is a map $U_{t_0} \to V_0$.

At this point, the map $\widetilde{h}$ satisfies the conditions required in Theorem 7.3.2 on the subset $W_0 \subseteq Z \times \mathbb{I}$. In particular, $\widetilde{h}$ is continuous (being the composition of continuous functions), and we have

$$\widetilde{h}(z,0) = (p_0^{-1} \circ h)(z,0) = p_0^{-1}(h(z,0)) = \widetilde{h_0}(z), \qquad (7.11)$$

where the equality $p_0^{-1}(h(z,0)) = \widetilde{h_0}(z)$ follows from the fact that $\widetilde{h_0}$ is a lift of h_0; more explicitly, since $p \circ \widetilde{h_0} = h_0$, we obtain $p_0(\widetilde{h_0}(z)) = p(\widetilde{h_0}(z)) = h_0(z) = h(z,0)$. Moreover, the map $\widetilde{h}$ so far is uniquely determined (by properties of the covering map).

Next we would like to extend the domain of definition of $\widetilde{h}$ to all of $Z \times \mathbb{I}$. Let $k \in \mathbb{N}$, $1 \le k \le r$, and suppose that we have constructed $\widetilde{h}$ so that it is defined on $W_0 \cup W_1 \cup \cdots \cup W_{k-1}$, satisfying the conditions of Theorem 7.3.2 on its domain. We extend the domain of definition for $\widetilde{h}$ to include W_k as follows. Consider $U_{t_k} \subseteq X$. We have $h(z, t_k) \subset U_{t_k}$, and $h[W_k] = h[N_{t_k} \times I_k] \subseteq U_{t_k}$. Now $W_{k-1} \cap W_k \ne \emptyset$ (since $I_{t_{k-1}} \cap I_{t_k} \ne \emptyset$ and both N_{k-1} and N_k contain at least the point z). Let $T = W_{k-1} \cap W_k$. Since $T \subseteq W_{k-1}$, the map $\widetilde{h}$ has already been defined on T, and we have $h[T] \subseteq U_{t_k}$. Let $V_k \subseteq p^{-1}[U_{t_k}] \subseteq E$ be an open set homeomorphic to U_{t_k} such that $\widetilde{h}[T] \subseteq V_k$ (which is possible to find because $p[\widetilde{h}[T]] = (p \circ \widetilde{h})[T] = h[T] \subseteq U_{t_k}$). Clearly, such V_k is unique (for the same reason V_0 was unique). Let $p_k = p|_{V_k}$, so that $p_k^{-1} : U_{t_k} \to V_k$ is a homeomorphism. Extend $\widetilde{h}$ by defining $\widetilde{h} = p_k^{-1} \circ h$ on W_k. Thus $\widetilde{h}$ is the unique map defined on $W_0 \cup W_1 \cup \cdots \cup W_k$ such that $h = p \circ \widetilde{h}$ and such that $\widetilde{h}$ restricts to $\widetilde{h_0}$ on the neighborhood of $Z \times \{0\}$ on which it is defined. After finitely many steps, $\widetilde{h}$ can be extended to all of $W_0 \cup W_1 \cup \cdots \cup W_r$.

Finally, since $z \in Z$ was arbitrary, we have in fact defined $\widetilde{h}$ on all of $Z \times \mathbb{I}$. Continuity of $\widetilde{h}$ follows directly from the construction: $\widetilde{h}$ is continuous on every neighborhood $N_t \times I_t$, and the values of the function must agree on every intersection of those neighborhoods owing to uniqueness of the construction. □

As a direct consequence, paths can be lifted uniquely to covering spaces.

Proposition 7.3.3 (Path Lifting Property). *Suppose $p : E \to X$ is a covering map.*

(a) Suppose $\gamma : \mathbb{I} \to X$ is a path with initial point $\gamma(0) = x_0$. If $e_0 \in E$ is any fixed preimage of x_0 with respect to p (i.e., $e_0 \in p^{-1}[x_0]$), then there is a unique path $\widetilde{\gamma} : \mathbb{I} \to E$ lifting γ and such that $\widetilde{\gamma}(0) = e_0$.

(b) If $\gamma \simeq \eta$ are paths in X and $\widetilde{\gamma}, \widetilde{\eta}$ are their lifts in E such that $\widetilde{\gamma}(0) = \widetilde{\eta}(0) = e_0$, then $\widetilde{\gamma} \simeq \widetilde{\eta}$ in E via a homotopy relative to e_0.

The proof of part (a) follows by letting $Z = \{z_0\}$ be any one-point space and applying Theorem 7.3.2 to γ regarded as a map $Z \times \mathbb{I} \approx \mathbb{I} \to X$. Part (b) follows by letting $Z = \mathbb{I}$. We are now ready to work out our first nontrivial fundamental group.

Fundamental Group of the Circle

For each integer n, consider the loop λ_n traversing the circle n times counterclockwise if $n \geq 0$, or $|n|$ times clockwise if $n < 0$.

$$\begin{aligned} \lambda_n &: \quad \mathbb{I} \to \mathbb{R}^2 \\ \lambda_n(s) &= (\cos 2\pi(ns), \sin 2\pi(ns)) \end{aligned} \tag{7.12}$$

Note that $\lambda_n(0) = \lambda_n(1) = (1, 0) = s_0$, which we take to be the basepoint of $\mathbb{S}^1$. The circle $\mathbb{S}^1$ itself can be thought of as the image of $\mathbb{I}$ under the map λ_1. Now the homotopy classes of the loops λ_n defined by (7.12) for various $n \in \mathbb{Z}$ are elements of the fundamental group of $\mathbb{S}^1$, so we can define another function,

$$\begin{aligned} \phi &: \quad \mathbb{Z} \to \pi_1(\mathbb{S}^1) \\ \phi(n) &= [\lambda_n]. \end{aligned} \tag{7.13}$$

In fact, ϕ is a group homomorphism, as it can be verified directly that $\lambda_n \cdot \lambda_m \simeq \lambda_{n+m}$ for all $n, m \in \mathbb{Z}$. That is,

$$\phi(n + m) = [\lambda_{n+m}] = [\lambda_m \cdot \lambda_n] = [\lambda_m] \cdot [\lambda_n] = \phi(n) \cdot \phi(m).$$

At this point we only know that there is a set of loops $[\lambda_n]$ in $\pi_1(\mathbb{S}^1)$ indexed by $\mathbb{Z}$, but we do not know whether two loops λ_n and λ_m may in fact be the same in $\pi_1(\mathbb{S}^1)$ (i.e., is ϕ *injective?*), nor can we be sure that every loop in $\pi_1(\mathbb{S}^1)$ is represented by one of the loops λ_n (i.e., is ϕ *surjective?*). These two questions will be addressed by finding an *inverse* function for ϕ. As you might imagine, this is the hard part of the argument, which will be taken up in the proof of Theorem 7.3.4 below. But first consider the function

$$\begin{aligned} p &: \quad \mathbb{R}^1 \to \mathbb{S}^1 \\ p(x) &= (\cos 2\pi x, \sin 2\pi x). \end{aligned} \tag{7.14}$$

The map p has the property that $p(n) = s_0$ for every $n \in \mathbb{Z}$. It may help to visualize p as taking the entire number line, coiling it up, and projecting all of the coils onto the circle $\mathbb{S}^1$, as shown in Figure 7.13. This map p is a covering. For example, if we let $U_1 = \lambda_1[(1/8, 7/8)]$ and $U_2 = \lambda_1[[0, 3/8) \cup (5/8, 1]]$, then $\mathbb{S}^1 = U_1 \cup U_2$, and we see that each of $p^{-1}[U_1]$ and $p^{-1}[U_2]$ is a disjoint union of open intervals in $\mathbb{R}^1$ (see Figure 7.14).

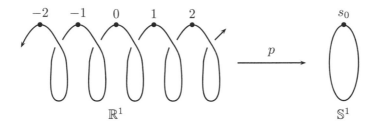

Figure 7.13: The "coiling" map $p : \mathbb{R}^1 \to \mathbb{S}^1$.

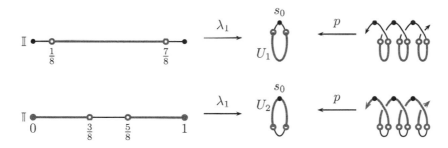

Figure 7.14: U_1 is the image of $(1/8, 7/8)$ in $\mathbb{S}^1$, and U_2 is the image of $[0, 3/8) \cup (5/8, 1]$ in $\mathbb{S}^1$, under the map $\lambda_1 : \mathbb{I} \to \mathbb{S}^1$. The inverse images $p^{-1}[U_1]$ and $p^{-1}[U_2]$ are each a disjoint union of (infinitely many) open intervals in $\mathbb{R}^1$.

Theorem 7.3.4. $\pi_1(\mathbb{S}^1) = \{[\lambda_n] \mid n \in \mathbb{Z}\} \cong \mathbb{Z}$

Proof. Suppose γ is a loop in $\mathbb{S}^1$ with basepoint s_0. Here we are regarding γ as a path $\mathbb{I} \to \mathbb{S}^1$ with $\gamma(0) = \gamma(1) = s_0$. Let $p : \mathbb{R}^1 \to \mathbb{S}^1$ be the covering map defined by (7.14). Now using p, Proposition 7.3.3(a) can be used to lift γ uniquely to a path $\widetilde{\gamma} : \mathbb{I} \to \mathbb{R}^1$ such that $\widetilde{\gamma}(0) = 0 \in \mathbb{R}^1$. Since $p^{-1}[\{s_0\}] = \mathbb{Z} \subseteq \mathbb{R}^1$, we know that $\widetilde{\gamma}(1)$ is equal to some integer $n(\gamma)$, as illustrated in Figure 7.15.

The loop $\lambda_{n(\gamma)}$ in $\mathbb{S}^1$ also lifts uniquely to a path $\widetilde{\lambda_{n(\gamma)}}$ in $\mathbb{R}^1$ having initial point 0 and terminal point $n(\gamma)$. Recall from Example 125 that there is only a single homotopy class of loops in any given Euclidean space $\mathbb{R}^n$. A similar argument shows that all paths having the same initial and terminal points are

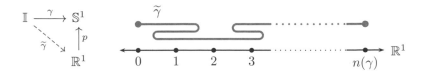

Figure 7.15: There is a unique lift $\widetilde{\gamma}$ of γ such that $\widetilde{\gamma}(0) = 0 \in \mathbb{R}^1$.

homotopic through an endpoint-preserving homotopy. Thus $\widetilde{\gamma} \simeq \widetilde{\lambda_{n(\gamma)}}$ via a homotopy $\widetilde{h} : \mathbb{I} \times \mathbb{I} \to \mathbb{R}^1$. Composing with the map p, we find $\gamma \simeq \lambda_{n(\gamma)}$ via the basepoint-preserving homotopy $p \circ \widetilde{h}$.

Next we must prove that the value of $n(\gamma)$ is *uniquely determined* by the homotopy class of γ. Suppose $\gamma \simeq \eta$ via a homotopy $h : \mathbb{I} \times \mathbb{I} \to \mathbb{S}^1$ (again veiwing the domain of the loops as $\mathbb{I}$ rather than $\mathbb{S}^1$). Then the lifts $\widetilde{\gamma}$ and $\widetilde{\eta}$ are homotopic via a unique homotopy $\widetilde{h} : \mathbb{I} \times \mathbb{I} \to \mathbb{R}^1$, preserving the initial point $0 \in \mathbb{R}^1$, by Proposition 7.3.3(b). The map $f(t) = \widetilde{h}(1, t)$ identifies the terminal points of each path $\widetilde{h}_t$. In particular, $f(0) = \widetilde{\gamma}(1) = n(\gamma)$. Noting that $p \circ \widetilde{h} = h$, and the fact that h is a basepoint-preserving homotopy, we find that the codomain of f is the discrete space $\mathbb{Z}$. The only continuous functions into a discrete space are the constant functions; therefore $f(1) = \widetilde{\eta}(1) = n(\gamma)$ as well. This shows that $n(\eta) = n(\gamma)$, which is what we needed to prove. (Observe that this argument also shows that $[\lambda_m] \neq [\lambda_n]$ if $n \neq m \in \mathbb{Z}$.)

The preceding discussion shows that $\pi_1(\mathbb{S}^1) = \{[\lambda_n] \, n \in \mathbb{Z}\}$, and provides a well-defined function, inverse to ϕ, as defined by (7.13):

$$\phi^{-1} \quad : \quad \pi_1(\mathbb{S}^1) \to \mathbb{Z},$$
$$\phi^{-1}([\lambda_n]) \quad = \quad n.$$

This proves that ϕ is bijective, and so must be a group isomorphism. $\qquad \square$

Brouwer Fixed-Point Theorem

We end this section with an important application of the fundamental group. Take two identical sheets of paper and place them on the table, one right on top of the other. Imagine that both sheets represent the same rectangle of the coordinate plane, so a point that is x centimeters from the left edge and y centimeters from the bottom edge would correspond to the point $(x, y) \in \mathbb{R}^2$ in both sheets. If the two sheets are aligned perfectly, then every point of the top sheet is directly over its corresponding point on the bottom piece. Now shift the top sheet 1 cm to the left (of course, this will cause a strip of the left side of the top sheet to fall outside the region of the bottom sheet). Since every x-value is decreased by 1, there are *no* points of the top sheet directly above their corresponding point on the bottom.

Now take the top sheet and fold it, twist it, and crumple it up in any way you want, so long as you don't tear it. Then drop it onto the bottom sheet, in such a way so that no part of the crumpled paper falls outside of the bottom sheet. Then there *must be at least one point* on the top sheet that is directly over its corresponding point on the bottom. Before we can prove this surprising result, we must rephrase it in a precise mathematical way.

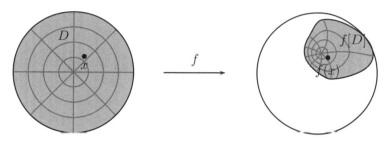

Figure 7.16: Any continuous map from a convex subset of the plane to itself must have a fixed point.

The sheet of paper represents a space X (in the above case, X is a closed rectangle subset of the plane). Folding, twisting, and crumpling are all operations that *preserve closeness*; in other words, the end result is the image[8] of a *continuous* function. The next key ingredient is to require the codomain of f to be X as well. This is what we meant when we required the crumpled sheet to be placed completely within the confines of the bottom sheet. With these requirements in place, we will argue that at least one point $x \in X$ has the property that $f(x)$ is "directly over" x. What we really mean here is that $f(x) = x$, and if such a point x exists, then it is called a **fixed point** for the function f (see Figure 7.16). Now we may state and prove a version of the **Brouwer Fixed-Point Theorem**. There are many generalizations of the theorem, but we will focus on the version for spaces that look like disks in the plane.

Theorem 7.3.5 (Brouwer Fixed-Point Theorem). *Let D be any space homeomorphic to a closed disk $\mathbb{D}^2 \subseteq \mathbb{R}^2$. Every continuous map $f : D \to D$ has a fixed point.*

Proof. The argument is based on a computation of fundamental groups. Since π_1 is a topological invariant, it is sufficient to prove the result when D is the closed unit disk centered at the origin. The boundary of D is the unit circle $\mathbb{S}^1$. Suppose $f : D \to D$ is continuous but has no fixed points. That is, $f(x) \neq x$ for every $x \in D$. Thinking of points in $\mathbb{R}^2$ as vectors, the vector function $\mathbf{v}(x) = x - f(x)$ must be nonzero for all $x \in D$, and since f is continuous in x, so is $\mathbf{v}$. Consider a

---

[8]Technically, after crumpling the paper, we mapped it down to the bottom sheet by a *projection* function. Since projection is continuous, the composition is also continuous.

function $g : D \to \mathbb{S}^1$ defined as follows: For each $x \in D$, construct a ray starting at $f(x)$ in the direction of $\mathbf{v}(x)$ (which is possible since $\mathbf{v}(x) \neq \mathbf{0}$). The ray must intersect the bounding circle in one point that is not the initial point $f(x)$; define $g(x)$ to be that point of intersection, as shown in Figure 7.17. Note that if x is already on the boundary circle, then x itself is the point of intersection, and $g(x) = x$ in that case.

We must establish that g is continuous. Let $\epsilon > 0$ and consider all points within ϵ distance of $g(x)$ as our *target*. For sufficiently small $\delta > 0$, both $x + \delta$ and $f(x + \delta)$ can be made arbitrarily close to x and $f(x)$, respectively (again, refering to Figure 7.17). Think of x as the tip of an archer's arrow, and $f(x)$ as the tail. Owing to real-world conditions (unpredictable changes in wind, arrow imperfections, slight jitters of the hands of the archer), the arrow may never be positioned *exactly* in line to hit the dead center of the target. But so long as the arrow is positioned *precisely enough*, it will hit somewhere within the target.[9] That is, δ can be chosen small enough so that the ray drawn from $f(x + \delta)$ to $x + \delta$ intersects $\mathbb{S}^1$ within ϵ distance of $g(x)$.

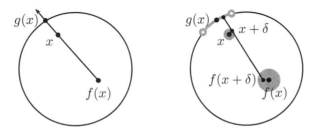

Figure 7.17: *Left*, definition of the map $g : D \to \mathbb{S}^1$. *Right*, verifying that g is continuous.

At this point we have defined a continuous function $g : D \to \mathbb{S}^1$ that has the additional feature that $g(x) = x$ for all $x \in \mathbb{S}^1$. Let $i : \mathbb{S}^1 \to D$ be the inclusion map (which is also continuous). Then the composition $g \circ i : \mathbb{S}^1 \to \mathbb{S}^1$ is the identity on $\mathbb{S}^1$. On the other hand, there is a big problem with the induced maps on fundamental groups.

$$
\begin{array}{ccc}
\mathbb{S}^1 \xrightarrow{\;i\;} D & & \pi_1(\mathbb{S}^1) = \mathbb{Z} \xrightarrow{\;i_*\;} \pi_1(D) = 1 \\
{\scriptstyle \text{id}}\searrow \;\; \downarrow {\scriptstyle g} & \Longrightarrow & {\scriptstyle \text{id}}\searrow \;\; \downarrow {\scriptstyle g_*} \\
\mathbb{S}^1 & & \pi_1(\mathbb{S}^1) = \mathbb{Z}
\end{array}
$$

---

[9]This illustration breaks down if we follow it too far. For ϵ on the order of a millimeter, perhaps no human archer can reasonably keep δ small enough to suffice. If ϵ is on the order of about 10^{-35} meters (the *Planck distance*), then no archer, human nor machine, could reduce their δ small enough to compensate for so-called *quantum fluctuations* that would perturb the motion of the arrow unpredictably.

According to the diagram, we should have $g_* \circ i_* = (\mathrm{id})_* = \mathrm{id} : \mathbb{Z} \to \mathbb{Z}$, the map that sends every element of $\mathbb{Z}$ to itself. However, the only possible map from $\mathbb{Z}$ to the trivial group 1 sends every element to the identity element $\mathbf{1} \in 1$. Hence the composition $g_* \circ i_*$ must send every element of $\mathbb{Z}$ to the identity element $0 \in \mathbb{Z}$. This contradiction proves that no fixed point free map $f : D \to D$ could possibly exist. $\square$

Exercises

1. Flesh out the proof of Proposition 7.3.3.

2. Construct a homotopy $\lambda_n \cdot \lambda_m \simeq \lambda_{n+m}$ for arbitrary $n, m \in \mathbb{Z}$.

3. Consider $U_1 = \lambda_1[(1/8, 7/8)]$ and $U_2 = \lambda_1[[0, 3/8) \cup (5/8, 1]]$ as defined above. Identify explicitly, as a disjoint union of open intervals, the preimage sets $p^{-1}[U_1]$ and $p^{-1}[U_2]$.

4. Mimic the proof of Theorem 7.3.4 to show that $\pi_1(\mathbb{T}) = \mathbb{Z} \times \mathbb{Z}$. (*Hint:* Use the covering map illustrated in Figure 7.11.)

5. Show that the Brouwer Fixed-Point Theorem does not apply to the torus by finding a fixed point free map $f : \mathbb{T} \to \mathbb{T}$.

7.4   Compact Surfaces and Knot Complements

Theorem 7.2.6 implies that if $\pi_1(X) \ncong \pi_1(Y)$, then X and Y are not homeomorphic. That is, the fundamental group is a topological invariant. In fact, π_1 is a more powerful invariant than some of the others we have encountered so far in this text. For example, some nonhomeomorphic surfaces have the same Euler characteristic but may differ in their fundamental groups. We explore two kinds of spaces in this section: compact surfaces (and surfaces-with-boundary) and knot complements.

Fundamental Group for a Plane Model

Figure 7.18 illustrates a few (nonbased) loops in the plane model of the torus. Since all four corners are identified as the same point of $\mathbb{T}$, the "edges" a and b are really loops. By deformation, we may verify that any loop in $\mathbb{T}$ is either homotopic to a constant loop or homotopic to a product of the loops a and b, but this verification is nontrivial.[10] Now consider all loops based at the "corner" point x_0. Since every $\gamma \in \mathrm{Loop}(\mathbb{T}, x_0)$ can be deformed to a loop involving only a and b, we say that $[a]$ and $[b]$ are the *generators* of the fundamental group

---

[10]For example, how can we be sure that *every* loop can be deformed in this way? There are strange pathologies to consider, such as *space-filling* curves – see Armstrong [Arm10], §2.3.

$\pi_1(\mathbb{T})$. By abuse of notation, we will use the letters a and b to stand for $[a]$ and $[b]$, respectively; context will make it clear whether a and b are specific loops or homotopy classes.

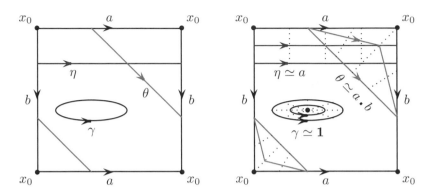

Figure 7.18: A few loops in $\mathbb{T}$: $[\gamma] = [\mathbf{1}]$, $[\eta] = [a]$, and $[\theta] = [a] \cdot [b]$.

We see that a commutes with b (i.e., $ab = ba$), as shown in Figure 7.19. Thus every loop in $\text{Loop}(\mathbb{T}, x_0)$ is homotopic to $a^m b^n$, for some $m, n \in \mathbb{Z}$. Moreover, the product of two such expressions is accomplished by simply adding exponents: $(a^{m_1} b^{n_1}) \cdot (a^{m_2} b^{n_2}) = a^{m_1 + m_2} b^{n_1 + n_2}$. It turns out that there are no further *relations* among the loops,[11] so $\pi_1(\mathbb{T})$ is isomorphic to the group $\mathbb{Z}^2$, with "product" operation given by adding pairs of integers: $(m_1, n_1) \cdot (m_2, n_2) = (m_1 + m_2, n_1 + n_2)$ (see also §7.3, Exercise 4). Note that $\mathbb{Z}^2$ is an *abelian* group, though typically the fundamental group is nonabelian. We may write:

$$\pi_1(\mathbb{T}) = \{a^m b^n \mid m, n \in \mathbb{Z}\} \cong (\mathbb{Z}^2, +). \tag{7.15}$$

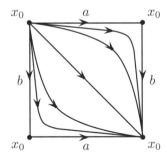

Figure 7.19: Here $ab = ba \in \pi_1(\mathbb{T})$.

---

[11] The fact that there are no other relations is a consequence of the **Seifert-van Kampen Theorem**, which is beyond the scope of this elementary text.

More precisely, we may give a presentation of the group (see §B.1 for more details). The fundamental group of $\mathbb{T}$ has the presentation

$$\pi_1(\mathbb{T}) = \langle a, b \mid ab = ba \rangle. \tag{7.16}$$

Note that $ab = ba$ is equivalent to $aba^{-1}b^{-1} = \mathbf{1}$, so that we could have written

$$\pi_1(\mathbb{T}) = \langle a, b \mid aba^{-1}b^{-1} = \mathbf{1} \rangle. \tag{7.17}$$

It is no coincidence that the expression $aba^{-1}b^{-1}$ is simply a *word* (in the sense of §5.2) describing the plane model for $\mathbb{T}$. If all vertices of the plane model are identified to a single point x_0, then every letter $a, b, c, \ldots$ labeling a particular edge is a loop. By deforming any given loop so that it gets pushed to the edges, there will always be some word in the letters that represents the homotopy class of the loop. Furthermore, the loop that traverses the outer edge of the plane model exactly once is homotopic to the constant loop by a homotopy that "shrinks" the loop down to the basepoint (see Figure 7.20), and it turns out that this defines the *only* relation in the fundamental group, while the distinct letters of the plane model are the generators. For example, $\pi_1(\mathbb{T})$ has two generators, a and b, representing the loops shown in Figure 5.12 (§5.2), while $\pi_1(2\mathbb{T})$ has four generators, $a_1, b_1, a_2,$ and b_2 (recall Figure 5.13).

The same idea can be used to determine fundamental groups for any space that has a plane model, including surfaces-with-boundary and other spaces like the *dunce cap* (Figure 5.14), though more care must be taken if the vertices are not all identified to a single point.

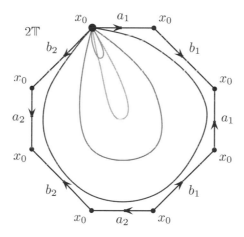

Figure 7.20: In the double torus, $a_1 b_1 a_1^{-1} b_1^{-1} a_2 b_2 a_2^{-1} b_2^{-1} \simeq \mathbf{1}$.

Example 126. Find the fundamental group of each space.

(a) $\mathbb{P}$ (b) $\mathbb{S}^2$ (c) Möbius strip (d) Dunce cap

Solution:

(a) A word for $\mathbb{P}$ is a^2 (see Figure 5.24). Since both vertices of the plane model represent the same point, we have $\pi_1(\mathbb{P}) = \langle a \mid a^2 = 1 \rangle$, which is isomorphic to the *cyclic* group $\mathbb{Z}/2\mathbb{Z}$ of two elements.

(b) The simplest plane model for $\mathbb{S}^2$ has the word aa^{-1}; however, a itself is not a loop in $\mathbb{S}^2$ because the two vertices of the plane model are distinct points in the sphere. The only loop involving the edges of the model would be the entire loop aa^{-1}, and so we have a relation $aa^{-1} = 1$, but by multiplying a on the right, we end up with the tautology $a = a$. Thus there are no generators and no relations in the presentation for this group. In fact, $\pi_1(\mathbb{S}^2) = 1$, the trivial group.

(c) Take $abcb$ as the word for the Möbius strip M. The vertices of the plane model are identified in pairs. Using v_0 as a basepoint, both ab and cb are loops in M. The relation $abcb = (ab)(cb) = 1$ implies that $cb = (ab)^{-1}$, and so there is really only one generator, $w = ab$, with no relations: $\pi_1(M) = \langle w \rangle \cong \mathbb{Z}$.

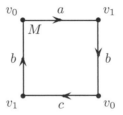

(d) The dunce cap has word a^3 (see Figure 5.14), and so its fundamental group is the cyclic group $\langle a \mid a^3 = 1 \rangle \cong \mathbb{Z}/3\mathbb{Z}$.

Knot Complements

Suppose $K : \mathbb{S}^1 \to \mathbb{R}^3$ is a knot in the sense of Definition 6.3.1. The *complement* of K in $\mathbb{R}^3$ is a noncompact 3-manifold $M_K = \mathbb{R}^3 \setminus K[\mathbb{S}^1]$ (see §6.4, Exercise 10). Imagine a solid three-dimensional block of wood extending indefinitely in all directions, but somewhere inside the block, a termite has eaten away a closed path through the wood in the shape of a knot. If the termite actually has no length, width, or height (zero dimensional), then the path would be one dimensional, hence an embedding of $\mathbb{S}^1$, and the remaining wood can be thought of as a knot complement. What is the fundamental group of M_K? Of course the answer to this question depends on the knot; if $K = \mathbb{U}$ (the *unknot*), then it shouldn't be too difficult to see that $\pi_1(M_K) \cong \mathbb{Z}$, generated by a single loop γ that links with $\mathbb{U}$, as illustrated in Figure 7.21. Note that we may assume that all knots avoid $\mathbf{0}$, and so we can take the origin to be the basepoint for all loops in M_K; then $\pi_1(M_K) = \pi_1(M_K, \mathbf{0})$ is a knot invariant called the *knot group* of K. In

Figure 7.21: Here $\pi_1(M_U) \cong \mathbb{Z} \cong \langle \gamma \rangle$.

fact, the knot group is an invariant of links as well, though the literature seems to avoid the term "link group."

Definition 7.4.1. The **knot group** of a knot or link K is the fundamental group of the complement $M_K = \mathbb{R}^3 \setminus K$.

There is a procedure for constructing the knot group based only on a plane diagram for the knot. We give the result here, called the **Wirtinger presentation**, without justifying all the details; for more information, see, for example, Armstrong [Arm10] (Chapter 10), or the classic *Introduction to Knot Theory* by Crowell and Fox [CF08].

- If not already specified, give orientations to all components of the link.

- Label each strand of the diagram by a unique letter (recall, from §6.3, that the strands are the connected components of the knot diagram). These letters are the generators of the knot group.

- Each crossing corresponds to a relation. Positive and negative crossings are handled a bit differently; see Figure 7.22.

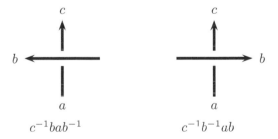

Figure 7.22: *Left to right*, relations in the Wirtinger presentation arising from negative and positive crossings.

Let's see how this works in the case of the right trefoil knot.

Example 127. An oriented right trefoil knot diagram is shown below with labels a, b, and c on the three strands. Labels are repeated near each crossing so that it's easier to see what goes into each relation.

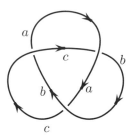

The upper left crossing contributes $a^{-1}c^{-1}bc$; upper right crossing, $b^{-1}a^{-1}ca$; and lower crossing, $c^{-1}b^{-1}ab$. Thus the knot group for the right trefoil has the presentation $\langle a, b, c \mid a^{-1}c^{-1}bc = b^{-1}a^{-1}ca = c^{-1}b^{-1}ab = \mathbf{1} \rangle$. Note that this presentation is not the most efficient way to describe the knot group. Let's explore the relations a bit more.

$$a^{-1}c^{-1}bc = \mathbf{1} \implies c^{-1}bc = a \implies bc = ca$$

Similarly, we obtain $ca = ab$ and $ab = bc$. But notice that only two relations are necessary, say, $ab = bc$ and $bc = ca$, since then the remaining one would follow from transitivity. Furthermore, if $ca = ab$, then $c = aba^{-1}$, which may be substituted for c in the other relations. After simplifying using the group axioms, we find the following.

$$ab = bc \quad \implies \quad ab = b(aba^{-1}) \implies aba = bab$$
$$bc = ca \quad \implies \quad b(aba^{-1}) = (aba^{-1})a \implies baba^{-1} = ab \implies bab = aba$$

Observe that we've reduced to the single relation, $aba = bab$. Thus the knot group for the right trefoil is: $\langle a, b \mid aba = bab \rangle$.

Now why does this procedure work? Let's take our oriented knot diagram and construct a certain model of the knot in $\mathbb{R}^3$. The diagram itself is two dimensional, and so all of the strands may embed into the plane $z = 0$ in $\mathbb{R}^3$. Place the diagram in a way so that it avoids the origin. For each undercrossing, connect the strands with an arc that goes below the plane (as illustrated in Figure 7.23 for the trefoil). Now for each strand a, draw a loop γ_a based at $\mathbf{0}$ encircling a – these are the generators of the knot group. For consistency, let each loop consist of an arc θ_a starting at the origin, leading to a small circle enclosing the strand (and no others) oriented so that it satisfies the *right-hand rule*[12] with respect to the strand, and then the reverse arc θ_a^{-1} back to the origin.

Each relation is a statement about how these loops interact near a crossing. For example, if we move the loop γ_a through an undercrossing, then the loop

---

[12]This idea should be familiar to those who have seen some physics. On your right hand, stick your thumb up and curl your fingers around. Then if your thumb points in the direction of the strand, your fingers will curl according to the proper orientation for the loop encircling the strand.

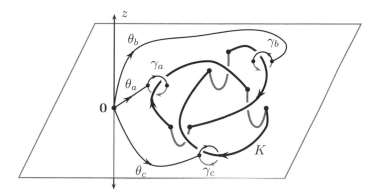

Figure 7.23: Embedding a knot diagram into $\mathbb{R}^3$ so that the strands embed into the plane $z = 0$. Gray curves go below the plane. The knot group is generated by loops that wind around each of the strands as shown.

now encircles another strand, say, strand c. There should be some relationship between γ_a and γ_c involving the over-strand at that crossing. Suppose the over-strand is b, and consider the loop γ_b. By carefully deforming the loops, it turns out that γ_c is simply a *conjugate*[13] of γ_a by γ_b; depending on the orientations involved, we have either $\gamma_c \simeq \gamma_b \cdot \gamma_a \cdot \gamma_b^{-1}$ or $\gamma_c \simeq \gamma_b^{-1} \cdot \gamma_a \cdot \gamma_b$. The case of a positive crossing is illustrated in Figure 7.24. Now for brevity we simply use a, b, c in place of the homotopy classes $[\gamma_a]$, $[\gamma_b]$, $[\gamma_c]$, which leads to the relations as displayed in Figure 7.22.

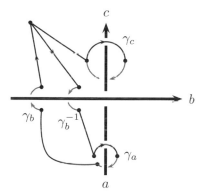

Figure 7.24: Moving a loop through a crossing results in a conjugate: $\gamma_c \simeq \gamma_b^{-1} \cdot \gamma_a \cdot \gamma_b$. Gray curves go below the plane.

The knot group is a fairly powerful knot invariant, but it has two major drawbacks. First, it cannot distinguish between mirror images. If $K \subseteq \mathbb{R}^3$,

---

[13]In group theory, an expression such as xyx^{-1} or $x^{-1}yx$ is called a **conjugate** of y.

then the homeomorphism $f : (x, y, z) \mapsto (x, y, -z)$ reflects K to its mirror image K^*. Thus, since the fundamental group is a topological invariant, we have $\pi_1(\mathbb{R}^3 \setminus K^*) \cong \pi_1(\mathbb{R}^3 \setminus K)$. Second, it is generally difficult to prove that two groups are the same or different based only on presentations for the groups.

Exercises

1. Determine the homotopy class of a (p, q)-torus knot (see Example 115 from §6.3), viewed as a loop in $\mathbb{T}$. Use a and b as generators for $\pi_1(\mathbb{T})$ as in (7.15).

2. Find the fundamental group of each surface, writing it in terms of generators and relations as in (7.17).

   (a) $3\mathbb{T}$ (b) $\mathbb{K}$ (c) $\mathbb{T} \# \mathbb{K}$ (d) $n\mathbb{P}$

3. Construct a space whose fundamental group is isomorphic to $\mathbb{Z}/4\mathbb{Z}$.

4. Compute the knot group of the left trefoil knot. Show that there is a labeling of the strands so that its group matches the knot group of the right trefoil, $G = \langle a, b \mid aba = bab \rangle$.

5. Find the Wirtinger presentation for the figure-eight knot (rightmost knot in Figure 6.33).

6. Show that the knot group of the Hopf link is the abelian group $\langle a, b \mid ab = ba \rangle \cong \mathbb{Z}^2$. Demonstrate this using two key rings linked together and a piece of string.

7.5   *Higher Homotopy Groups

You may have been wondering what the "1" signifies in the notation $\pi_1(X)$ for the fundamental group. It turns out that $\pi_1(X)$ is one of an infinite family of topological invariants. Thinking of $\pi_1(X)$ as the set of homotopy classes of maps $\mathbb{S}^1 \to X$, it is natural to consider homotopy classes of maps $\mathbb{S}^n \to X$ for any integer $n \geq 0$. Taken together, $\pi_n(X)$ for $n \geq 0$ are called the **homotopy groups** of X. Although the higher homotopy groups tend to be much harder to compute, they are extremely important because, taken together, they come close to being a *complete invariant* for homotopy equivalence, in a limited sense of the term anyway. For a large class of spaces, those that are homotopy equivalent to a *CW complex*[14] if there is a map $f : X \to Y$ that induces isomorphisms on every homotopy group (i.e., $f_* : \pi_n(X) \to \pi_n(Y)$ is an isomorphism $\pi_n(X) \cong \pi_n(Y)$ for every integer $n \geq 0$), then it follows that $X \simeq Y$. This important result is

---

[14]CW complexes include the *cell complexes* mentioned in Chapter 5 as well as *infinite-dimensional* spaces that are built up from cell complexes in a certain way.

known as **Whitehead's Theorem**.[15] Conversely, if two spaces X and Y differ at even a single homotopy group (i.e., $\pi_n(X) \not\cong \pi_n(Y)$ for some n), then X cannot be homotopic to Y (hence X and Y cannot be homeomorphic either).

Maps of Spheres

Recall Definition 7.1.3; two maps γ_0 and γ_1 from $\mathbb{S}^n$ to X are homotopic ($\gamma_0 \simeq \gamma_1$) if there is a map $h : \mathbb{S}^n \times \mathbb{I} \to X$ such that $\gamma_0 = h(s, 0)$ and $\gamma_1 = h(s, 1)$ for all $s \in \mathbb{S}^n$. Suppose $\mathbb{S}^n$ is represented by the unit sphere in $\mathbb{R}^{n+1}$, and let $s_0 = (1, 0, \ldots, 0)$ be the chosen basepoint for $\mathbb{S}^n$; let x_0 be the chosen basepoint for X. The homotopy h is *basepoint preserving* if $h(s_0, t) = x_0$ for all $t \in \mathbb{I}$.

> **Definition 7.5.1.** For each integer $n \geq 0$, define the nth **homotopy group** of X (with respect to a chosen basepoint $x_0 \in X$) by
>
> $$\pi_n(X) = \{[\gamma] \mid \gamma : (\mathbb{S}^n, s_0) \to (X, x_0)\},$$
>
> where the equivalence class $[\gamma]$ is defined by basepoint-preserving homotopy.

Consider the simplest case, $n = 0$. Since $\mathbb{S}^0 = \{-1, 1\}$ is nothing more than a two-point discrete set, every function $\gamma : \mathbb{S}^0 \to X$ is continuous, and may be described by the pair of points $\gamma(1) = x_0$ and $\gamma(-1) = x_1$ in X. Furthermore, since $\gamma(1)$ *must* be the basepoint x_0, we only need to know x_1 to describe γ completely. In other words, the map $\gamma : (\mathbb{S}^0, s_0) \to (X, x_0)$ is equivalent to a choice of a single point $x_1 \in X$. We defined $\pi_1(X)$ in §7.2 for arc-connected spaces to avoid ambiguities with different choices of basepoints; however, when X is arc-connected, $\pi_0(X)$ is *trivial*. This is because if there is an arc $\eta : \mathbb{I} \to X$ such that $\eta(0) = x_1 = \gamma(-1)$ and $\eta(1) = x_0$, then the following homotopy $h : \mathbb{S}^0 \times \mathbb{I} \to X$ can be defined.

$$h(s, t) = \begin{cases} x_0, & s = 1 \\ \eta(t), & s = -1 \end{cases}$$

Note that $h(s_0, t) = h(1, t) = x_0$ for all $t \in \mathbb{I}$ (so h is basepoint-preserving); $h(-1, 0) = \eta(0) = x_1$ (so $h(s, 0) = \gamma(s)$); and $h(-1, 1) = \eta(1) = x_0$ (so $h(s, 1)$ is the *constant* map $\mathbb{S}^0 \to X$ sending both points to the basepoint x_0). Thus $\gamma \simeq \mathbf{1}$, and so $\pi_0(X)$ has only one element.

The situation is a little more interesting for spaces X that are not arc-connected. Suppose $X = X_1 \cup X_2 \cup \ldots \cup X_r$, where each subspace X_k is a distinct arc-connected component of X. Then it's not too difficult to prove that $\pi_0(X)$ has exactly r elements.[16] Figure 7.25 may give a clue as to how to show this.

---

[15] For a proof of Whitehead's Theorem, see Hatcher [Hat02].

[16] The set $\pi_0(X)$, however, cannot generally be given a natural group structure unless the space X itself carries a group structure.

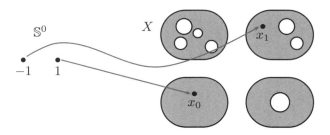

Figure 7.25: A map $\gamma : \mathbb{S}^0 \to X$, in which X has four connected components.

The second homotopy group of a space, $\pi_2(X)$, identifies nontrivial maps of the sphere $\mathbb{S}^2$ into X. For example, Figure 7.26 demonstrates a trivial and nontrivial element of $\pi_2(\mathbb{R}^3\backslash\{\mathbf{0}\})$. It is often very challenging to imagine different ways in which a sphere could "wrap" itself around within a given topological space, but there are certain tools that can be used to compute the homotopy groups.[17]

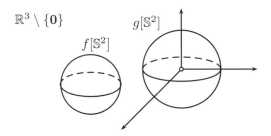

Figure 7.26: Two maps $f, g : \mathbb{S}^2 \to \mathbb{R}^3 \setminus \{\mathbf{0}\}$. Here $f \simeq \mathbf{1}$ because the image of the sphere can be shrunk to a point, but since $g[\mathbb{S}^2]$ encloses the missing point at the origin, the image of g cannot be shrunk continuously to a point. In fact, $[g]$ is a generator of $\pi_2(\mathbb{R}^3 \setminus \{\mathbf{0}\}) \cong \mathbb{Z}$.

For $n \geq 2$, when $\pi_n(X)$ is *trivial* (in other words, consists of a single homotopy class), we denote this by $\pi_n(X) = 0$. Later we will show that $\pi_n(X)$ is an abelian group for $n \geq 2$, and so you can think of 0 as the trivial abelian group with a single element, the additive identity element: $0 = \{\mathbf{0}\}$.

Example 128. For all integers $r \geq 0$, $n \geq 2$, it can be shown that $\pi_n(\mathbb{R}^r) = 0$. The proof is a straightforward extension of Example 125 (see §7.1, Exercise 1).

---

[17]These kinds of tools fall well outside the scope of this elementary textbook.

Group Structure

The homotopy groups are called *groups* because they all (except for $\pi_0(X)$) possess a product function. It was relatively easy to define the product of two loops, because we all have the intuitive understanding that a point can travel along one loop for a finite time and then get on another loop and travel, as easily as you can transfer from one subway line to another. There is a natural idea of time progression in a map $\gamma : \mathbb{I} \to X$; start at $\gamma(0)$ and end at $\gamma(1)$, and this time progression still makes sense when the endpoints of $\mathbb{I}$ are identified so that γ becomes a map $\mathbb{S}^1 \to X$. But what does it mean to "travel" along the map $\gamma : \mathbb{S}^n \to X$ when $n \geq 2$? The key is to flatten out $\mathbb{S}^n$ and then identify one direction of travel as a *time* dimension. Recall that there is a homeomorphism $\mathbb{S}^n \approx \mathbb{I}^n / \partial \mathbb{I}^n$. When $n \geq 1$, we may write $\mathbb{I}^n = \mathbb{I} \times \mathbb{I}^{n-1}$ and identify the first component $\mathbb{I}$ as *time*. In other words, if $s = (s_1, s_2, \ldots, s_n) \in \mathbb{I}^n$, then regard the first component s_1 as time; the components $s_2, \ldots, s_n$ play no substantial role in the product definition below. The product of two maps $\gamma_0, \gamma_1 : \mathbb{I}^n \to X$ can thus be defined.

$$\gamma_0 \cdot \gamma_1 \quad : \quad \mathbb{I}^n = \mathbb{I} \times \mathbb{I}^{n-1} \to X$$
$$(\gamma_0 \cdot \gamma_1)(s) \;\; = \;\; \begin{cases} \gamma_0(2s_1, s_2, \ldots, s_n), & s_1 \in [0, 1/2] \\ \gamma_1(2s_1 - 1, s_2, \ldots, s_n), & s_1 \in (1/2, 1] \end{cases} \tag{7.18}$$

Any map $\gamma : \mathbb{S}^n \to X$ can be regarded as a map from $\mathbb{I}^n$ to X via the quotient map $\mathbb{I}^n \to \mathbb{I}^n / \partial \mathbb{I}^n \approx \mathbb{S}^n$. Such a map would necessarily send all of the boundary $\partial \mathbb{I}^n$ to the basepoint x_0 of X. We should verify that if γ_0 and γ_1 are maps whose domains are both $\mathbb{S}^n$, then their product can be viewed[18] as a map of $\mathbb{S}^n$ too. This is an easy exercise. Moreover, the proof that the product map actually gives a valid group structure on $\pi_n(X)$ is essentially the same as the proof of Theorem 7.2.2 and discussion preceding that theorem.

Formula (7.18) reduces to Definition 7.1.6 when $n = 1$, since then $\mathbb{I}^{n-1} = \mathbb{I}^0$ is a single point. Figure 7.27 illustrates how the product works on maps from $\mathbb{S}^2$ to a space X. Regarding a map $\gamma : \mathbb{S}^2 \to X$ as a map $\mathbb{I}^2 \to X$ (whose value on the boundary is x_0), we compose two such maps, γ_0 and γ_1, by compressing along the first component of the domain.

$$(\gamma_0 \cdot \gamma_1)(s_1, s_2) = \begin{cases} \gamma_0(2s_1, s_2), & s_1 \in [0, 1/2] \\ \gamma_1(2s_1 - 1, s_2), & s_1 \in (1/2, 1] \end{cases}$$

It turns out that for the $n \geq 2$, the group structure of $\pi_n(X)$ is commutative. Essentially, there is enough room for maps to "slide past" one another when the domain has dimension two or higher. The sequence of diagrams in Figure 7.28 suggests a homotopy $\gamma_0 \cdot \gamma_1 \simeq \gamma_1 \cdot \gamma_0$ (for $\gamma_1, \gamma_2 : \mathbb{S}^2 \to X$ in particular), although a precise proof is not given here. The domain of each map γ_i $(i = 0, 1)$ is shrunk small enough so that those regions can interchange positions. The gray area

---

[18]In more precise language, we must verify that the map $\gamma_0 \cdot \gamma_1$ as defined in (7.18) *factors through* $\mathbb{I}^n / \partial \mathbb{I}^n$.

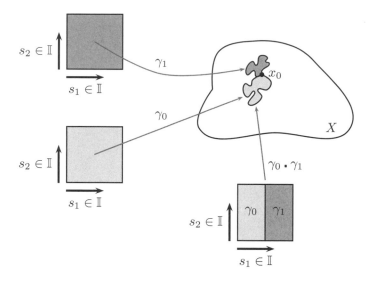

Figure 7.27: The product $\gamma_0 \cdot \gamma_1$.

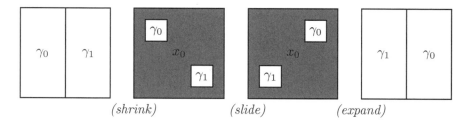

Figure 7.28: Shrinking and sliding the domain of γ_0 past that of γ_1, showing how $[\gamma_0] + [\gamma_1] = [\gamma_1] + [\gamma_0]$.

indicates regions of the domain that get sent to the basepoint $x_0 \in X$. Since the boundary of each domain for γ_i must map to x_0 anyway, there is continuity in each frame of the homotopy. It takes a more careful argument, though, to ensure continuity of the homotopy on all of $\mathbb{S}^n \times \mathbb{I}$ (perhaps the diligent reader can formalize the proof). The reason this same idea doesn't work to show that $\pi_1(X)$ is commutative is that the domain of the maps γ_i must be at least two-dimensional to accommodate such an interchange. Two regions of the one-dimensional segment $\mathbb{I}$ cannot be made to slide past one another, no matter how much those regions are shrunk.

Because of commutativity, we typically use the notation $+$ for the group operation. If $n \geq 2$, and $[\gamma], [\eta] \in \pi_n(X)$, then $[\gamma] + [\eta] = [\eta] + [\gamma]$.

An important property of fundamental groups that is shared by the higher homotopy groups is the existence of induced homomorphisms. Suppose $f : X \to Y$. Then there is an induced group homomorphism $f_* : \pi_n(X) \to \pi_n(Y)$ for each integer $n \geq 0$. The homomorphism f_* is defined in precisely the same way is in Definition 7.2.4, by $f_*([\gamma]) = [f \circ \gamma]$, and satisfies the same properties as in Theorem 7.2.5. As a consequence, the homotopy groups π_n for $n \in \mathbb{N} \cup \{0\}$ are topological invariants.

Homotopy Groups of the Spheres

We end this section with something of a mystery. While it is quite easy to obtain complete information about $\pi_n(\mathbb{S}^1)$, there's no such luck for $\pi_n(\mathbb{S}^r)$ for any $r \geq 2$. Here are some of the things we do know.[19]

- If $n < r$, then $\pi_n(\mathbb{S}^r) = 0$. Intuitively, this means that there is no nontrivial way to map a lower-dimensional sphere into a higher-dimensional sphere. Every such map is homotopic to the map sending all points of $\mathbb{S}^n$ to the basepoint of $\mathbb{S}^r$.

- We have $\pi_3(\mathbb{S}^2) = \mathbb{Z}$. It may come as a surprise that there are nontrivial ways in which the three-dimensional sphere can "wrap around" a two-dimensional sphere. There is a map $\eta : \mathbb{S}^3 \to \mathbb{S}^2$ called the **Hopf fibra-tion**.[20] The preimage of any point $x \in \mathbb{S}^2$ is a circle in $\mathbb{S}^3$, and for this reason we often represent the fibration as a sequence,

$$\mathbb{S}^1 \longrightarrow \mathbb{S}^3 \longrightarrow \mathbb{S}^2.$$

  We leave further discussion of the Hopf fibration to more advanced texts. Even visualizing η is quite a challenge.[21]

- The vast majority of the groups $\pi_n(\mathbb{S}^r)$ are known to be finite, the only exceptions being $\pi_n(\mathbb{S}^n)$ and $\pi_{4n-1}(\mathbb{S}^{2n})$ for $n \in \mathbb{N}$.

- It's known that $\pi_n(\mathbb{S}^3) = \pi_n(\mathbb{S}^2)$ for all $n \geq 3$. This curious result is actually a by-product of the Hopf fibration and the fact that $\pi_n(\mathbb{S}^1)$ is trivial for $n \geq 2$.

- For all $r \geq n + 2$, it is known that $\pi_{r+n}(\mathbb{S}^r)$ is **stable**, in the sense that $\pi_{r_1+n}(\mathbb{S}^{r_1}) = \pi_{r_2+n}(\mathbb{S}^{r_2})$ for every pair of integers $r_1, r_2 \geq n + 2$. This is called the **Freudenthal Suspension Theorem**. For example, when $n = 0$, we have $\pi_2(\mathbb{S}^2) = \pi_3(\mathbb{S}^3) = \pi_4(\mathbb{S}^4) = \cdots$. It can be shown that $\pi_2(\mathbb{S}^2) = \mathbb{Z}$; hence $\pi_n(\mathbb{S}^n) = \mathbb{Z}$ for all $n \in \mathbb{Z}$. (Recall that in §7.3 we proved the result for $n = 1$.) The so-called **stable homotopy groups** of the spheres are defined for each $n \in \mathbb{N}$ by $\pi_n^s = \pi_{r+n}(\mathbb{S}^r)$ for $r \geq n + 2$. While determining the stable homotopy groups is generally regarded as a more tractable problem than determining all the **unstable** groups

---

[19]For discussions and proof of these results, see Hatcher [Hat02], May [May99], etc.

[20]We say a map $f : E \to X$ is a *fibration* if f satisfies the Homotopy Lifting Property (Theorem 7.3.2), and so all covering maps are examples of fibrations.

[21]Niles Johnson has done a remarkable job of this; see `http://nilesjohnson.net/hopf.html`.

(the various groups $\pi_n(\mathbb{S}^r)$ for $n, r \in \mathbb{N}$), topologists are still far from a complete answer. As the following table suggests,[22] no straightforward pattern emerges for π_n^s.

n	0	1	2	3	4	5	6	7	8	9	10	11	12
Size of π_n^s	∞	2	2	24	0	0	2	240	4	8	6	504	0

Exercises

1. Suppose $X = X_1 \cup X_2 \cup \ldots \cup X_r$, where each X_k is a distinct arc-connected component of X. Show that $\pi_0(X)$ has exactly r elements by giving an explicit bijection between $\pi_0(X)$ and $\{X_k \mid k = 1, 2, \ldots r\}$.

2. Verify that for maps $\gamma_0, \gamma_1 : (\mathbb{S}^n, s_0) \to (X, x_0)$, the product map $\gamma_0 \cdot \gamma_1 : \mathbb{I}^n \to X$ defined in (7.18) sends all of $\partial \mathbb{I}^n$ to x_0; hence can be interpreted as a map $\mathbb{S}^n \to X$.

Supplemental Reading

- Crowell and Fox [CF08], Chapter VI for knot groups.

- Hatcher [Hat02], Chapter 1, for the fundamental group; Chapter 4 for higher homotopy groups.

- Massey [Mas91], especially Chapter IV for methods for finding fundamental groups of plane models of surfaces.

- May [May99], various chapters, for a concise treatment of homotopy theory using more advanced techniques.

- Munkres [Mun00], Chapters 9 and 12.

- Spanier [Spa94], Chapters 1, 2, and 7.

---

[22]Larger tables of unstable and stable homotopy groups can be found at `https://en.wikipedia.org/wiki/Homotopy_groups_of_spheres`, for example.

Chapter 8

Introduction to Homology

Chapter 7 introduced us to an important and powerful set of invariants called the fundamental group and the higher homotopy groups. Taken together, these groups, $\pi_n(X)$ for $n \in \mathbb{N}$, provide detailed information about the topological structure of a space X, but they are very difficult to compute in practice – at the time of this writing, we don't even know all of the homotopy groups of the sphere $\mathbb{S}^2$ (recall §7.5). However, there is a related family of invariants called the *homology* groups that are typically easier to calculate and yet can carry a wealth of information about the topological properties of the space.

When you think of a torus, what's the first thing that comes to mind? It's the *hole* in the middle, right? In fact, the spaces $n\mathbb{T}$ for various n can easily be distinguished by the number of holes. Now let's be careful though. The hole is not really *part of* the torus any more than the space inside a room is part of the material structure of the house containing that room (instead, the room emerges as a *feature* of the house because of the way the walls, ceilings, and floors are arranged). While we sometimes think of a hole as *what's missing*, a much better way of identifying it is to describe how the space *surrounds* it. We call a hole n-dimensional if a sphere $\mathbb{S}^n$ can surround it in a nontrivial[1] way.

Let's take a motivating example. Consider the object shown in Figure 8.1. It consists of the surface of a sphere together with a disk through the equator and an arc attached at two points. We label the top hemisphere A, bottom hemisphere B, the equatorial disk D, and the arc C. A circle $\mathbb{S}^1$ surrounds the empty space "within" the arc handle C, illustrating a one-dimensional hole. The disk D partitions the interior of the sphere into two "chambers," so there are two enclosed regions: the upper chamber bounded by $A \cup D$, and the lower chamber bounded by $B \cup D$. Indeed, $A \cup D$ and $B \cup D$ each represent nontrivial embeddings of $\mathbb{S}^2$ into the space. But what about the sphere itself? Clearly the sphere $A \cup B$ is nontrivial, but we say it is not *independent* of the two spheres $A \cup D$ and $B \cup D$. In some sense, there should be an *addition* operation such

---

[1]We are using the word *nontrivial* here to mean that the surrounding sphere is unable to be deformed within the space to a single point.

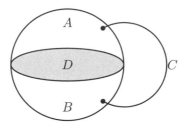

Figure 8.1: This space has a one-dimensional hole (encircled partly by C and partly by a path on the sphere) and two two-dimensional holes (enclosed by $A \cup D$ and $B \cup D$).

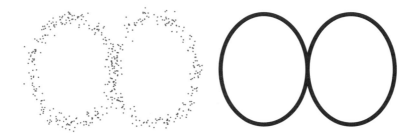

Figure 8.2: The point cloud on the left may be a noisy random sampling of the space on the right.

that *sum of two chambers is itself a chamber* (but we must be vague at this point until introducing the required mathematical machinery).

So homology is basically about identifying "holes" of various dimensions in a space (of course there's much more to the story, as we shall see in this chapter). Moreover, owing to the relative flexibility in how homology can be computed and interpreted (relative to homotopy groups, that is), homology turns out to be a surprisingly powerful tool in the field of *data analysis*. Take, for example, the data set pictured on the left in Figure 8.2. It is clear that the data are clustered in a pattern that has two "holes," and the distribution of points reflects an essentially one-dimensional skeleton, leading one to conjecture that the ideal data set should look something like the figure-eight pictured on the right in the figure. The data set, or *point cloud*, itself then would be interpreted as a finite random sampling from the ideal space with *noise* or *error* in the sampling. While the precise methods used in *topological data analysis* are beyond the scope of this text, it is interesting to note that homology theory (in the form of so-called *persistence homology*) plays a major role.[2]

---

[2]See the wonderfully readable article by Gunnar Carlsson posted online at `http://www.ayasdi.com/blog/bigdata/why-topological-data-analysis-works/`. See also [Car09].

8.1 Rational Homology

Before defining homology in this text, it is important to understand that there is not just one theory of homology. Different homology theories[3] exist, applying to many different kinds of topological spaces. In this chapter we develop the *rational*[4] homology of triangulated spaces, but let's first try to understand what homology should be measuring in the easy case of one-dimensional cell complexes, or *graphs* (recall Chapter 6 for definitions).

Homology of Graphs

We start by defining a rational homology theory for graphs. This allows us to introduce essential terminology and concepts without too much complication. The only kinds of "holes" we see in graphs are those surrounded by closed walks.[5] Such a closed walk is called a **1-cycle** of the graph (though the terminology is much broader than the graph-theoretic definition of cycle from Chapter 6).

How many 1-cycles could a graph have? This seems like an easy question to answer in practice – simply look at the graph and count them, right? However, this task is impossible in general since *absolutely every* closed walk surrounding a hole in the graph qualifies, including walks that circle around the same set of edges an arbitrary number of times. Rather than attempting to list all the 1-cycles (a fruitless task), homology presents the information in the most efficient format – in terms of a *cycle basis*, a minimal set of 1-cycles such that every 1-cycle in the graph can be expressed unambiguously as a combination of those basis cycles.

Thus far in this text, graphs have been **undirected**, in the sense that there is no indicated orientation (direction) on any edge. To properly define the homology of a graph, we need to specify the orientation of each edge; in other words, to consider **directed** graphs. It turns out that the specification is immaterial; different choices of orientations lead to isomorphic homology groups.

Let's locate a few 1-cycles in the graph G shown in Figure 8.3. We identify the cycles in a peculiar way: by listing a *sum* of edges, where the sign of an edge indicates the orientation, and allowing terms to commute and cancel when appropriate (in contrast to loops as defined in §7.1). For example, let's describe the triangular cycle with vertices u, v, and w. Starting at vertex u and proceeding directly to v, the edge c is traversed in the same direction as its arrow; this contributes $+c$ to the sum. Then from v to w with the flow, another term $+d$ is added. Finally $-b$ is added to the sum corresponding to the journey from w

---

[3]Just as a set $\mathscr{T} \subseteq \mathcal{P}(X)$ must satisfy a certain set of axioms to be called a topology on X, a *homology theory* must satisfy a specific list of conditions called the **Eilenberg-Steenrod axioms** [ES45] on a specified class of topological spaces.

[4]Here *rational* refers to the rational numbers $\mathbb{Q}$, not as a synonym for *reasonable* or *sane* – though the case can be made, tongue-in-cheek of course, that there are some incredibly insane homology theories out there.

[5]This includes loops (walks of length 1), pairs of parallel edges (length 2), circuits and cycles (length ≥ 3), as well as more general closed walks that may backtrack along some edges.

back to u, against the flow. Thus there is a cycle $(+c)+(+d)+(-b) = -b+c+d$ in the graph. Of course there are many more, including $-a+b$, f, $-a+c+d$, and even $b+(-d)+e+f+f+f+(-e)+(-c)+a+(-b) = a-c-d+3f$.

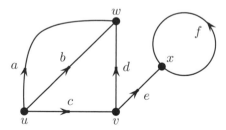

Figure 8.3: A directed graph G with vertex set $\{u, v, w, x\}$ and edge set $\{a, b, c, d, e, f\}$.

What makes each of these expressions a 1-cycle? Each directed edge $e \in E(G)$ has a tail and a head vertex – say, v_0 and v_1, respectively – and we may write $e = [v_0, v_1]$. The term $+e$ represents a journey *from* v_0 *to* v_1. Similarly, $-e$ is a journey *from* v_1 *to* v_0 (equivalently, $-[v_0, v_1] = [v_1, v_0]$). Imagine each word *from* is replaced by a minus sign, and each word *to* is replaced by a plus sign. The key is that each vertex involved in the sum of edges shows up as both a tail and a head, and so can "cancel" completely. For example, the 1-cycle $-b+c+d$ could be broken down as follows.

$$-b = -[u, w] = [w, u] \quad \mapsto \quad \text{from } w, \text{ to } u \quad \mapsto \quad (-w) + u$$
$$+c = [u, v] \quad \mapsto \quad \text{from } u, \text{ to } v \quad \mapsto \quad (-u) + v$$
$$+d = [v, w] \quad \mapsto \quad \text{from } v, \text{ to } w \quad \mapsto \quad (-v) + w$$

When all six terms are combined, the result is 0. This is the essence of what it means to be a 1-cycle.

Another feature of 1-cycles is the ability to add them to each other to produce new 1-cycles. For example, the 1-cycles $\alpha = -b+c+d$ and $\beta = -a+b$ sum to $\alpha + \beta = -a+c+d$. Cycles can even be added to themselves (e.g., $\alpha + \alpha = 2\alpha = 2c + 2d - 2b$) or negated (e.g., $-\beta = a - b$). Intuitively, a multiple of a cycle should mean traversing it that many times, and the opposite of a cycle should mean traversing it the opposite direction; however, in *rational homology* we would also like to include more general coefficients (e.g., $\frac{3}{4}\alpha - \frac{2}{7}\beta$ qualifies as a rational 1-cycle), and so a 1-cycle may not always be realizable as a walk in the graph.

At this point we must introduce tools and terminology from *linear algebra*. Refer to §B.2 or any standard elementary text (e.g., Lay et al. [LLM15], Poole and Lipsett [PL14]) for more information. Recall that if $A = \{a_1, a_2, \ldots, a_n\}$ is a set, then a (formal) **linear combination** of A is any expression of the form

$$\sum_{k=1}^{n} c_k a_k = c_1 a_1 + c_2 a_2 + \cdots + c_n a_n, \tag{8.1}$$

in which each coefficient c_k is a **scalar**. In this section, we take all scalars to be in the set of rational numbers, $\mathbb{Q}$, though the methods presented here work over any field ($\mathbb{R}, \mathbb{C}$, etc.), but not the set of integers $\mathbb{Z}$ – more about this point in §8.2. That is, we assume for now that for each $k \in \{1, 2, \ldots, n\}$, we have $c_k \in \mathbb{Q}$ in (8.1).

Definition 8.1.1. Let G be a directed graph.

- A **0-chain** in G is any linear combination of the vertices of the graph. The set of all 0-chains of G is denoted $C_0(G; \mathbb{Q})$.

- A **1-chain** in G is any linear combination of the edges of the graph. The set of all 1-chains of G is denoted $C_1(G; \mathbb{Q})$.

Recall the set of all linear combinations of elements from a set A is called the **span** of A, and may be written $\mathrm{Span}_R A$, where R is the set of scalars over which we are working. So $C_0(G; \mathbb{Q}) = \mathrm{Span}_{\mathbb{Q}} V(G)$, and $C_1(G; \mathbb{Q}) = \mathrm{Span}_{\mathbb{Q}} E(G)$. If $x \in C_0(G; \mathbb{Q})$, then we say x has **degree** 0; similarly, $x \in C_1(G; \mathbb{Q})$ has degree 1. (This use of the term *degree* is different than the graph-theoretic notion of degree from §6.1.) Define two functions from the set of edges to the set of vertices, called **boundary functions**.

$$\partial_0(e) = \partial_0([v_0, v_1]) \quad = \quad v_1 \quad (= \text{head vertex})$$
$$\partial_1(e) = \partial_1([v_0, v_1]) \quad = \quad v_0 \quad (= \text{tail vertex})$$

Then encode the ideas of *to* and *from* using signs. We define a function d_1 on the set of edges by

$$d_1(e) = d_1([v_0, v_1]) = \partial_0([v_0, v_1]) - \partial_1([v_0, v_1]) = v_1 - v_0. \quad (8.2)$$

Extend d_1 to all of $C_1(G; \mathbb{Q})$ by *linearity*. That is, if $E(G) = \{e_1, e_2, \ldots, e_n\}$, then define

$$d_1 \quad : \quad C_1(G; \mathbb{Q}) \to C_0(G; \mathbb{Q})$$
$$d_1 \left(\sum_{i=1}^{n} \lambda_i e_i \right) \quad = \quad \sum_{i=1}^{n} \lambda_i d_1(e_i).$$

The map d_1 is an example of a **boundary homomorphism**, and it will be used to determine the 1-cycles of a graph.

Definition 8.1.2. Let G be a directed graph. A **1-cycle** in G is any 1-chain α such that $d_1(\alpha) = \mathbf{0}$.

To develop a more intuitive feel for how Definition 8.1.2 works, refer to Figure 8.4. Each edge in a given 1-cycle contributes a positive and negative boundary vertex, which cancel with one another, resulting in a sum of $\mathbf{0}$.

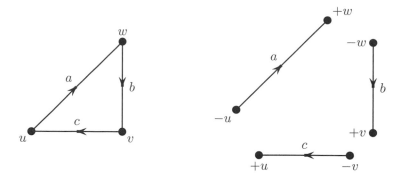

Figure 8.4: *Left*, here $\alpha = a + b + c$ is a 1-cycle in the graph. *Right*, verifying that $d_1(\alpha) = \mathbf{0}$, $d_1(a + b + c) = d_1(a) + d_1(b) + d_1(c) = \partial_0(a) - \partial_1(a) + \partial_0(b) - \partial_1(b) + \partial_0(c) - \partial_1(c) = w - u + v - w + u - v = \mathbf{0}$.

Now we may define the *degree*-1 *homology*[6] of a graph.

> **Definition 8.1.3.** The first **homology group** $H_1(G;\mathbb{Q})$ of a graph G is defined as the set of all rational 1-cycles in G. The **first Betti number** of G is the dimension of $H_1(G;\mathbb{Q})$.

Example 129. Describe the homology group $H_1(G;\mathbb{Q})$ for the graph shown in Figure 8.3. What is the first Betti number?

Solution: We are looking for every 1-chain α that is a 1-cycle (i.e., $d_1(\alpha) = \mathbf{0}$). Since a general 1-cycle is nothing more than a linear combination of the edges, and the function d_1 is linear, we can set up a *matrix* to represent the action of d_1. Each column of the matrix corresponds to an edge and each row to a vertex. Now for an edge $e_i = [u_0, u_1]$, we have $d_1(e_i) = u_1 - u_0$; encode this in the matrix by adding 1 to the entry corresponding to u_1 and adding -1 to the entry corresponding to u_0 in the column for e_i. If $u_0 = u_1$, then the net effect is to add $\mathbf{0}$.

$$
\begin{array}{c}
u \\ v \\ w \\ x
\end{array}
\begin{pmatrix}
\begin{array}{cccccc}
a & b & c & d & e & f
\end{array} \\
\begin{array}{cccccc}
-1 & -1 & -1 & 0 & 0 & 0 \\
0 & 0 & 1 & -1 & -1 & 0 \\
1 & 1 & 0 & 1 & 0 & 0 \\
0 & 0 & 0 & 0 & 1 & 0
\end{array}
\end{pmatrix}
$$

Solve to find the *null space* of the matrix. This can be done by hand, but there are also numerous options for doing the work by computer.[7] We obtain three

---

[6] **Caution**: This particular definition of homology applies *only* to graphs.

[7] For example, in Sagemath, you can use the command `right_kernel()` on the matrix. The result is a matrix whose rows represent a basis for the null space.

independent basis 1-cycles, $a - c - d$, $b - c - d$, and f. All other 1-cycles are linear combinations of these three (e.g., $(a - c - d) - (b - c - d) = a - b$ is also a cycle). Thus the first Betti number is 3, and

$$H_1(G; \mathbb{Q}) = \mathrm{Span}_{\mathbb{Q}}\{a - c - d,\, b - c - d,\, f\} \cong \mathbb{Q}^3.$$

Now that we've gotten our feet wet, let's expand our definition of homology to include some higher-dimensional spaces. In addition to vertices and edges, we also now consider triangles and their higher-dimensional analogs called *simplices*.

Simplices

Consider the triangle $\triangle ABC$ defined by three points in the plane, say, $A = (-1, -1)$, $B = (2, -3)$, and $C = (1, 3)$. You might recall that the *centroid* (or *center of mass*) M of the triangle is found by averaging the coordinates:

$$M = \frac{A + B + C}{3} = \left(\frac{-1 + 2 + 1}{3}, \frac{-1 + (-3) + 3}{3} \right) = \left(\frac{2}{3}, -\frac{1}{3} \right).$$

The point M could also have been expressed as $M = \frac{1}{3}A + \frac{1}{3}B + \frac{1}{3}C$, which is a type of *weighted* average. A **weighted average** is simply a linear combination in which the coefficients are nonnegative and sum to 1. It's a fairly well-known fact of analytic geometry that any point on the boundary or interior of $\triangle ABC$ is a weighted average of the vertices A, B, and C (see Figure 8.5):

$$\triangle ABC = \{w_0 A + w_1 B + w_2 C \mid w_0 + w_1 + w_2 = 1;\ w_k \geq 0,\ \forall k\}.$$

We call the triangle a 2-*simplex* because it is essentially two dimensional (except on its boundary, of course). Note that if the points A', B', and C' happened to lie on a single line (as shown in Figure 8.5), then the "triangle" formed from these points is not a triangle at all, but a line segment – in this situation, we call the simplex *degenerate*.

An n-*simplex* (plural: *simplices*[8]) is the analog of a triangle in arbitrary dimensions. In the definition below, we use the term **hyperplane**. An r-dimensional hyperplane H of a Euclidean space $\mathbb{R}^n$ is a subset $H \subseteq \mathbb{R}^n$ having the geometrical properties of $\mathbb{R}^r$ (flatness, infinite extent, etc.). For example, any line in $\mathbb{R}^n$ is a one-dimensional hyperplane of $\mathbb{R}^n$. Planes are two-dimensional hyperplanes, and so on.

---

[8] *Simplices* is the plural of *simplex*, just as *vertices* is the plural of *vertex*, *matrices* is the plural of *matrix*, and *indices* is the plural of *index*. Isn't Latin fun?

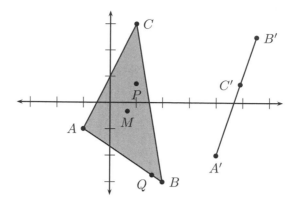

Figure 8.5: $\triangle ABC$ is a 2-simplex defined as the set of weighted averages of A, B, C; e.g., $M = \frac{1}{3}A + \frac{1}{3}B + \frac{1}{3}C = (2/3, -1/3)$, $P = \frac{1}{7}A + \frac{2}{7}B + \frac{4}{7}C = (1, 5/7)$, $Q = (0.13)A + (0.87)B + (0)C = (1.61, -2.74)$. On the other hand, points A', B', and C' are collinear, and so $A'C'B'$ is a degenerate 2-simplex. Ignoring the point C', we could say that $\overline{A'B'}$ is a 1-simplex.

Definition 8.1.4. Let $n \geq 0$ be an integer. An n-dimensional **simplex** (or n-**simplex**) Δ^n is a subset of a Euclidean space defined as the set of weighted averages,

$$\Delta^n = \left\{ \sum_{k=0}^{n} w_k v_k \;\middle|\; \sum_{k=0}^{n} w_k = 1; \; w_k \geq 0, \; \forall k \right\}.$$

The simplex may be denoted by the bracket $[v_0, v_1, v_2, \ldots, v_n]$ of $n + 1$ points, called the **vertices** of the simplex. If the vertices all lie in the same $(n-1)$-dimensional hyperplane, then the simplex is called **degenerate**; otherwise, **nondegenerate**.

The definition of simplex given above is equivalent to the following, which you might encounter in other texts:

> Δ^n is the *convex hull* of $n + 1$ points, $v_0, v_1, \ldots, v_n$, in a Euclidean space, and the simplex is nondegenerate if and only if the $n + 1$ points are in *general position*.

Just as the direction along an edge matters in defining homology for a graph, the ordering of the vertices of a simplex matters. The order of vertices matters in the bracket notation, and must be specified before any calculations are carried out. The 2-simplex $\triangle ABC$ from Figure 8.5 may be represented in six different permutations:

$$[A, B, C], [A, C, B], [B, A, C], [B, C, A], [C, A, B], [C, B, A],$$

but $[A, B, C] \neq [A, C, B]$ as the order of vertices is not the same. The **orientation** of the simplex is implied by the order of its vertices – swapping the position of any pair of vertices in the bracket notation changes the sign of the orientation. If we say that $[A, B, C]$ has positive orientation, then so do $[B, C, A]$ and $[C, A, B]$, while $[A, C, B], [B, A, C]$, and $[C, B, A]$ have negative orientations.

An n-simplex may be viewed as a cell complex (recall Definition 5.4.2 of §5.4) whose cells are lower-dimensional simplices. For example, a 3-simplex is a tetrahedron, which has four 0-cells, six 1-cells, four 2-cells, and a single 3-cell, as shown in Figure 8.6. Note that each k-cell is simply a $(k+1)$-element subset of the set of vertices of the 3-simplex.

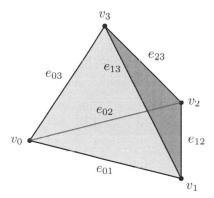

Figure 8.6: Cell structure of a 3-simplex. There are four vertices (0-cells), v_0, v_1, v_2, and v_3, that serve as the endpoints for six edges (1-cells), labeled $e_{01} = [v_0, v_1]$, $e_{02} = [v_0, v_2]$, $e_{03} = [v_0, v_3]$, $e_{12} = [v_1, v_2]$, $e_{13} = [v_1, v_3]$, and $e_{23} = [v_2, v_3]$. The edges bound four faces (2-cells), $f_{012} = [v_0, v_1, v_2]$, $f_{013} = [v_0, v_1, v_3]$, $f_{023} = [v_0, v_2, v_3]$, and $f_{123} = [v_1, v_2, v_3]$ (not labeled in the figure), which comprise the boundary of the simplex $\sigma = [v_0, v_1, v_2, v_3]$ (3-cell).

Boundaries

The boundary of an n-simplex Δ^n consists of the $n+1$ distinct $(n-1)$-dimensional faces, which are in fact $(n-1)$-simplices. Moreover, all of these faces can be determined in a *combinatorial* manner by deleting each vertex in turn from the bracket notation $\Delta^n = [v_0, \ldots, v_n]$. Define a set of **boundary functions**, ∂_k for $0 \leq k \leq n$, that act on abstract brackets via:

$$\partial_0([v_0, v_1, v_2, v_3, \ldots, v_n]) = [v_1, v_2, v_3, \ldots, v_n]$$
$$\partial_1([v_0, v_1, v_2, v_3, \ldots, v_n]) = [v_0, v_2, v_3, \ldots, v_n]$$
$$\partial_2([v_0, v_1, v_2, v_3, \ldots, v_n]) = [v_0, v_1, v_3, \ldots, v_n]$$
$$\vdots$$
$$\partial_n([v_0, v_1, v_2, v_3, \ldots, v_n]) = [v_0, v_1, v_2, v_3, \ldots, v_{n-1}].$$

It is convenient to use the "hat" notation: any element with a hat in the bracket is meant to be omitted. So $[\widehat{u}, v] = [v]$ and $[u, \widehat{v}] = [u]$. Thus we may write for each $0 \leq k \leq n$,

$$\partial_k([v_0, \ldots, v_n]) = [v_0, \ldots, \widehat{v_k}, \ldots, v_n]. \tag{8.3}$$

When $n = 0$, the formula seems to break down. A 0-simplex is a single vertex, $[v_0]$, and according to (8.3), $\partial_0([v_0]) = [\,]$ is an *empty* bracket. But a single vertex should not have a "boundary," so we will define the boundary function on 0-simplices by $\partial_0([v_0]) = \mathbf{0}$. For $n \geq 1$, every boundary simplex of a given n-simplex σ is identified with $\partial_k(\sigma)$ for some $0 \leq k \leq n$. We define the **boundary homomorphism** as an alternating sum of boundary functions, d_n for each $n \in \mathbb{N} \cup \{0\}$, acting on abstract n-simplices (brackets) by:

$$d_n = \sum_{k=0}^{n} (-1)^k \partial_k. \tag{8.4}$$

For example, when $n = 1$, we have $d_1 = \partial_0 - \partial_1$, in agreement with (8.2). For $n = 2$, we have $d_2 = \partial_0 - \partial_1 + \partial_2$ or, more explicitly,

$$
\begin{aligned}
d_2([v_0, v_1, v_2]) &= \partial_0([v_0, v_1, v_2]) - \partial_1([v_0, v_1, v_2]) + \partial_2([v_0, v_1, v_2]) \\
&= [v_1, v_2] - [v_0, v_2] + [v_0, v_1].
\end{aligned}
$$

Now if a topological space X has a triangulation, then for each whole number n the space of n-**chains** $C_n(X; \mathbb{Q})$ will be defined as the (rational) span of the n-simplices in the triangulation. Since d_n is defined on abstract brackets, d_n makes sense when applied to the particular n-simplices of X. Extend (8.4) to all of $C_n(X; \mathbb{Q})$ by linearity. Elements of the form $d_n(\alpha) \in C_n(X; \mathbb{Q})$ for some $\alpha \in C_{n+1}(X; \mathbb{Q})$ are called n-**boundaries**, since these are n-chains that bound an $(n + 1)$-chain. Elements $\alpha \in C_n(X; \mathbb{Q})$ such that $d_n(\alpha) = \mathbf{0}$ are called n-**cycles**. We shall prove that *every boundary is a cycle*. That is, if $n \geq 1$ and if $\alpha = d_{n+1}(\beta) \in C_n(X; \mathbb{Q})$ for some $\beta \in C_{n+1}(X; \mathbb{Q})$, then $d_n(\alpha) = \mathbf{0} \in C_{n-1}(X; \mathbb{Q})$. First we prove the following useful identity.

Proposition 8.1.5. *If $k < \ell$, then $\partial_k \circ \partial_\ell = \partial_{\ell-1} \circ \partial_k$.*

Proof. Suppose $k < \ell$ and consider an abstract bracket of the form

$$\alpha = [v_0, \ldots, v_k, \ldots, v_\ell, \ldots, v_n].$$

Applying $\partial_k \circ \partial_\ell$, we obtain:

$$
\begin{aligned}
\partial_k(\partial_\ell(\alpha)) &= \partial_k([v_0, \ldots, v_k, \ldots, \widehat{v_\ell}, \ldots, v_n]) \\
&= [v_0, \ldots, \widehat{v_k}, \ldots, \widehat{v_\ell}, \ldots, v_n]. \tag{8.5}
\end{aligned}
$$

Next, consider $\partial_{\ell-1} \circ \partial_k$ applied to α. Since $k < \ell$, removing v_k first affects the position of every entry following, so that v_ℓ would now be entry $\ell - 1$.

$$
\begin{aligned}
\partial_{\ell-1}\left(\partial_k(\alpha)\right) &= \partial_{\ell-1}\left([v_0, \ldots, \widehat{v_k}, \ldots, v_\ell, \ldots, v_n]\right) \\
&= [v_0, \ldots, \widehat{v_k}, \ldots, \widehat{v_\ell}, \ldots, v_n] \qquad (8.6)
\end{aligned}
$$

Clearly (8.6) is identical to (8.5); hence $\partial_k \circ \partial_\ell = \partial_{\ell-1} \circ \partial_k$. $\qquad \square$

Theorem 8.1.6. *For each $n \in \mathbb{N} \cup \{0\}$, we have $(d_n \circ d_{n+1})(\alpha) = \mathbf{0}$ for all $\alpha \in C_{n+1}(X; \mathbb{Q})$.*

This property of the boundary homomorphism, which is often abbreviated to the expression $d^2 = 0$, is a launching point for the study of *homological algebra*.

Proof. We must show that if $\alpha = [v_0, \ldots, v_{n+1}]$ is arbitrary, then $d_n\left(d_{n+1}(\alpha)\right) = 0$. The proof hinges on Proposition 8.1.5 and a clever partition of a double sum.

$$
\begin{aligned}
d_n\left(d_{n+1}(\alpha)\right) &= \sum_{k=0}^{n} (-1)^k \partial_k \left(\sum_{\ell=0}^{n+1} (-1)^\ell \partial_\ell(\alpha)\right) \\
&= \sum_{k=0}^{n} \sum_{\ell=0}^{n+1} (-1)^{k+\ell} \partial_k \partial_\ell(\alpha) \\
&= \sum_{k<\ell} (-1)^{k+\ell} \partial_k \partial_\ell(\alpha) + \sum_{k \geq \ell} (-1)^{k+\ell} \partial_k \partial_\ell(\alpha) \\
&= \sum_{k<\ell} (-1)^{k+\ell} \partial_{\ell-1} \partial_k(\alpha) + \sum_{k \geq \ell} (-1)^{k+\ell} \partial_k \partial_\ell(\alpha)
\end{aligned}
$$

The first sum is over all k, ℓ such that $0 \leq k < \ell \leq n+1$, which is equivalent to: $0 \leq k \leq \ell - 1 \leq n$. The second sum, which is over all k, ℓ such that $0 \leq \ell \leq k \leq n$, has exactly the same number of terms. Reindex the first sum with $k' = \ell - 1$ and $\ell' = k$.

$$
\begin{aligned}
d_n\left(d_{n+1}(\alpha)\right) &= \sum_{k' \geq \ell'} (-1)^{k'+\ell'+1} \partial_{k'} \partial_{\ell'}(\alpha) + \sum_{k \geq \ell} (-1)^{k+\ell} \partial_k \partial_\ell(\alpha) \\
&= \mathbf{0}
\end{aligned}
$$

$\qquad \square$

Rational Homology of Triangulated Spaces

The main idea of homology is to locate those n-cycles that are *not* n-boundaries. That is, we want to find the n-dimensional holes of X. As in graph homology, we cannot simply count up all the n-cycles – instead we should determine a *basis* of independent cycles. Let $Z_n(X;\mathbb{Q})$ be the subset of $C_n(X;\mathbb{Q})$ consisting only of cycles. That is,

$$Z_n(X;\mathbb{Q}) = \{\alpha \in C_n(X;\mathbb{Q}) \mid d_n(\alpha) = \mathbf{0}\}. \tag{8.7}$$

Now if $\alpha \in Z_n(X;\mathbb{Q})$ is just the *boundary* of some chain in $C_{n+1}(X;\mathbb{Q})$, then α is not really a hole and should not be counted in the homology of X. How can we get rid of these kinds of cycles? In terms of set theory, we may set up an equivalence relation that identifies every boundary $\beta = d_{n+1}(\alpha)$ with the null chain $\mathbf{0}$. However, we also require that the equivalence relation ($\sim$) be compatible with the vector space structure on $C_n(X;\mathbb{Q})$, in the sense that if $\beta \sim \mathbf{0}$, then $\alpha + \beta \sim \alpha$ for all $\alpha \in Z_n(X;\mathbb{Q})$. The proper definition is given below in Definition 8.1.7. Let $B_n(X;\mathbb{Q})$ be the subset of $C_n(X;\mathbb{Q})$ consisting of all the boundaries:

$$B_n(X;\mathbb{Q}) = \{d_{n+1}(\alpha) \in C_n(X;\mathbb{Q}) \mid \alpha \in C_{n+1}(X;\mathbb{Q})\}. \tag{8.8}$$

It follows from Theorem 8.1.6 that $B_n(X;\mathbb{Q})$ as defined above is actually a subset of $Z_n(X;\mathbb{Q})$. Both $Z_n(X;\mathbb{Q})$ and $B_n(X;\mathbb{Q})$ are vector spaces over $\mathbb{Q}$ for each n, and we assume that they are finite-dimensional spaces, since we are working only with finitely triangulated spaces.

> **Definition 8.1.7.** Two n-cycles, $\alpha, \beta \in Z_n(X;\mathbb{Q})$, are said to be **homologous** if their difference is a boundary, that is, if
>
> $$\alpha - \beta \in B_n(X;\mathbb{Q}).$$

Note that if $\alpha \in B_n(X;\mathbb{Q})$, then α is homologous to $\mathbf{0}$ since $\alpha - \mathbf{0} = \alpha \in B_n(X;\mathbb{Q})$. Finally, the *rational homology* of X is defined by:

> **Definition 8.1.8.** $H_n(X;\mathbb{Q}) = Z_n(X;\mathbb{Q})/\sim$, where $\alpha \sim \beta$ if and only if α and β are homologous.

It can be shown that $H_n(X;\mathbb{Q})$ thus defined also has a (finite-dimensional) vector space structure over $\mathbb{Q}$, and so there is a basis. The number of basis elements (dimension) of $H_n(X;\mathbb{Q})$ is called the nth **Betti number**, $\beta_n(X)$, of the space. That is, $\beta_n(X)$ is the whole number that satisfies:

$$H_n(X;\mathbb{Q}) \approx \mathbb{Q}^{\beta_n(X)}.$$

For now, we will find out how to use the *rank* of matrices to determine the Betti numbers of a triangulated space. More detailed computations of homology must wait until §8.2. Consider a linear map $f : \mathbb{Q}^q \to \mathbb{Q}^p$. The map f can be represented by a $p \times q$ matrix F, whose columns record the action of f on each basis element of the domain $\mathbb{Q}^q$ in terms of the basis of the codomain $\mathbb{Q}^p$. Thus every boundary map d_n can be encoded as a matrix (recall that we encoded the map d_1 from Example 129 in this way). Now consider the chain groups in degrees $n-1$, n, and $n+1$ for a space X, together with the boundary maps d_n and d_{n+1} as shown below. Here we have identified the dimension of each space $C_n(X; \mathbb{Q})$ by r_n and the representing matrix for each d_n by D_n.

$$C_{n-1}(X; \mathbb{Q}) \xleftarrow{\;d_n\;} C_n(X; \mathbb{Q}) \xleftarrow{\;d_{n+1}\;} C_{n+1}(X; \mathbb{Q}) \tag{8.9}$$
$$\| \qquad\qquad \| \qquad\qquad \|$$
$$\mathbb{Q}^{r_{n-1}} \xleftarrow{\;\;D_n\;\;} \mathbb{Q}^{r_n} \xleftarrow{\;\;D_{n+1}\;\;} \mathbb{Q}^{r_{n+1}}$$

From linear algebra, it is known that the rank of a matrix is the same as the dimension of the image of the map. Thus, with (8.8) in mind, we have

$$\dim(B_n(X; \mathbb{Q})) = \operatorname{rank}(D_{n+1}). \tag{8.10}$$

Equation (8.7) defines the space $Z_n(X; \mathbb{Q})$ as the *kernel* of the boundary map d_n (more about that term in the next section), which corresponds to the null space of the representing matrix D_n. Now using the *Rank Theorem* (Theorem B.2.11),

$$\dim(Z_n(X; \mathbb{Q})) = r_n - \operatorname{rank}(D_n). \tag{8.11}$$

Finally, as a result of Definition 8.1.8 and well-known results about vector spaces,[9] we have $\dim(H_n(X; \mathbb{Q})) = \dim(Z_n(X; \mathbb{Q})) - \dim(B_n(X; \mathbb{Q}))$, which leads to a computational tool for computing Betti numbers:

$$\beta_n(X) = \dim(H_n(X; \mathbb{Q})) = r_n - \operatorname{rank}(D_n) - \operatorname{rank}(D_{n+1}). \tag{8.12}$$

Example 130. The boundary of a 3-cell serves as a triangulation for the sphere $\mathbb{S}^2$. Use this triangulation to compute the Betti numbers of $\mathbb{S}^2$.

Solution: First let's identify the chains in each dimension. Let $\Delta^3 = [v_0, v_1, v_2, v_3]$, but since we are only interested in the boundary of the simplex, there are no 3-simplices.

$$C_2(\mathbb{S}^2; \mathbb{Q}) = \operatorname{Span}_{\mathbb{Q}}\{[v_1, v_2, v_3], [v_0, v_2, v_3], [v_0, v_1, v_3], [v_0, v_1, v_2]\} \approx \mathbb{Q}^4$$
$$C_1(\mathbb{S}^2; \mathbb{Q}) = \operatorname{Span}_{\mathbb{Q}}\{[v_0, v_1], [v_0, v_2], [v_0, v_3], [v_1, v_2], [v_1, v_3], [v_2, v_3]\} \approx \mathbb{Q}^6$$
$$C_0(\mathbb{S}^2; \mathbb{Q}) = \operatorname{Span}_{\mathbb{Q}}\{[v_0], [v_1], [v_2], [v_3]\} \approx \mathbb{Q}^4$$

We have a sequence of three nontrivial chain groups and two nontrivial boundary homomorphisms.

$$C_0(\mathbb{S}^2; \mathbb{Q}) \xleftarrow{\;d_1\;} C_1(\mathbb{S}^2; \mathbb{Q}) \xleftarrow{\;d_2\;} C_2(\mathbb{S}^2; \mathbb{Q})$$
$$\| \qquad\qquad \| \qquad\qquad \|$$
$$\mathbb{Q}^4 \xleftarrow{\;\;D_1\;\;} \mathbb{Q}^6 \xleftarrow{\;\;D_2\;\;} \mathbb{Q}^4$$

---

[9]In linear-algebraic terms, the homology $H_n(X; \mathbb{Q}) = Z_n(X; \mathbb{Q})/B_n(X; \mathbb{Q})$ is a *quotient space*. There is a well-known formula for the dimension of any quotient space: $\dim(V/W) = \dim(V) - \dim(W)$.

The representing matrices are as follows.

$$
D_1 = \begin{array}{c} \\ [v_0] \\ [v_1] \\ [v_2] \\ [v_3] \end{array}
\begin{array}{c} \begin{array}{cccccc} [v_0,v_1] & [v_0,v_2] & [v_0,v_3] & [v_1,v_2] & [v_1,v_3] & [v_2,v_3] \end{array} \\
\left(\begin{array}{cccccc}
-1 & -1 & -1 & 0 & 0 & 0 \\
1 & 0 & 0 & -1 & -1 & 0 \\
0 & 1 & 0 & 1 & 0 & -1 \\
0 & 0 & 1 & 0 & 1 & 1
\end{array} \right) \end{array}
$$

$$
D_2 = \begin{array}{c} \\ [v_0,v_1] \\ [v_0,v_2] \\ [v_0,v_3] \\ [v_1,v_2] \\ [v_1,v_3] \\ [v_2,v_3] \end{array}
\begin{array}{c} \begin{array}{cccc} [v_1,v_2,v_3] & [v_0,v_2,v_3] & [v_0,v_1,v_3] & [v_0,v_1,v_2] \end{array} \\
\left(\begin{array}{cccc}
0 & 0 & 1 & 1 \\
0 & 1 & 0 & -1 \\
0 & -1 & -1 & 0 \\
1 & 0 & 0 & 1 \\
-1 & 0 & 1 & 0 \\
1 & 1 & 0 & 0
\end{array} \right) \end{array}
$$

With the help of a computer algebra system, or by reducing the matrices by hand, we find $\operatorname{rank}(D_1) = \operatorname{rank}(D_2) = 3$.

$$
\begin{aligned}
\beta_0(\mathbb{S}^2) &= 4 - \operatorname{rank}(D_1) = 4 - 3 = 1 \\
\beta_1(\mathbb{S}^2) &= 6 - \operatorname{rank}(D_1) - \operatorname{rank}(D_2) = 6 - 3 - 3 = 0 \\
\beta_2(\mathbb{S}^2) &= 4 - \operatorname{rank}(D_2) = 4 - 3 = 1 \\
\beta_n(\mathbb{S}^2) &= 0, \quad \text{if } n \geq 3
\end{aligned}
$$

Recall the definition of Euler characteristic given in §5.4, as the alternating sum of the number of cells in each dimension for a particular cell decomposition of the space. This definition leaves a lot to be desired. Does the definition depend on the choice of cell decomposition? For compact surfaces, this question can be answered by appealing to triangulations, but what about in higher dimensions, or in spaces that are not manifolds? It turns out that the Euler characteristic is nothing more than the alternating sum of the dimensions of homology groups, and each homology group is a topological invariant.[10]

Definition 8.1.9. The Euler characteristic of a triangulated space X is equal to the alternating sum of its Betti numbers.

$$
\chi(X) = \sum_{n \geq 0} (-1)^n \beta_n(X)
$$

Theorem 8.1.10. *Definition 8.1.9 coincides with Definition 5.4.3 when X has a triangulation.*

---

[10]See Hatcher [Hat02] for details. The proof requires the definition of *singular homology*, which does not rely in any way on triangulations or cell decompositions of the space; however, this powerful machinery falls outside the scope of this text.

Proof. Suppose X has a triangulation with r_n n-simplices in dimension n, for $n = 1, 2, \ldots, N$. Using (8.12), and the fact that there are no simplices of dimension smaller than 0 or larger than N,

$$
\begin{aligned}
\sum_{n \geq 0} (-1)^n \beta_n(X) &= \sum_{n \geq 0} (-1)^n (r_n - \operatorname{rank}(D_n) - \operatorname{rank}(D_{n+1})) \\
&= r_0 - \operatorname{rank}(D_1) \\
&\quad -r_1 + \operatorname{rank}(D_1) + \operatorname{rank}(D_2) \\
&\quad +r_2 - \operatorname{rank}(D_2) - \operatorname{rank}(D_3) \\
&\quad \vdots \\
&\quad +(-1)^N r_N - (-1)^N \operatorname{rank}(D_N) \\
&= \sum_{n=1}^{N} (-1)^n r_n \\
&= \chi(X).
\end{aligned}
$$

Note how the sum above *telescopes*, yielding the Euler characteristic formula as shown in Definition 5.4.3. $\qquad \square$

Exercises

1. Show that the property of being homologous is an equivalence relation.

2. Find a basis for $H_1(G; \mathbb{Q})$ and the first Betti number for each of the following graphs (see Chapter 6).

   (a) K_3

   (b) K_6

   (c) G_1 from Figure 6.1

   (d) G_4 from Figure 6.1

   (e) P from Figure 6.8

   (f) C_n for any $n \geq 3$

3. Show that a properly embeded graph G and its dual G^* have isomorphic cycle spaces. (*Hint:* Both G and G^* define a cell structure for the manifold on which they are embedded.)

4. Let $c_r(n)$ be the number of r-cells in the simplex Δ^n. For each fixed $n \geq 0$, consider the sequence $(c_r(n))_{r \geq 0}$ of cell counts by dimension. For example, Figure 8.6 illustrates that

$$
(c_r(3))_{r \geq 0} = (4, 6, 4, 1, 0, 0, \ldots).
$$

   (a) Build a table of the nonzero values of $c_r(n)$ for $n = 0, 1, 2, 3, 4$.

   (b) Use the combinatorial description of Δ^5 to find $(c_r(5))_{r \geq 0}$. Append the nonzero values to your table from part (a).

(c) Do the numbers in your table look familiar? Conjecture a formula for $c_r(n)$. Prove your formula. (*Hint:* Each k-cell is a $(k+1)$-element subset $[v_{i_0}, \ldots, v_{i_k}]$ of the n-simplex $[v_0, \ldots, v_n]$.)

5. Verify that $(d_3 \circ d_4)([v_0, v_1, v_2, v_3, v_4]) = \mathbf{0}$.

6. Compute the Betti numbers for each space. (*Hint:* First find a triangulation for each surface, using the plane model.)

    (a) $\mathbb{T}$ (b) $\mathbb{K}$ (c) $\mathbb{P}$ (d) $\mathbb{S}^1 \times \mathbb{I}$

7. Construct a triangulation of $\mathbb{S}^2$ having more than four vertices. Use your triangulation to compute the Betti numbers for the sphere, and verify that you get the same answers as in Example 130.

8. Compute all of the nonzero Betti numbers for Δ^n, for $n = 0, 1, 2, 3$. What do you guess the answer would be for $n \in \mathbb{N}$ in general?

9. Using the homeomorphism $\mathbb{S}^n \approx \partial \Delta^{n+1}$, compute all of the nonzero Betti numbers for $\mathbb{S}^n$, for arbitrary $n \in \mathbb{N}$.

8.2   Integral Homology

What makes rational homology easy to calculate is the ability to use linear algebra to analyze the boundary maps. But there is a trade-off. Rational homology doesn't *see* the difference between the closed disk $\mathbb{D}^2$, the projective plane $\mathbb{P}^2$, and the dunce cap (see Figure 5.14 in Chapter 5), for example. It's as if we are trying to determine what object is under our living room chair just by reaching down and poking it with our fingers. We might be able to tell it's a coin (as opposed to a key or cell phone), but have a hard time telling whether it's a penny or nickel. With another tool, a flashlight, we could look under the chair and distinguish the coin by its color. In this section we will see how to compute homology with different *coefficients*, specifically the integers $\mathbb{Z}$, and with tools from linear algebra we can develop powerful invariants that shed light on features of topological spaces that may be missed by rational homology.

Modules and Torsion

Integral homology is defined in much the same way as rational homology, as the equivalence classes of cycles with respect to the boundaries; however, certain *extra* features emerge when working over $\mathbb{Z}$ rather than $\mathbb{R}$ or $\mathbb{Q}$. Consider, for example, all scalar multiples of the vector $\mathbf{x} = (2, 1)$, that is, the span of $\{\mathbf{x}\}$, but with scalars restricted to $\mathbb{Z}$. As Figure 8.7 illustrates, $\mathrm{Span}_{\mathbb{Z}}\{\mathbf{x}\}$ is a discrete subset of $\mathbb{R}^2$. Vectors like $(8, 4)$ and $(-6, -3)$ are in the span, but not $(1, 1/2)$ or $(2\pi, \pi)$.

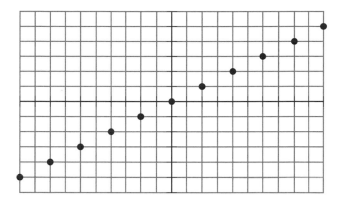

Figure 8.7: $\mathrm{Span}_{\mathbb{Z}}\{(2,1)\}$.

From this point, we assume all vectors and matrices must have integer entries, and the scalars are restricted to $\mathbb{Z}$. In fact, it helps to think of $\mathbb{Z}^r$ as something like a vector space over $\mathbb{Z}$, which we call a $\mathbb{Z}$-**module**.[11] The upshot is that certain matrix methods can still be used.

At the same time, every $\mathbb{Z}$-module is also an abelian group under vector addition, which is commutative and associative, with the vector **0** as identity. Conversely, every abelian group G may also be regarded as a $\mathbb{Z}$-module, with scalar multiplication defined for $n \in \mathbb{Z}$ and $g \subset G$ by:

$$
n \cdot g = \begin{cases} \underbrace{g + g + \cdots + g}_{n}, & n > 0; \\ \mathbf{0}, & n = 0; \\ \underbrace{(-g) + (-g) + \cdots + (-g)}_{|n|}, & n < 0 \end{cases}
$$

One feature of $\mathbb{Z}$-modules not enjoyed by vector spaces is the existence of nontrivial **torsion** elements. An element $m \in M$ is a torsion element if there is some nonzero $n \in \mathbb{Z}$ such that $n \cdot m = 0$. If n is the smallest positive number such that $n \cdot m = 0$, then m is said to have **order** n. For example, consider the *cyclic group* $\mathbb{Z}/5\mathbb{Z} = \{0, 1, 2, 3, 4\}$ (see Examples 149 and 156 in the appendices). Every nonzero element of $\mathbb{Z}/5\mathbb{Z}$ is a torsion element with order 5.

A $\mathbb{Z}$-module M is called **free** if M is isomorphic (as an abelian group) to $\mathbb{Z}^r$ for some $r \in \mathbb{N}$. If M is free as the span of an independent set $A = \{a_1, a_2, \ldots, a_n\}$, then we say that A is a **basis** for M. The **dimension** or **rank** of a free $\mathbb{Z}$-module M is the number of basis elements. If M contains torsion, then M cannot be free because there would be a nontrivial relation among any set of spanning elements.

---

[11]Although we do not formally define *module* in this text, we use the term because module theory forms the bedrock of much of algebraic topology.

A function $f : M \to N$ between $\mathbb{Z}$-modules is called a **homomorphism** if f is a group homomorphism. Briefly, every homomorphism $f : M \to N$ is a *linear map*, in the sense that it satisfies both of the following properties.

1. $f(m_1 + m_2) = f(m_1) + f(m_2)$, $\forall m_1, m_2 \in M$

2. $f(n \cdot m) = nf(m)$, $\forall n \in \mathbb{Z}, m \in M$

Definition 8.2.1. If $f : M \to N$ is a homomorphism, the **image**, $\mathrm{im}(f)$, and **kernel**, $\ker(f)$, are defined as follows.

$$
\begin{aligned}
\mathrm{im}(f) &= \{f(m) \mid m \in M\} \subseteq N \\
\ker(f) &= \{m \in M \mid f(m) = \mathbf{0} \in N\} \subseteq M
\end{aligned}
$$

Integral Homology of Delta Complexes

In §8.1, we defined the space $C_n(X; \mathbb{Q})$ of n-chains for a given triangulated space X as the rational span of n-simplices in the triangulation. Similarly, one could define integral n-chains as the integral span of n-simplices; however, there is a more efficient way to proceed. A **delta complex** (or Δ-**complex**) structure on a space X is a collection of maps $\sigma_k : \Delta^{n(k)} \to X$, where $n(k) \in \mathbb{N} \cup \{0\}$ is the dimension of the domain simplex, such that:

- The restriction $\sigma_k\big|_{\mathrm{int}(\Delta^{n(k)})}$ is a homeomorphism for all k.

- If $D \subseteq \Delta^{n(k)}$ is a single face of the simplex, then $\sigma_k\big|_D = \sigma_\ell$ for some map $\sigma_\ell : \Delta^{n(k)-1} \to X$ in the collection (recall that each face of an n-simplex is an $(n-1)$-simplex).

- Any subset $Y \subseteq X$ is open (in X) if and only if $\sigma_k^{-1}[Y] \subseteq \Delta^{n(k)}$ is open for each k.

The maps σ_k are nothing more than *attaching maps* for cells in a certain kind of *cell decomposition* for X in which the n-skeleton is identified with the images of σ_k such that $n(k) \leq n$ (compare Definitions 5.4.1 and 5.4.2). While these conditions may seem esoteric at first, they imply that X has something like a triangulation, except that the boundary faces of any given n-simplex may be identified to other faces in less restrictive ways. The advantage to working with Δ-complexes is that they are often much less cumbersome than triangulations while at the same time admitting a straightforward boundary homomorphism. First we define the $\mathbb{Z}$-module of n-**chains** as follows:

$$C_n(X; \mathbb{Z}) = \mathrm{Span}_{\mathbb{Z}}\{\sigma \mid \sigma \text{ is an } n\text{-simplex in a delta decomposition of } X\}.$$

The boundary functions ∂_k and homomorphisms $d_n : C_n(X; \mathbb{Z}) \to C_{n-1}(X; \mathbb{Z})$ are defined in a similar way as in (8.3) and (8.4), though we must be careful as simplices are now no longer uniquely identifiable by their list of vertices alone. In order to work with a Δ-complex, first give all vertices, edges, and higher k-simplices unique labels and then make the identifications when defining the chain groups (via a *quotient* of modules – see below). The *cycles* and *boundaries* can now be defined in module-theoretic terms:

> **Definition 8.2.2.** For each $n \in \mathbb{N} \cup \{0\}$,
>
> $$n\text{-}\mathbf{cycles}: \quad Z_n(X;\mathbb{Z}) \;=\; \ker(d_n).$$
> $$n\text{-}\mathbf{boundaries}: \quad B_n(X;\mathbb{Z}) \;=\; \mathrm{im}(d_{n+1}).$$

As before, we have $d^2 = 0$, which implies that $B_n(X;\mathbb{Z}) \subseteq Z_n(X;\mathbb{Z})$. The previous definition of *homologous* cycles still applies (see Definition 8.1.7) in the more general setting. If N is any submodule of a module M, the **quotient module** M/N may be defined by $M/N = M/\sim$, where $m_1 \sim m_2 \iff m_1 - m_2 \in N$. Then M/N inherits an $\mathbb{Z}$-module structure from M (see Exercise 2). Each homology group is then defined to be the *quotient modules* of the cycles with respect to the boundaries.

> **Definition 8.2.3.** The **integral homology** of X is defined for each $n \in \mathbb{N} \cup \{0\}$ by:
>
> $$H_n(X;\mathbb{Z}) - \ker(d_n)/\mathrm{im}(d_{n+1}) = Z_n(X;\mathbb{Z})/B_n(X;\mathbb{Z}).$$

Let's see if we can make sense of all of these new definitions by finding the homology of the projective plane. If the techniques displayed in the following example seem a bit ad hoc, do not despair. There are more algorithmic methods available in general.

Example 131. The projective plane $\mathbb{P}$ has a plane model with word a^2. Let's decompose $\mathbb{P}$ into a Δ-complex by putting an extra vertex in the center along with two extra edges as shown in Figure 8.8.

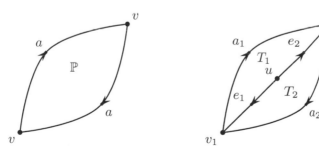

Figure 8.8: $\mathbb{P}$ as a Δ-complex with identified edges $a = a_1 = a_2$ and vertices $v = v_1 = v_2$.

We will begin the procedure for finding the integral homology for this space. There are only two unique vertices (0-simplices), u and v (since $v = v_1 = v_2$), so

$$C_0(\mathbb{P}; \mathbb{Z}) = \text{Span}_{\mathbb{Z}}\{u, v\} \cong \mathbb{Z}^2.$$

The edges (1-simplices) are e_1, e_2, and a (since $a = a_1 = a_2$).

$$C_1(\mathbb{P}; \mathbb{Z}) = \text{Span}_{\mathbb{Z}}\{e_1, e_2, a\} \cong \mathbb{Z}^3$$

There are two distinct "triangles" (2-simplices): T_1, having edges a_1, $-e_2$ and e_1; and T_2, having edges a_2, $-e_1$ and e_2, in clockwise order.

$$C_2(\mathbb{P}; \mathbb{Z}) = \text{Span}_{\mathbb{Z}}\{T_1, T_2\} \cong \mathbb{Z}^2$$

There are no higher-dimensional chains, so $C_n(\mathbb{P}; \mathbb{Z}) = 0$ for $n \geq 3$, and in particular, $d_3 = 0$. For convenience, we introduce d_0 as a map from $C_0(\mathbb{P}; \mathbb{Z})$ to $C_{-1}(\mathbb{P}; \mathbb{Z}) = 0$, even though the concept of "(-1)-simplex" does not make much geometric sense. The result is that $d_0 = 0$, so that the kernel of d_0 is all of $C_0(\mathbb{P}; \mathbb{Z})$. The chain groups and boundary homomorphisms are summarized in the diagram below.

$$C_{-1}(\mathbb{P}; \mathbb{Z}) \xleftarrow{d_0} C_0(\mathbb{P}; \mathbb{Z}) \xleftarrow{d_1} C_1(\mathbb{P}; \mathbb{Z}) \xleftarrow{d_2} C_2(\mathbb{P}; \mathbb{Z}) \xleftarrow{d_3} C_3(\mathbb{P}; \mathbb{Z})$$

$$0 \xleftarrow{\quad 0 \quad} \mathbb{Z}^2 \longleftarrow \mathbb{Z}^3 \longleftarrow \mathbb{Z}^2 \xleftarrow{\quad 0 \quad} 0$$

Now let's compute the integral homology, $H_n(\mathbb{P}; \mathbb{Z})$, which involves finding kernels and images of the boundary homomorphisms. We have $d_1(e_1) = d_1([u, v_1]) = v_1 - u = v - u$ and, similarly, $d_1(e_2) = v - u$. To find $d_1(a)$, use either representative a_1 or a_2, so $d_1(a_1) = v_1 - v_2 = v - v = \mathbf{0}$. This implies that $\text{im}(d_1)$ is spanned by a single boundary cycle, $v - u$. Thus, in degree $n = 0$, we have

$$H_0(\mathbb{P}; \mathbb{Z}) = \ker(d_0)/\text{im}(d_1) = \text{Span}_{\mathbb{Z}}\{u, v\}/\text{Span}_{\mathbb{Z}}\{v - u\}.$$

This means that $H_0(\mathbb{P}; \mathbb{Z})$ is the quotient module spanned by u and v but with the relation that $v - u \sim 0$ or, equivalently, $v \sim u$. Since v and u are homologous, either one may serve as a representative, and so we may write:

$$H_0(\mathbb{P}; \mathbb{Z}) \cong \mathbb{Z}, \text{ generated by } v.$$

Let's move on to $H_1(\mathbb{P}; \mathbb{Z})$. Since $d_1(e_1) = d_1(e_2)$, we have $d_1(e_1 - e_2) = \mathbf{0}$, implying $e_1 - e_2 \in \ker(d_1)$. Also, $a \in \ker(d_1)$, since $d_1(a) = \mathbf{0}$. There are no other indepedent elements in the kernel since, otherwise, $d_1 = 0$ on $\mathbb{Z}^3$. What about the image of d_2? Using bracket notation, we may write $T_1 = [v_1, v_2, u]$

and $T_2 = [v_2, v_1, u]$. Now using $d_2 = \partial_0 - \partial_1 + \partial_2$, and identifying $a = a_1 = a_2$, we have:

$$d_2(T_1) = [v_2, u] - [v_1, u] + [v_1, v_2] = (-e_2) - (-e_1) + a_1 = e_1 - e_2 + a$$
$$d_2(T_2) = [v_1, u] - [v_2, u] + [v_2, v_1] = (-e_1) - (-e_2) + a_2 = -e_1 + e_2 + a.$$

Notice though that $d_2(T_1 + T_2) = (e_1 - e_2 + a) + (-e_1 + e_2 + a) = 2a$. So $2a \in \mathrm{im}(d_2)$. It can be shown that $a \notin \mathrm{im}(d_2)$, so the element a becomes *torsion* of order 2 in the homology group. Furthermore, since $-e_1 + e_2 + a \in \mathrm{im}(d_2)$, it follows that $a \sim e_1 - c_2$ in the quotient module. Thus

$$\begin{aligned} H_1(\mathbb{P}; \mathbb{Z}) &= \ker(d_1)/\mathrm{im}(d_2) \\ &= \mathrm{Span}_{\mathbb{Z}}\{e_1 - e_2, a\}/(2a \sim \mathbf{0}, a \sim e_1 - e_2) \\ &\cong \mathrm{Span}_{\mathbb{Z}}\{a\}/\mathrm{Span}_{\mathbb{Z}}\{2a\} \\ &\cong \mathbb{Z}/2\mathbb{Z}, \text{ generated by } a. \end{aligned}$$

Finally, we compute $H_2(\mathbb{P}; \mathbb{Z})$. Note that $\mathrm{im}(d_3) = 0$, so it all comes down to finding $\ker(d_2)$. Suppose that $d_2(m_1 T_1 + m_2 T_2) = 0$ for some $m_1, m_2 \in \mathbb{Z}$. Then

$$\begin{aligned} 0 &= m_1 d_2(T_1) + m_2 d_2(T_2) \\ 0 &= m_1 e_1 - m_1 e_2 + m_1 a - m_2 e_1 + m_2 e_2 + m_2 a \\ 0 &= (m_1 - m_2)e_1 - (m_1 - m_2)e_2 + (m_1 + m_2)a. \end{aligned}$$

This implies both $m_1 - m_2 = 0$ and $m_1 + m_2 = 0$. But the only solution to this system is $(m_1, m_2) = (0, 0)$. Thus $\mathbf{0}$ is the only element in $\ker(d_2)$. The degree 2 homology is therefore trivial.

$$H_2(\mathbb{P}; \mathbb{Z}) = \ker(d_2)/\mathrm{im}(d_3) = 0$$

Example 132. Below is a delta complex structure for torus using only two triangles. Simply draw a diagonal in the square plane model for $\mathbb{T}$.

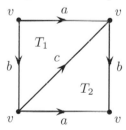

With respect to this decomposition, we have the following.

$$\begin{aligned} C_0(\mathbb{T}; \mathbb{Z}) &= \mathrm{Span}_{\mathbb{Z}}\{v\} \cong \mathbb{Z} \\ C_1(\mathbb{T}; \mathbb{Z}) &= \mathrm{Span}_{\mathbb{Z}}\{a, b, c\} \cong \mathbb{Z}^3 \\ C_2(\mathbb{T}; \mathbb{Z}) &= \mathrm{Span}_{\mathbb{Z}}\{T_1, T_2\} \cong \mathbb{Z}^2 \end{aligned}$$

The boundary functions are defined as follows.

$$d_2(T_1) \;=\; b - a + c$$
$$d_2(T_2) \;=\; c - a + b$$
$$d_1(a) = d_1(b) = d_1(c) \;=\; v - v = \mathbf{0}$$

Since $d_1 = 0$, we obtain $H_0(\mathbb{T};\mathbb{Z}) \cong \mathbb{Z}$ with generator $\{v\}$. Furthermore,

$$H_1(\mathbb{T};\mathbb{Z}) = \ker(d_1)/\mathrm{im}(d_2) = \mathrm{Span}_{\mathbb{Z}}\{a,b,c\}/(b-a+c \sim \mathbf{0}, c-a+b \sim \mathbf{0}).$$

Both relations imply that $c \sim a - b$, but there are no further relations between a and b, so we get $H_1(\mathbb{T};\mathbb{Z}) \cong \mathbb{Z}^2$ on generators $\{a,b\}$. Finally, since $d_2(T_1 - T_2) = (b - a + c) - (c - a + b) = \mathbf{0}$, certainly $T_1 - T_2 \in Z_2(\mathbb{T};\mathbb{Z})$. There are no other independent cycles, so $H_2(\mathbb{T};\mathbb{Z}) \cong \mathbb{Z}$. The higher-dimensional homology is trivial.

Note that the procedure for finding homology of a delta complex can be automated using advanced techniques from linear algebra, including the *Smith normal form*. Unfortunately, we cannot explore the topic further in this text.

Exercises

1. Let $u = (2,1)$ and $v = (1,3)$. Find all elements of $\mathrm{Span}_{\mathbb{Z}}\{u,v\}$ that are within the region $[-6,6] \times [-6,6]$ in the plane. Plot each element as a point.

2. Suppose $N \subseteq M$ are both $\mathbb{Z}$-modules. Show that the quotient M/N is a $\mathbb{Z}$-module, where $[m_1]+[m_2] = [m_1+m_2]$ and $n \cdot [m] = [nm]$ on equivalence classes.

3. The notation $\mathbb{Z}/n\mathbb{Z}$ used in this text for the cyclic groups suggests that $\mathbb{Z}/n\mathbb{Z}$ is really a quotient module. Explain how the group $\mathbb{Z}/5\mathbb{Z}$ is a quotient of $\mathbb{Z}$ by $5\mathbb{Z}$.

4. Let X be the dunce cap, which has a plane model with word a^3. Find $H_n(X;\mathbb{Z})$ for $n \in \mathbb{N} \cup \{0\}$.

5. Using Example 132 as a guide, find a simple delta complex structure on the Klein bottle $\mathbb{K}$ and determine both $H_n(\mathbb{K};\mathbb{Q})$ and $H_n(\mathbb{K};\mathbb{Z})$ for all $n \in \mathbb{N} \cup \{0\}$.

Supplemental Reading

- Hatcher [Hat02], Chapters 2–3.

Appendix A

Review of Set Theory and Functions

Mathematicians typically use the language of **set theory** in order to make abstract definitions precise. This appendix serves as a brief review of the concepts and notations of set theory that will be used specifically in our study of topology, as well as basic proof structure. It is not our purpose to delve into an axiomatic treatment of sets, but rather to develop what is called *naive set theory*, in which we define and reason about sets using everyday English language. The reader is encouraged to explore the suggested readings at the end of this appendix if a more detailed treatment of set theory is desired.

A.1 Sets and Operations on Sets

In this section, we define the basic constructions of set theory. We also informally introduce proof techniques, including proof by contradiction and proof by induction.

Sets and Elements

We think of sets both as "containers" for objects and as "classifiers" of objects. A set may *contain* specific objects, just as my desk drawer contains pens, pencils, and coins, but a set may also *define* its elements; for example, the set of all students who will receive a passing grade in next year's calculus course (the elements of this set are not known yet, but will be defined in a year).

> **Definition A.1.1.** A **set** is a collection of objects, called **elements**, with no regard to order or multiplicity. The notation $A = \{a, b, c, \ldots\}$ may be used to describe a set A as having elements a, b, c, etc.

The elements of a set may be anything, including numbers, vectors, points, functions, or even other sets.[1] For example, the set $B = \{4, \mathbf{i}, f(x) = x^2, \mathbb{R}^2, \{2, 3, 4\}\}$ contains the number 4, the vector $\mathbf{i}$, a function f defined by the rule $f(x) = x^2$, the Cartesian plane $\mathbb{R}^2$, and the set $\{2, 3, 4\}$. Note that by definition all of the following are the exact same set, the set containing the elements a and b and nothing else:

$$\{a, b\} = \{b, a\} = \{a, a, a, a, a, a, a, b\} = \{b, a, b, a, \ldots\}.$$

We use the notation $x \in A$ (or $A \ni x$) to denote that x is an element of A, while $y \notin A$ (or $A \not\ni y$) denotes that y is not an element of A. Thus we may write $1 \in \{1, 2\}$, $2 \in \{1, 2\}$, $3 \notin \{1, 2\}$.

A set may have any number of elements, even infinitely many. Consider the set $\mathbb{N} = \{1, 2, 3, 4, \ldots\}$ ($\mathbb{N}$ stands for the set of **natural numbers**). Clearly $1 \in \mathbb{N}$, $2 \in \mathbb{N}$, and since the dots "..." indicate continuation of the pattern forever, we can be sure that $1,000,000 \in \mathbb{N}$ and $10^{100} - 1 \in \mathbb{N}$. But what does it mean to say that a pattern continues *forever*? Surely we cannot list or even imagine an infinite number of elements.[2] Putting aside the discussion of whether infinity makes *philosophical* sense, let us agree that it makes *mathematical* sense. In practice, we usually deal with sets (infinite or not) by considering elements only *one at a time*. Indeed, there is a definition for the set $\mathbb{N}$ that avoids those pesky three dots altogether.

1. $1 \in \mathbb{N}$.

2. If $n \in \mathbb{N}$, then $n + 1 \in \mathbb{N}$.

3. There are no elements in $\mathbb{N}$ except those defined by rules 1 and 2.

This type of definition is called **recursive** or **inductive**. Now let's prove that $\mathbb{N}$ has an **infinite** number of elements. The proof is by **contradiction**.

Example 133. Prove there is no largest integer. *Solution:* Assume *to the contrary* that there *is* a largest integer $N \in \mathbb{Z}$. Consider $K = N + 1$. K is clearly an integer, by rule 2. Now since $1 > 0$, we have $N + 1 > N + 0 = N$, or $K > N$, *contradicting* the fact that N is largest. Thus there is no largest integer.

The natural numbers have a property called **well ordering**, which we will assume without proof.

---

[1] For technical reasons, in this text, we will never permit a set to be an element of *itself.*

[2] We may ask if there even is such a thing as *infinity* if we, as human beings, can only hold finitely many things in our minds and have only finitely much time to ponder them. In fact, a professor of mine once expressed this quandary to the class by asking, in a tone that I might call mathematical sermonizing, "Do you *believe* in those dots? You have to *believe* in those dots..." (The professor was Robert Young at Oberlin College. The class was Real Analysis. The quote is reconstructed from memory, and so may not be verbatim.)

> **Proposition A.1.2** (Well-Ordering Principle). *Every nonempty set of natural numbers contains a smallest element.*

Suppose now that $P(n)$ is a mathematical statement involving an arbitrary natural number n, whose truth we want to determine. For example,

$$P(n) = \sum_{i=1}^{n} i = \frac{n(n+1)}{2}. \tag{A.1}$$

For any given $n \in \mathbb{N}$, the statement $P(n)$ may be verified. Thus $P(3)$ is the statement $\sum_{i=1}^{3} i = \frac{3(3+1)}{2}$, and we can see that $P(3)$ is true because $\sum_{i=1}^{3} i = 1 + 2 + 3 = 6$ and $\frac{3(3+1)}{2} = \frac{12}{2} = 6$. You might verify $P(n)$ is true for other values of n, but how can you tell whether it is *always* true? The proof technique we require for this job is called **mathematical induction**.

> **Proposition A.1.3** (Principle of Mathematical Induction). *Let $n_0 \in \mathbb{N}$ (typically, $n_0 = 1$). Let $P(n)$ be a mathematical statement, and suppose the following:*
>
> - *[Base case] $P(n_0)$ is true.*
>
> - *[Inductive step] For any integer $k > n_0$, if $P(k)$ is true, then $P(k+1)$ is true.*
>
> *Then the statement $P(n)$ is true for all $n \geq n_0$.*

In fact, the principle of mathematical induction is equivalent to the well-ordering principle (see Exercise 5), and we shall assume both without proof in this text. In mathematical induction, the inductive step requires the truth of $P(k)$ to be *assumed* when proving $P(k+1)$. We say that $P(k)$ is the **inductive hypothesis**. This may seem strange at first, because logically we are never allowed to simply assume a statement is true before we have proven it, but this is something different. We are saying that *if $P(k)$ is true, then $P(k+1)$ would follow. It's as if each statement $P(n)$ is a domino in an infinite line of dominoes.[3] There are two conditions that must be met if we expect all of the dominoes to be knocked over.

- There is an initial domino that can easily be knocked over (*base case*).

- All of the other dominoes are close enough together that if any one domino (k) is knocked over, then it is certain that the next domino ($k+1$) will also be knocked over (*inductive step*).

---

[3]By the way, there is an art and a science to *domino toppling*. Very impressive configurations with hundreds of thousands of dominos have been created and knocked down with the push of a single domino. See `http://www.domino-play.com/TopplingBasic.htm`.

Example 134.   Use mathematical induction to prove (A.1) for all $n \in \mathbb{N}$.

Solution: **Base case.** ($n = 1$) The statement $P(1)$ is true, since

$$\sum_{i=1}^{1} i = 1, \quad \text{and} \quad \frac{1(1+1)}{2} = \frac{2}{2} = 1.$$

Inductive step. Let $k \geq 1$ and *assume* that $P(k)$ is true. In other words, assume the *inductive hypothesis*,

$$\sum_{i=1}^{k} i = \frac{k(k+1)}{2}. \tag{A.2}$$

We want to show that $P(k+1)$ is true; that is, we need to prove that $\sum_{i=1}^{k+1} i = \frac{(k+1)([k+1]+1)}{2}$. Start with the lefthand side (**Caution:** Never begin this part of the proof with the full statement of $P(k+1)$, as this *begs the question*. We do not yet know if $P(k+1)$ is true, so we cannot assume it.), so that

$$\begin{aligned}
\sum_{i=1}^{k+1} i &= \left(\sum_{i=1}^{k} i \right) + (k+1) \\
&= \frac{k(k+1)}{2} + (k+1), \quad \text{(by inductive hypothesis)} \\
&= \frac{k^2 + 3k + 2}{2} = \frac{(k+1)(k+2)}{2} = \frac{(k+1)([k+1]+1)}{2}.
\end{aligned}$$

The last line shows that $\sum_{i=1}^{k+1} i = \frac{(k+1)([k+1]+1)}{2}$; in other words, $P(k+1)$ is true. Thus, by induction, $\sum_{i=1}^{n} i = \frac{n(n+1)}{2}$ for all $n \in \mathbb{N}$.

Now let's get back to set theory. Another way to define a set is by using *set-builder notation*. Set-builder notation has the form $A = \{x \in S \mid P(x)\}$, where S is a predefined set, and $P(x)$ is some mathematical statement involving x. If $P(x)$ is true, then $x \in A$; if $P(y)$ is false, then $y \notin A$. If the predefined set S is omitted, then assume elements are taken from the largest set that makes sense in that context.

Example 135.   Write the set $C = \{x \mid x < 10 \text{ and } x \text{ is prime}\}$ by listing all the elements.

Solution: Since *prime* may only refer to natural numbers in this context, we have to assume $C = \{x \in \mathbb{N} \mid x < 10 \text{ and } x \text{ is prime}\}$. Therefore $C = \{2, 3, 5, 7\}$.

Certain sets, such as the integers and real numbers, are taken for granted in this text, since it would be outside our scope to attempt to construct these from the foundations. Below we list the notations and descriptions of five important sets of numbers.

Notation	Name	Description
$\mathbb{N}$	natural numbers	$\{1, 2, 3, 4, \ldots\}$
$\mathbb{Z}$	integers	$\{\ldots, -3, -2, -1, 0, 1, 2, 3, \ldots\}$
$\mathbb{Q}$	rational numbers	$\left\{\frac{m}{n} \mid m \in \mathbb{Z},\, n \in \mathbb{N}\right\}$
$\mathbb{R}$	real numbers	all points on the number line
$\mathbb{C}$	complex numbers	$\{a + bi \mid a, b \in \mathbb{R}\}$, where $i^2 = -1$

Just as it is useful to have a concept of zero in arithmetic, it is useful to define a set consisting of no elements at all.

> **Definition A.1.4.** The **empty set** is a set having no elements, and it is given the notation $\emptyset$ or $\{\}$.

Two sets are considered equal (*equivalent* in the realm of set theory) if they have exactly the same elements. Therefore there is only one empty set.

Subsets

A subset is simply a collection of some (or all) of the elements of a given set.

> **Definition A.1.5.** A set A is a **subset** of a set B, written $A \subseteq B$, if every element of A is also an element of B; that is, if $x \in A$, then $x \in B$.

We may also write $B \supseteq A$ to indicate that A is a subset of B or, equivalently, B is a **superset** of A. Note that the statements $A \subseteq A$ and $A \supseteq A$ are true for any set A. (We do not make use of the notations $\subset$ and $\supset$ in this text.) The notation $A \nsubseteq B$ means A is not a subset of B, which would be the case if there is at least one element $x \in A$ such that $x \notin B$. It is essential to understand the distinction between the notations $\in$ and $\subseteq$.

Example 136. Suppose $A = \{1, 2, \{3\}, \{4, 5\}\}$. A has four elements, $1 \in A$, $2 \in A$, $\{3\} \in A$, and $\{4, 5\} \in A$, but it's not correct to write $3 \in A$. We have subsets $\{1\} \subseteq A$, $\{2\} \subseteq A$, $\{1, 2\} \subseteq A$, but $\{3\} \nsubseteq A$ (since $\{3\}$ is an element of A, not a subset of A); in fact, it is true that $\{\{3\}\} \subseteq A$ and $\{\{4, 5\}\} \subseteq A$.

In order to prove that $A \subseteq B$ when the sets A and B are infinite, it is typically necessary to frame the argument in terms of arbitrary elements – that is, we show that any *arbitrarily chosen* element $x \in A$ also must be an element of B, *using only properties common to all elements of A and B.*

Example 137. Let $A = \{n \in \mathbb{Z} \mid n = 6k + 4, \text{for some } k \in \mathbb{Z}\}$, and $B = \{n \in \mathbb{Z} \mid n = 3k + 1, \text{for some } k \in \mathbb{Z}\}$. Prove that $A \subseteq B$.

Solution: Let $n \in A$ be arbitrary. Then there exists an integer k such that $n = 6k + 4$. Rewriting this expression, we find that $n = 3(2k) + (3 + 1) = 3(2k + 1) + 1$. Now since $2k + 1 \in \mathbb{Z}$, the number n is in B. Thus $A \subseteq B$.

Certain subsets of $\mathbb{R}$ called **intervals** play a significant role in analysis and topology. Standard interval notation is used in this textbook, as illustrated in the table below.

Notation	Set Description	Visualization on the Number Line
(a, b)	$\{x \mid a < x < b\}$	
$[a, b)$	$\{x \mid a \leq x < b\}$	
$(a, b]$	$\{x \mid a < x \leq b\}$	
$[a, b]$	$\{x \mid a \leq x \leq b\}$	
(a, ∞)	$\{x \mid x > a\}$	
$[a, \infty)$	$\{x \mid x \geq a\}$	
$(-\infty, b)$	$\{x \mid x < b\}$	
$(-\infty, b]$	$\{x \mid x \leq b\}$	
$(-\infty, \infty)$	$\mathbb{R}$ *(all real numbers)*	

The set of all subsets of a given set is called the *power set.*

Definition A.1.6. Given a set A, the **power set** of A, written $\mathcal{P}(A)$, is the set of all subsets of A.

Caution: $\mathcal{P}(A)$ is, by definition, a set of sets. Note in particular that $\emptyset \in \mathcal{P}(A)$ and $A \in \mathcal{P}(A)$ for any set A.

Example 138. List all the elements in the power set of $\{1, 2, 3, 4\}$.

Solution: There are 16 elements:

$$\mathcal{P}(\{1,2,3,4\}) \;=\; \{\emptyset, \{1\}, \{2\}, \{3\}, \{4\},$$
$$\{1,2\}, \{1,3\}, \{1,4\}, \{2,3\}, \{2,4\}, \{3,4\},$$
$$\{1,2,3\}, \{1,2,4\}, \{1,3,4\}, \{2,3,4\}, \{1,2,3,4\}\}.$$

Logical Notations

As we look toward more formal proofs involving subsets and set equality, it helps to introduce some useful notations from logic. A logical **implication** is a statement of the form "if P, then Q," often notated more concisely by $P \implies Q$. In order to prove the implication, we first *assume* the hypothesis P, and then use this to *derive* the conclusion Q. Looking back to Example 137, this is exactly what we did to show the subset relationship, $n \in A \implies n \in B$.

A logical *equivalence*, or **biconditional**, is a statement of the form "P if and only if Q." This type of statement, often written $P \iff Q$, is equivalent to the statement "$P \implies Q$ and $Q \implies P$." Thus the definitions for subset, set equality, and power set may be compactly expressed using logical notation.

$$A \subseteq B \quad \iff \quad x \in A \implies x \in B$$
$$A = B \quad \iff \quad A \subseteq B \text{ and } A \supseteq B$$
$$B \in \mathcal{P}(A) \quad \iff \quad B \subseteq A$$

Another useful pair of notations are the *quantifiers*.

- **Universal quantifier**, $\forall$. "For all" or "for each."

- **Existential quantifier**, $\exists$. "There exists" or "for some."

For example, the mathematical statement "$\forall n \in \mathbb{N}, n^2 \in \mathbb{N}$" means that for *any natural number at all*, its square must also be a natural number. On the other hand, "$\exists n \in \mathbb{N}$ such that $n^2 = 4$" means that there is (at least one) natural number whose square is equal to 4. Both statements are true, of course.

Union, Intersection, and Difference Set

Given two sets, A and B, we define two operations, *union* and *intersection*.

Definition A.1.7. Let A and B be sets.

- The **union** of A and B, written $A \cup B$, is the set whose elements are all of those from A or from B, that is,

$$A \cup B = \{x \mid x \in A \text{ or } x \in B\}.$$

- The **intersection** of A and B, written $A \cap B$, is the set whose elements are all of those that are in both A and B, that is,

$$A \cap B = \{x \mid x \in A \text{ and } x \in B\}.$$

Example 139. Let $P = \{n \in \mathbb{N} \mid n = 5k \text{ for some } k \in \mathbb{N}\}$, and $D = \{n \in \mathbb{N} \mid n = 2k \text{ for some } k \in \mathbb{N}\}$. Describe $P \cup D$ and $P \cap D$.

Solution: First, let's write out some of the elements of each set to get a better sense for what they are.

$$P = \{5(1), 5(2), 5(3), 5(4), 5(5), 5(6), \ldots\} = \{5, 10, 15, 20, 25, 30, \ldots\}$$
$$D = \{2(1), 2(2), 2(3), 2(4), 2(5), 2(6), \ldots\} = \{2, 4, 6, 8, 10, 12, \ldots\}$$

It should be clear that P is the set of positive multiples of 5, and D is the set of positive multiples of 2. The union will have all multiples of 5 or 2:

$$P \cup D = \{2, 4, 5, 6, 8, 10, 12, 14, 15, 16, 18, 20, \ldots\}.$$

The intersection will contain only numbers that are multiples of both 2 and 5, which implies only multiples of 10:

$$P \cap D = \{10, 20, 30, 40, 50, 60, \ldots\} = \{n \in \mathbb{N} \mid n = 10k \text{ for some } k \in \mathbb{N}\}.$$

It is interesting to note that $P \cap D$ can be written in set-builder notation in a similar way as P and D were given (as multiples of a single number), but $P \cup D$ does not have a similar description.

Definition A.1.8. The **difference** between A and B, written $A \setminus B$, is the set of all elements of A that are not elements of B, that is,

$$A \setminus B = \{x \in A \mid x \notin B\}.$$

Note, by definition, $A \setminus B \subseteq A$ and $(A \setminus B) \cap B = \emptyset$. For any sets S and T, we say that S and T are **disjoint** if $S \cap T = \emptyset$. Hence for any sets A and B, the difference $A \setminus B$ is disjoint from B. If the union of two or more pairwise-disjoint sets $S_1, S_2, \ldots, S_n$ is equal to a given set S, then we say S is **partitioned** by $S_1, S_2, \ldots, S_n$. (Unless otherwise specified, we do not insist that any S_k in a partition be nonempty.) For example, $A = (A \setminus B) \cup (A \cap B)$ is a partition of the set A into two subsets based on whether the element is also in B or not. There is also a useful partition of the union of A and B into three logical subsets:

$$A \cup B = (A \setminus B) \cup (A \cap B) \cup (B \setminus A).$$

Definition A.1.8 defines the difference $A \setminus B$ as the set of elements of A that are not in B. But what about the set of *all* elements that are not in B? It turns out that this concept is not well defined. For example, if $B = \{1, 2, 3\}$, then what is *not* in B? Well, certainly $4 \notin B$, $0 \notin B$, and $32 \notin B$. But also $1.5 \notin B$, $\pi \notin B$, $\sqrt{-1} \notin B$, $(1, 2) \notin B$, $f(x) = x^2 \notin B$, Tuesday $\notin B$, and $B \notin B$. The concept of "*everything that is not in B*" is not well defined[4] because we have no well-defined notion of "*everything.*" In practice, when we want to know what is *not* in a set, we usually have some *universal set* U in mind already. For example, if we consider $U = \mathbb{N}$, then the set of elements not in $B = \{1, 2, 3\}$ would be the set $\{n \in \mathbb{N} \mid n \geq 4\}$. If instead we were working in $U = \mathbb{R}$, then the set of elements not in B would be $\{x \in \mathbb{R} \mid x \neq 1, 2, 3\}$. This leads to a definition for the *complement* of a set.

Definition A.1.9. Suppose $A \subset U$, where U is a given **universal set**. The **complement** of a set A (in U) is the set A^c defined as the difference set,

$$A^c = U \setminus A.$$

The Venn diagram in Figure A.1 may help to visualize the relationships between two sets A and B in a universal set U.

Cartesian Products

When you think of the xy-plane (also called the *Cartesian plane*), you may think of all pairs of numbers (x, y) such that both x and y are real numbers. In fact, this is a construction known as the *Cartesian product* of $\mathbb{R}$ with itself.

Definition A.1.10. The **Cartesian product** of two sets A and B is the set of *ordered pairs*,

$$A \times B = \{(x, y) \mid x \in A, \, y \in B\}.$$

Ordered pair means that the order matters. On the plane, $(1, 2)$ and $(2, 1)$ are distinct points. The Cartesian product of any finite number of sets is the

---

[4]For more about this conundrum, see Russell's Paradox: https://en.wikipedia.org/wiki/Russell's_paradox.

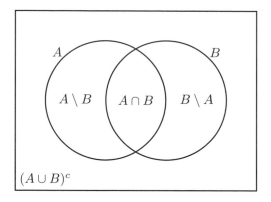

Figure A.1: Venn diagram for two sets, A and B, both subsets of a universal set U. A and B are represented by the two disk regions, and U is the entire rectangular region. $A \cup B = (A \setminus B) \cup (A \cap B) \cup (B \setminus A)$.

set of all *ordered n-tuples*:

$$A_1 \times A_2 \times \cdots \times A_n = \{(x_1, x_2, \ldots, x_n) \mid x_k \in A_k \text{ for each } k = 1, 2, \ldots, n\}. \quad \text{(A.3)}$$

Example 140. Describe the following sets.

(a) $\{4, 5, 6\} \times \{1, 5\}$

(b) $\mathbb{N} \times \{a, b\}$

(c) $\underbrace{\mathbb{R} \times \mathbb{R} \times \cdots \times \mathbb{R}}_{n}$

(d) $[4, 6] \times [1, 5]$

(e) $\mathbb{R}^n \times \mathbb{R}^m$

(f) $\mathbb{R} \times \mathbb{Z}$

Solution:

(a) There are six elements in the product, choosing one from $\{4, 5, 6\}$ and one from $\{1, 5\}$ for each pair:

$$\{4, 5, 6\} \times \{1, 5\} = \{(4, 1), (5, 1), (6, 1), (4, 5), (5, 5), (6, 5)\}.$$

(b) While we cannot list every element, we can describe the set in a nice way as a union.

$$\mathbb{N} \times \{a, b\} = \{(n, a) \mid n \in \mathbb{N}\} \cup \{(n, b) \mid n \in \mathbb{N}\}$$

(c) This is another way to interpret $\mathbb{R}^n$:

$$\underbrace{\mathbb{R} \times \mathbb{R} \times \cdots \times \mathbb{R}}_{n} = \mathbb{R}^n = \{(x_1, x_2, \ldots, x_n) \mid x_k \in \mathbb{R}\}.$$

(d) The product of intervals is a rectangular region of the plane.

$$[4,6] \times [1,5] = \{(x,y) \in \mathbb{R}^2 \mid 4 \leq x \leq 6, \text{ and } 1 \leq y \leq 5\}$$

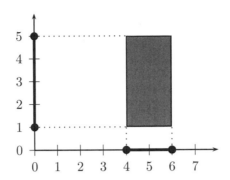

(e) Considering $\mathbb{R}^n = \mathbb{R} \times \cdots \times \mathbb{R}$ (n times), and $\mathbb{R}^m = \mathbb{R} \times \cdots \times \mathbb{R}$ (m times), we have a natural identification $\mathbb{R}^n \times \mathbb{R}^m = \mathbb{R}^{n+m}$. Explicitly, the element $\big((x_1, x_2, \ldots, x_n), (y_1, y_2, \ldots, y_m)\big) \in \mathbb{R}^n \times \mathbb{R}^m$ may be identified with the point $(x_1, x_2, \ldots, x_n, y_1, y_2, \ldots, y_m) \in \mathbb{R}^{n+m}$.

(f) As a set, $\mathbb{R} \times \mathbb{Z} = \{(x,n) \mid x \in \mathbb{R}, \ n \in \mathbb{Z}\}$. It may be useful to interpret this set within $\mathbb{R}^2$ as the union of all horizontal lines at integer heights.

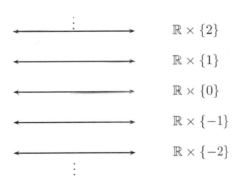

Sometimes we use the Cartesian product symbol $\times$ between a set X and a single element $a \in Y$. In such cases we interpret $X \times a = X \times \{a\}$.

Example 141. Let $m, n \geq 1$. Then $\mathbb{R}^m \times \mathbf{0} \subseteq \mathbb{R}^{m+n}$, where $\mathbf{0} \in \mathbb{R}^n$. Moreover, $\mathbf{0} \times \mathbb{R}^n \subseteq \mathbb{R}^{m+n}$, where $\mathbf{0} \in \mathbb{R}^m$. By abuse of notation, we typically identify $\mathbb{R}^m = \mathbb{R}^m \times \mathbf{0}$ and $\mathbb{R}^n = \mathbf{0} \times \mathbb{R}^n$ so that we can say $\mathbb{R}^m \subseteq \mathbb{R}^{m+n}$ and $\mathbb{R}^n \subseteq \mathbb{R}^{m+n}$. (These are particular *embeddings* of $\mathbb{R}^m$ and $\mathbb{R}^n$ into $\mathbb{R}^{m+n}$. There are many other ways to embed these spaces, though.)

Arbitrary Unions, Intersections, and Products

The definitions for union and intersection given above apply only to two sets at a time. Using a *recursive definition*, we extend to n sets, for $n > 2$, as follows.

$$A_1 \cup A_2 \cup \cdots \cup A_n = (A_1 \cup A_2 \cup A_3 \cup \cdots \cup A_{n-1}) \cup A_n \qquad (A.4)$$

$$A_1 \cap A_2 \cap \cdots \cap A_n = (A_1 \cap A_2 \cap A_3 \cap \cdots \cap A_{n-1}) \cap A_n \qquad (A.5)$$

Recall that when we want to abbreviate the notation for a sum of many terms, $a_1 + a_2 + \cdots + a_n$, we might use *sigma notation*,

$$\sum_{k=1}^{n} a_k = a_1 + a_2 + \cdots + a_n.$$

An analogous notation exists for unions and intersections:

$$\bigcup_{k=1}^{n} A_k = A_1 \cup A_2 \cup \cdots \cup A_n, \quad \text{and} \quad \bigcap_{k=1}^{n} A_k = A_1 \cap A_2 \cap \cdots \cap A_n.$$

Example 142.   Let $S_k = \{k, k+1\}$ for each $k \in \mathbb{N}$. Find $\displaystyle\bigcup_{k=1}^{4} S_k$.

Solution:

$$\bigcup_{k=1}^{4} S_k = S_1 \cup S_2 \cup S_3 \cup S_4 = \{1,2\} \cup \{2,3\} \cup \{3,4\} \cup \{4,5\} = \{1,2,3,4,5\}.$$

Both operations, $\cap$ and $\cup$, are **associative** and **commutative**. That is, so long as there is only one operation (only $\cap$ or only $\cup$) in an expression, then the order and grouping of the individual sets are not important.

Example 143.   Let $A = \{x \in \mathbb{N} \mid x \text{ is prime}\}$, $B = \{x \in \mathbb{N} \mid x < 30\}$, and $C = \{x \in \mathbb{N} \mid x \geq 20\}$. Determine $A \cap B \cap C$.

Solution: Interpret $A \cap B \cap C$ as $(B \cap C) \cap A$, since $B \cap C$ is easy to determine and is a finite set.

$$
\begin{aligned}
(B \cap C) \cap A &= \{x \in \mathbb{N} \mid x < 30 \text{ and } x \geq 20\} \cap A \\
&= \{x \in \mathbb{N} \mid 20 \leq x < 30\} \cap A \\
&= \{x \in \mathbb{N} \mid 20 \leq x < 30 \text{ and } x \text{ is prime}\} \\
&= \{23, 29\}
\end{aligned}
$$

Thinking again back to sums of terms, it is useful to have a concept of the sum of infinitely many terms. In sigma notation, this looks like:

$$\sum_{k=1}^{\infty} a_k = a_1 + a_2 + a_3 + \cdots$$

There is no guarantee that such a sum evaluates to a well-defined numeric value. On the other hand, it is true that the union or intersection of infinitely many sets is again a set.

Definition A.1.11. Suppose that for each $k \in \mathbb{N}$, there is a set A_k. The union of the sets $A_1, A_2, A_3, \ldots$ is the set whose elements are all of those from any set A_k, that is,

$$\bigcup_{k \in \mathbb{N}} A_k = \bigcup_{k=1}^{\infty} A_k = A_1 \cup A_2 \cup A_3 \cup \cdots = \{x \mid x \in A_k \text{ for some } k \in \mathbb{N}\}.$$

The intersection of the sets $A_1, A_2, A_3, \ldots$ is the set whose elements are all of those that are common to every set A_k, that is,

$$\bigcap_{k \in \mathbb{N}} A_k = \bigcap_{k=1}^{\infty} A_k = A_1 \cap A_2 \cap A_3 \cap \cdots = \{x \mid x \in A_k \text{ for all } k \in \mathbb{N}\}.$$

Example 144. For each $k \in \mathbb{N}$, let

$$M_k = \{n \in \mathbb{N} \mid n = km \text{ for some } m \in \mathbb{N} \text{ such that } m > 1\}.$$

Describe $\displaystyle\bigcup_{k=2}^{\infty} M_k$ and $\displaystyle\bigcap_{k=2}^{\infty} M_k$.

Solution: Let's first get an idea for what each set M_k looks like.

$$M_2 = \{n \in \mathbb{N} \mid n = 2 \cdot m \text{ for some } m > 1\} = \{4, 6, 8, 10, 12, 14, \ldots\}$$
$$M_3 = \{n \in \mathbb{N} \mid n = 3 \cdot m \text{ for some } m > 1\} = \{6, 9, 12, 15, 18, 21, \ldots\}$$
$$M_4 = \{n \in \mathbb{N} \mid n = 4 \cdot m \text{ for some } m > 1\} = \{8, 12, 16, 20, 24, 28, \ldots\}$$

So each $M_k \subseteq \mathbb{N}$ is the set of multiples of k greater than the number k itself. So the union contains every natural number that can be written in the form km, where both k and m are at least 2. This is the set of all *composite* numbers, that is, all natural numbers greater than 1 that are not prime:

$$\bigcup_{k=2}^{\infty} M_k = \mathbb{N} \setminus \{1, 2, 3, 5, 7, 11, 13, 17, 19, \ldots\}.$$

Any number in the intersection must be a multiple of *every* number greater than 1. However, no such number exists, since every natural number is the product of only finitely many primes. In other words, the intersection is empty.

$$\bigcap_{k=2}^{\infty} M_k = \emptyset$$

Definition A.1.12. Suppose that for each $k \in \mathbb{N}$, there is a set A_k. The product of the sets $A_1, A_2, A_3 \ldots$ is the set of infinite tuples:

$$\prod_{k \in \mathbb{N}} A_k = \prod_{k=1}^{\infty} A_k = A_1 \times A_2 \times A_3 \times \cdots = \{(x_1, x_2, x_3, \ldots) \mid x_k \in A_k\}.$$

If all of the sets are equal, say, $A_k = A$ for all $k \in \mathbb{N}$, then we write $\prod_{k=1}^{\infty} A = A^{\infty}$.

Example 145. Let C be any set. A **sequence** in C is an infinite ordered tuple, $(c_1, c_2, c_3, \ldots)$. The set of all sequences in C is the infinite product C^{∞}. In particular,

- $\mathbb{Z}^{\infty}$ is the set of all integer sequences.

- $\mathbb{R}^{\infty}$ is the set of all real number sequences (or "infinite-dimensional" vectors).

Finally, in topology we may encounter unions or intersections of sets A_k, where k ranges over all of the elements of a given *index set* rather than just $\mathbb{N}$.

Definition A.1.13. Let $\mathcal{I}$ be a set (called the **index set**). Suppose that for each $k \in \mathcal{I}$, there is a set A_k. The union of A_k over $\mathcal{I}$ is:

$$\bigcup_{k \in \mathcal{I}} A_k = \{x \mid x \in A_k \text{ for some } k \in \mathcal{I}\}.$$

The intersection of A_k over $\mathcal{I}$ is:

$$\bigcap_{k \in \mathcal{I}} A_k = \{x \mid x \in A_k \text{ for all } k \in \mathcal{I}\}.$$

Example 146. Example 140(d) shows the set $\mathbb{R} \times \mathbb{Z}$ geometrically as a union of lines. The indexing set is $\mathbb{Z}$.

$$\mathbb{R} \times \mathbb{Z} = \bigcup_{n \in \mathbb{Z}} (\mathbb{R} \times \{n\})$$

Less obviously, we could break up the set over each real number and interpret $\mathbb{R} \times \mathbb{Z}$ as a *continuum* of integer sets:

$$\mathbb{R} \times \mathbb{Z} = \bigcup_{x \in \mathbb{R}} (\{x\} \times \mathbb{Z}).$$

Set Identities

Certain relationships exist among unions, intersections, and set differences, a number of which are collected here for reference. Exercise 10 asks you to prove each one.

Theorem A.1.14 (basic set identities). *For all sets, A, B, and C, the following identities hold.*

(a) $A \cup B = B \cup A$, *and* $A \cap B = B \cap A$

(b) $(A \cup B) \cup C = A \cup (B \cup C)$, *and* $(A \cap B) \cap C = A \cap (B \cap C)$

(c) $A \cup A = A$, *and* $A \cap A = A$

(d) $A \cup \emptyset = A$, *and* $A \cap \emptyset = \emptyset$

(e) $A \setminus \emptyset = A$, *and* $A \setminus A = \emptyset$

(f) $(A \cup B) \cap C = (A \cap C) \cup (B \cap C)$, *and* $(A \cap B) \cup C = (A \cup C) \cap (B \cup C)$

The following theorem, known as De Morgan's Law, is of particular importance. It ties together all three of the basic operations of union, intersection, and set difference.

Theorem A.1.15 (De Morgan's Law). *Let A, B, and C be sets.*

$$A \setminus (B \cup C) = (A \setminus B) \cap (A \setminus C) \qquad and$$
$$A \setminus (B \cap C) = (A \setminus B) \cup (A \setminus C)$$

We will prove the first half of the De Morgan Law, leaving the other for an exercise. Let's first develop a *game plan*. In order to prove two sets S and T are equal, it is necessary to prove both set inclusions, $S \subseteq T$ and $T \subseteq S$. Thus the proof below is separated into two parts. The parenthetical remarks are for the benefit of the reader and may be omitted in a formal proof.

Proof. **Part I.** To show: $A \setminus (B \cup C) \subseteq (A \setminus B) \cap (A \setminus C)$.

Let $x \in A \setminus (B \cup C)$. Then (by definition of *difference*), $x \in A$ and $x \notin B \cup C$. But if $x \notin B \cup C$, then certainly $x \notin B$ and $x \notin C$ (since if x were in either set B or C, then by definition of *union*, $x \in B \cup C$). Now we have $x \in A$ and $x \notin B$, so $x \in A \setminus B$, and $x \in A$ and $x \notin C$, so $x \in A \setminus C$. Since x is in both sets, we have $x \in (A \setminus B) \cap (A \setminus C)$ (by definition of *intersection*).

Part II. To show: $(A \setminus B) \cap (A \setminus C) \subseteq A \setminus (B \cup C)$.

Let $x \in (A \setminus B) \cap (A \setminus C)$. Then (by definition of *intersection*), $x \in A \setminus B$ and $x \in A \setminus C$. The former condition implies $x \in A$ and $x \notin B$, while the latter implies $x \in A$ and $x \notin C$ (by definition of *difference*). Since $x \notin B$ and $x \notin C$, then $x \notin B \cup C$ (since if x were in $B \cup C$, then either $x \in B$ or $x \in C$, by definition of *union*). Now we have $x \in A$ and $x \notin B \cup C$, which gives $x \in A \setminus (B \cup C)$ (by definition of *difference*).

Parts I and II imply that $A \setminus (B \cup C) = (A \setminus B) \cap (A \setminus C)$. $\square$

Exercises

1. Prove there is no smallest positive rational number.

2. Prove that if $x \in \mathbb{Q}$ and $y \in \mathbb{Q}$, then $x + y \in \mathbb{Q}$ and $x - y \in \mathbb{Q}$.

3. Prove that if $x \in \mathbb{Q}$ and $y \in \mathbb{R} \setminus \mathbb{Q}$, then $x + y \in \mathbb{R} \setminus \mathbb{Q}$. (*Hint:* Use Exercise 2 and a proof by contradiction.

4. Using induction, prove the formula for the sum of a finite geometric series:
   If $r \neq 1$, then $\displaystyle\sum_{k=0}^{n} r^k = \frac{1 - r^{n+1}}{1 - r}$ for all $n \in \mathbb{N}$.

5. Prove that the principle of mathematical induction is equivalent to the well-ordering principle. (*Hints:* To show that induction implies well-ordering, let S be a set of natural numbers having no minimum element, and arrive at a contradiction. To show that well-ordering implies induction, let S be the set of all $n \in \mathbb{N}$ such that $P(n)$ is not true, and show that $S = \emptyset$.)

6. Suppose $A = \{1, 4, \{1\}, \{3\}, \{1, 3, 4\}\}$. Identify which of the following statements are *true* or *false*.

(a) $1 \in A$	(e) $\{1\} \subseteq A$	(i) $\{1, 3, 4\} \in A$
(b) $3 \in A$	(f) $\{1\} \in A$	(j) $\{1, 3, 4\} \subseteq A$
(c) $4 \in A$	(g) $1 \subseteq A$	(k) $\{3\} \subseteq A$
(d) $\{4\} \in A$	(h) $\{1, 4\} \subseteq A$	(l) $\emptyset \subseteq A$

7. Suppose A is a finite set. How many elements are in its power set, $\mathcal{P}(A)$? How many elements are in $\mathcal{P}(\mathcal{P}(A))$?

8. Consider the sets M_k as defined in Example 144. Describe the following sets:

(a) $M_2 \cup M_3$	(c) $M_6 \cap M_{15}$	(e) $M_2 \times M_3 \times M_4$
(b) $M_2 \cap M_3$	(d) $M_2 \times M_3$	(f) $\prod_{k=2}^{\infty} M_k$

9. Show that if $A \cap B \neq \emptyset$ and $C \cap D \neq \emptyset$, then $(A \times C) \cap (B \times D) \neq \emptyset$. (*Hint:* Produce at least one element in $(A \times C) \cap (B \times D)$.)

10. Prove all of the set identities listed in Theorem A.1.14.

11. Prove the other half of De Morgan's Law, $A \setminus (B \cap C) = (A \setminus B) \cup (A \setminus C)$.

12. Prove De Morgan's Law for arbitrary unions and intersections:

$$A \setminus \bigcup_{k \in \mathcal{I}} B_k = \bigcap_{k \in \mathcal{I}} (A \setminus B_k) \qquad \text{and}$$

$$A \setminus \bigcap_{k \in \mathcal{I}} B_k = \bigcup_{k \in \mathcal{I}} (A \setminus B_k).$$

13. Prove a generalized distributive law for the intersection of two arbitrary unions,

$$\left(\bigcup_{k \in \mathcal{I}} A_k \right) \cap \left(\bigcup_{\ell \in \mathcal{J}} B_\ell \right) = \bigcup_{k \in \mathcal{I}, \ell \in \mathcal{J}} (A_k \cap B_\ell). \qquad (A.6)$$

14. Prove the following.

    (a) If $A \subseteq B$ and $B \subseteq C$, then $A \subset C$ (*transitivity* of $\subseteq$).
    (b) $(A^c)^c = A$.
    (c) If $A \subseteq B$, then $A \setminus B = \emptyset$.
    (d) If $C \subseteq A \cap B$, then $C \subseteq A$ and $C \subseteq B$.
    (e) $(A \setminus B) \setminus C = A \setminus (B \cup C)$.
    (f) If $A \subseteq B$, then $A \times C \subseteq B \times C$.
    (g) $A \times (B \cap C) = (A \times B) \cap (A \times C)$.
    (h) $(A \cap B) \times (C \cap D) - (A \times C) \cap (B \times D)$.

A.2 Relations and Functions

In this section, we define relations and functions using set theory.

Relations

> **Definition A.2.1.** Let X and Y be sets. A **relation** R between X and Y is a subset $R \subseteq X \times Y$. If $(x, y) \in R$, then we may write xRy and say "x is related to y via R."

Example 147. The following are examples of relations.

- Define a relation $S \subseteq \mathbb{N} \times \mathbb{R}$ by $S = \{(x, y) \in \mathbb{N} \times \mathbb{R} \mid y = \sqrt{x}\}$. The elements of S include $(1, 1)$, $(2, \sqrt{2})$, $(3, \sqrt{3})$, $(4, 2)$, etc.

- Define a relation $L \subseteq \mathbb{R} \times \mathbb{R} = \mathbb{R}^2$ by $L = \{(x, y) \in \mathbb{R}^2 \mid x \leq y\}$. That is, $xLy \iff x \leq y$. The set L may be visualized as the region of the plane $\mathbb{R}^2$ above and including the line $y = x$.

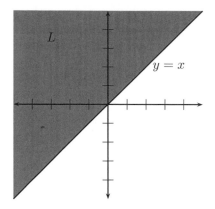

In this text we will mostly be concerned with two specific kinds of relation: *equivalence relations* and *functions*.

Definition A.2.2. An **equivalence relation** on a set X is a subset $E \subseteq X \times X$ satisfying the three properties below. We use the notation $x \sim y$ to indicate $(x, y) \in E$.

(i) [Reflexivity] $\forall x \in X, x \sim x$.

(ii) [Symmetry] $\forall x, y \in X, (x \sim y) \implies (y \sim x)$.

(iii) [Transitivity] $\forall x, y, z \in X, (x \sim y)$ and $(y \sim z) \implies (x \sim z)$.

Example 148.   Neither relation in Example 147 is an equivalence relation. For instance, S fails reflexivity since $(2, 2) \notin S$. The relation L does satisfy reflexivity since $x \leq x$ for all $x \in \mathbb{R}$, and transitivity follows from a well-known property: if $x \leq y$ and $y \leq z$, then $x \leq z$. However, L is not an equivalence relation because symmetry does not hold: $3 \leq 5$ but $5 \not\leq 3$, for example.

When there is an equivalence relation on a set, then the set can be **partitioned** into mutually disjount subsets called **equivalence classes**. We use the notation $[x]$ to stand for the equivalence class of X in which x is a member, that is,

$$[x] = \{z \in X \mid z \sim x\}.$$

Clearly, $x \sim y$ if and only if $[x] = [y]$. If X is a set with an equivalence relation $\sim$ defined on its elements, then the **quotient** of X with respect to $\sim$ is the set of all equivalence classes:

$$X/\sim = \{[x] \mid x \in X\}.$$

Example 149. The following are examples of equivalence relations.

- For any set X, let $E = \{(x, x) \mid x \in X\}$. This is the most restrictive equivalence relation that may be defined on a set: $x \sim y \iff x = y$. Certainly reflexivity is satisfied. Symmetry follows since if $x = y$, then $y = x$. Transitivity is also trivial: If $x = y$ and $y = z$, then $x = z$. The equivalence classes are the singleton sets, $[x] = \{x\}$. This relation is called the *equality* relation.

- Fix a natural number $d \geq 1$. Define an equivalence relation on $\mathbb{Z}$ as follows: $n \sim m \iff n - m$ is a multiple of d. For example, if $d = 5$, then we have $2 \sim 7 \sim 102 \sim -23 \sim \cdots$. You will prove in an exercise that this is indeed an equivalence relation, but let's explore what the equivalence classes look like. Each equivalence class is an arithmetic sequence, extending infinitely negative and positive. We have chosen *representatives*, 0, 1, 2, 3, 4, in each class, but any single element could have been chosen as representative (e.g., [5] instead of [0]). The process of determining which representative corresponds to a given number $n \in \mathbb{Z}$ is called **reduction modulo** d.

$$
\begin{aligned}
[0] &= \{\ldots, -15, -10, -5, 0, 5, 10, 15, 20, \ldots\} = \{5n \mid n \in \mathbb{Z}\} \\
[1] &= \{\ldots, -14, -9, -4, 1, 6, 11, 16, 21, \ldots\} = \{1 + 5n \mid n \in \mathbb{Z}\} \\
[2] &= \{\ldots, -13, -8, -3, 2, 7, 12, 17, 22, \ldots\} = \{2 + 5n \mid n \in \mathbb{Z}\} \\
[3] &= \{\ldots, -12, -7, -2, 3, 8, 13, 18, 23, \ldots\} = \{3 + 5n \mid n \in \mathbb{Z}\} \\
[4] &= \{\ldots, -11, -6, -1, 4, 9, 14, 19, 24, \ldots\} = \{4 + 5n \mid n \in \mathbb{Z}\}
\end{aligned}
$$

In this way, we have partitioned the infinite set $\mathbb{Z}$ into five classes. The standard notation for this set of equivalence classes is $\mathbb{Z}/5\mathbb{Z}$ (and $\mathbb{Z}/d\mathbb{Z}$ in general), and we may write $\mathbb{Z}/5\mathbb{Z} = \{[0], [1], [2], [3], [4]\}$ or $\{0, 1, 2, 3, 4\}$. With a little more work, it is possible to define the operations of addition and multiplication on these equivalence classes. These satisfy $[a] + [b] = [a+b]$ and $[a] \cdot [b] = [ab]$, as one might expect, and define *modular arithmetic* on the set $\mathbb{Z}/d\mathbb{Z}$.

Functions

A *function* is a type of relation $F \subseteq X \times Y$ in which all pairs satisfy the additional condition:

- For every $x \in X$, there is *exactly one* pair $(x, y) \in F$.

This condition implies that every $x \in X$ is related to a uniquely determined $y \in Y$. In other words, F defines a *rule* associating each x with a particular y. From now on, we will write functions in a more traditional way that may be familiar from calculus or algebra: $f(x) = y$.

> **Definition A.2.3.** Let D and C be sets. A **function** f from D to C, written symbolically as $f : D \to C$, is a rule that assigns to each element $x \in D$, a single element $y \in C$, and we may write $y = f(x)$. The set D is called the **domain** and C is called the **codomain** of f.

In algebra and calculus, the domain and codomain are usually not included in the definition of the function itself. Instead the domain is assumed to be the largest subset of $\mathbb{R}$ that makes sense for the function, while the codomain is rarely mentioned at all but understood to be $\mathbb{R}$. For example, when you see $f(x) = x^2$, it is usually understood that $f : \mathbb{R} \to \mathbb{R}$, and the rule is: "square the given real number." By our definition, f must be *defined* for every x in the domain set; however, we do not require all points of the codomain to be outputs for the function. Indeed, negative values, such as $y = -1$, cannot be *hit* by $f(x) = x^2$ (and yet we still may say $\mathbb{R}$ is the codomain).

When speaking about individual elements, we may use the notation $f : x \mapsto y$ or $x \mapsto y$ (*read:* "x maps to y") when the function is clear by context. In fact, we may even *define* the function by its rule, $x \mapsto f(x)$, when the domain and codomain are clear by context. Thus $x \mapsto x^2$ is another way to write $f(x) = x^2$. Note that in this small example we find multiple x-values that map to the same $f(x)$-value, for instance, $2 \mapsto 4$ and $-2 \mapsto 4$. It is important to understand that a function may have multiple inputs mapping to the same output, but may **not** have an input mapping to multiple outputs.

It is often helpful to represent a function by a *bubble diagram*, especially when the function has finite domain and codomain. Figure A.2 shows a few different functions from $D = \{1, 2, 3\}$ to $C = \{a, b, c, d\}$, while Figure A.3 demonstrates bubble diagrams that do not correspond to functions.

Suppose we are given a function $f : D \to C$ and $A \subseteq D$. Certainly f is defined on A since it is defined on all of D – we say that f **restricts** to A, and we write $f\big|_A$ to denote the function whose domain is the subset $A \subseteq D$ and has the same rule as f (i.e., $x \mapsto f(x)$). The restricted function is technically a different function because the domain is A rather than D; nevertheless, we often abuse notation, writing $f : A \to C$ when we really mean $f\big|_A : A \to C$.

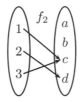 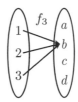

Figure A.2: Three functions. The rule for f_1 is: $f_1(1) = b$, $f_1(2) = d$, $f_1(3) = a$. The rule for f_2 is: $f_2(1) = c$, $f_2(2) = d$, $f_3(3) = c$. The function f_3 is a *constant* function, $f_3(1) = f_3(2) = f_3(3) = b$.

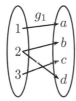 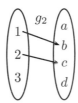 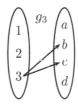

Figure A.3: Three nonfunctions. The diagram for g_1 provides two outputs, b and d, for the input value 2 (we say that $g_1(2)$ is not *well defined*). The diagram for g_2 lacks an output value corresponding to 3 (we say that $g_2(3)$ is *undefined*). The diagram for g_3 shows both defects ($g_3(1)$ and $g_3(2)$ are undefined and $g_3(3)$ is not well defined).

Injectivity, Surjectivity, and Bijectivity

Certain properties of functions will become important in this and later chapters of this book.

Definition A.2.4. Let $f : D \to C$ be a function.

- f is called **injective** (or **one-to-one**) if whenever $x_1, x_2 \in D$ and $x_1 \neq x_2$, then $f(x_1) \neq f(x_2)$. Equivalently, f is injective if $f(x_1) = f(x_2)$ implies $x_1 = x_2$.

- f is called **surjective** (or **onto**) if for every $y \in C$, there is at least one $x \in D$ such that $f(x) = y$.

- f is called **bijective** if f is both injective and surjective.

Figure A.4 illustrates injective, surjective, and bijective functions in terms of bubble diagrams.

Example 150. In this example, we consider the rule $x \mapsto x^2$ with respect to different domains and codomains. Refer to Figure A.5.

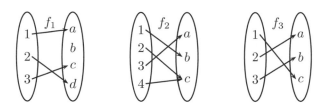

Figure A.4: Here f_1 is injective but not surjective, and f_2 is surjective but not injective. On the other hand, f_3 is both injective and surjective; hence it is a bijection.

(a) The function $f : \mathbb{R} \to \mathbb{R}$ defined by $f(x) = x^2$ is neither injective nor surjective. We see that f is not injective because $f(-2) = 4 = f(2)$; that is, with $x_1 = -2$ and $x_2 = 2$, we have $x_1 \neq x_2$, but $f(x_1) = f(x_2)$. We find that f is not surjective because $-1 \in \mathbb{R}$ (the codomain), but there is no $x \in \mathbb{R}$ (the domain) such that $f(x) = -1$.

(b) Next consider the restriction $f|_{[0,\infty)} : [0, \infty) \to \mathbb{R}$. This function is injective since now the inputs are restricted to nonnegative real numbers, but $f|_{[0,\infty)}$ is not surjective, as there is still no $x \in [0, \infty)$ such that $f(x) = -1$.

(c) Suppose we define $g : [0, \infty) \to [0, \infty)$ by $g(x) = x^2$. Notice that g has the same rule as f, but the domain and codomain of g are both restricted. Since g is both injective and surjective, g is a bijection.

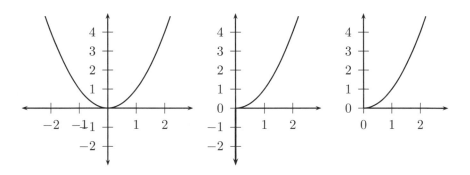

Figure A.5: *Left*, graph of f. *Center*, graph of $f|_{[0,\infty)}$. *Right*, graph of g.

Compositions and Inverses

The effect of applying one function followed by a second function is called *composition*. To be precise:

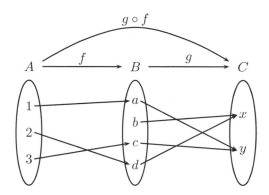

Figure A.6: Here $f : A \to B$, and $g : B \to C$. The composition $g \circ f$ is a function from A to C. According to this diagram, $(g \circ f)(1) = y$, $(g \circ f)(2) = x$, and $(g \circ f)(3) = y$.

Definition A.2.5. Suppose $f : A \to B$ and $g : B \to C$ are functions. The **composition** of f and g is a function $g \circ f : A \to C$ defined by the rule

$$(g \circ f)(x) = g(f(x)), \quad \text{for every } x \in A.$$

It is important to understand that composition of functions is not *commutative*. Even when the domains and codomains are given so that both $f \circ g$ and $g \circ f$ are defined, we have in general $f \circ g \neq g \circ f$. Also note that the definition of composition *requires* that the codomain of f be the same as the domain of g; otherwise, the composition is not defined. We may represent the composition of functions $f . A \to B$ and $g . B \to C$ using bubble diagrams as in Figure A.6.

Example 151. Let $\mathbb{R}^+$ represent the set of positive real numbers, $\{x \in \mathbb{R} \mid x > 0\}$, and let $\mathbb{R}^-$ represent the set of negative real numbers, $\{x \in \mathbb{R} \mid x < 0\}$. Define function $f : \mathbb{N} \to \mathbb{R}^+$ and $g : \mathbb{R}^+ \to \mathbb{R}^-$ by the following rules:

$$f(n) = 2^n, \qquad g(x) = -\frac{1}{x^4}.$$

Describe the composite function, $g \circ f$.

Solution: The domain of $g \circ f$ is the domain of f – that is, $\mathbb{N}$ – and the codomain is the codomain of g – that is, $\mathbb{R}^-$. To find the "rule" that defines $g \circ f$, simply work out $g(f(n))$.

$$g \circ f \quad : \quad \mathbb{N} \to \mathbb{R}^-$$
$$(g \circ f)(n) \quad = \quad g(f(n)) = g(2^n) = -\frac{1}{(2^n)^4} = -\frac{1}{2^{4n}}$$

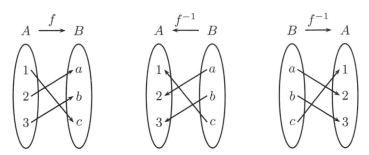

Figure A.7: Here f is bijective, so the inverse exists. Note that f^{-1} may be regarded as *reversing all arrows of f*. The third picture is f^{-1} displayed with domain on the left as normally depicted.

Whenever $f : A \to B$ is bijective, there is a way to "reverse" the direction of the arrow to get a function we call the **inverse**. See Figure A.7 for a bubble diagram interpretation of the inverse function.

Definition A.2.6. Suppose $f : A \to B$ is a bijective function. The **inverse** of f is a function $f^{-1} : B \to A$ defined by

$$f^{-1}(b) = a, \quad \text{if and only if} \quad b = f(a).$$

If f is not bijective, then f^{-1} does not exist.

Proposition A.2.7. *Functions $f : A \to B$ and $g : B \to A$ are inverses of one another if and only if $(g \circ f)(x) = x$ for all $x \in A$ and $(f \circ g)(y) = y$ for all $y \in B$.*

The phrase "f and g are inverses of one another" means both $g = f^{-1}$ and $f = g^{-1}$. We will only be concerned with the former, as the latter can be derived analogously. In order to prove an "if and only if," we must prove both directions. That is, we must first assume the statement on the left of the "if and only if," using it to derive the statement on the right, and then we must assume the right to derive the left. The proof given below may seem long at first, but it provides a model for the kind of formal proof that you may see or write on your way toward understanding topology.

Proof. **Part I.** Suppose $g = f^{-1}$.

Consider $(g \circ f)(x)$. By definition of composition, $(g \circ f)(x) = g(f(x))$. Now with any $b \in B$, and by definition of inverse functions, we have: $g(b) = x$ if and only if $b = f(x)$. Choosing now $b = f(x)$, we have

$$g(f(x)) = x \iff f(x) = f(x). \tag{A.7}$$

The right side of (A.7) is true (for all $x \in A$); therefore the left side is also true. This proves that $(g \circ f)(x) = x$.

Now consider $(f \circ g)(y)$. Again by definition of composition, $(f \circ g)(y) = f(g(y))$. For any $a \in A$, we have $g(y) = a$ if and only if $y = f(a)$. Choosing $a = g(y)$, we have

$$f(g(y)) = y \iff g(y) = g(y). \tag{A.8}$$

The right side of (A.8) is true (for all $y \in B$); therefore the left side is also true. This proves that $(f \circ g)(y) = y$.

Part II. Suppose both $(f \circ g)(x) = x$ for all $x \in A$ and $(g \circ f)(y) = y$ for all $y \in B$. We have to prove that $g = f^{-1}$. First, and most importantly, we have to show that f^{-1} exists, which is done by proving that f is bijective.

Injectivity of f: Consider two elements, $x_1, x_2 \in A$. Suppose $f(x_1) = f(x_2)$.

$$
\begin{aligned}
f(x_1) &= f(x_2) \\
g(f(x_1)) &= g(f(x_2)) \\
(g \circ f)(x_1) &= (g \circ f)(x_2) \\
x_1 &= x_2 \qquad \text{by hypothesis}
\end{aligned}
$$

Thus f is shown to be injective.

Surjectivity of f: Now consider any $y \in B$. We must show there is an $x \in A$ such that $f(x) = y$.

$$
\begin{aligned}
(f \circ g)(y) &= y \qquad \text{by hypothesis} \\
f(g(y)) &= y
\end{aligned}
$$

Therefore, for $x = g(y)$, we have $f(x) = y$. Thus f is surjective.

Now that we see f is bijective, we know that f^{-1} exists. All that remains is to show that g fits the definition to be f^{-1}. Let $a \in A$ and $b \in B$. We have to show both implications, $g(b) = a \implies b = f(a)$ and $b = f(a) \implies g(b) = a$.

$$
\begin{aligned}
g(b) &= a, \qquad \text{assumed} \\
f(g(b)) &= f(a) \\
(f \circ g)(b) &= f(a) \\
b &= f(a), \qquad \text{by hypothesis}
\end{aligned}
$$

This proves $g(b) = a \implies b = f(a)$. The argument showing $b = f(a) \implies g(b) = a$ is similar and is left as an exercise. Thus we have $g(b) = a$ if and only if $b = f(a)$, which shows that $g = f^{-1}$. $\qquad \square$

Example 152. Suppose $A = \{0, 1, 2, 3\}$ and $B = \{2, 3, 5, 7\}$, and let $f : A \to B$ be defined by the bubble diagram below. Determine the values of $f^{-1}(2)$, $f^{-1}(3)$, $f^{-1}(5)$, and $f^{-1}(7)$.

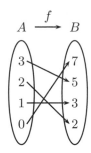

Solution: Simply follow the arrows backward to find $f^{-1}(2) = 2$, $f^{-1}(3) = 1$, $f^{-1}(5) = 3$, $f^{-1}(7) = 0$.

Example 153. Let $f : \mathbb{R} \to \mathbb{R}$ be defined by $f(x) = x^2$. Let $g : [0, \infty) \to [0, \infty)$ be defined by the same rule, $g(x) = x^2$. Determine the inverse function of each, if it exists.

Solution: The function f is neither injective nor surjective (see Example 150); therefore the inverse function f^{-1} does not exist. On the other hand, with restricted domain and codomain, g is bijective (see Example 150). To find g^{-1}, simply solve algebraically for x in $y = g(x)$:

$$y = x^2 \quad \Longrightarrow \quad \sqrt{y} = x \quad \Longrightarrow \quad g^{-1}(x) = \sqrt{x}.$$

Note that Proposition A.2.7 may be used to verify that $g^{-1}(x) = \sqrt{x}$.

$$
\begin{aligned}
(g^{-1} \circ g)(x) &= \sqrt{(x^2)} = x, \quad \text{since } x \geq 0 \\
(g \circ g^{-1})(x) &= \left(\sqrt{x}\right)^2 = x, \quad \text{since } x \geq 0
\end{aligned}
$$

Why doesn't this algebraic method work to show that the inverse of f could also be $\sqrt{x}$?

Forward and Inverse Image

Even when a function f fails to be injective or surjective, we may still want to find the x-values that solve $f(x) = y$. However, without an inverse function, we cannot in general find a unique result of the form $x = f^{-1}(y)$. Just think of $f(x) = x^2$ defined over all real numbers. When we solve $x^2 = 4$, we get two answers, $x = \pm 2$. On the other hand, the solution to $x^2 = -1$ is empty. The proper way to "reverse" a nonbijective function is to examine how f maps *subsets* to *subsets*.

Definition A.2.8. Suppose $f : A \to B$ is a function.

- Let $A_0 \subseteq A$. The **forward image** (or simply **image**) of A_0 under f is a subset of B defined by

$$f[A_0] = \{b \in B \mid b = f(a) \text{ for some } a \in A_0\} = \{f(a) \mid a \in A_0\}.$$

- Let $B_0 \subseteq B$. The **inverse image** (or **preimage**) of B_0 under f is a subset of A defined by

$$f^{-1}[B_0] = \{a \in A \mid f(a) \in B_0\}.$$

There is an unfortunate ambiguity in notation in Definition A.2.8. When working with the inverse image, $f^{-1}[B_0]$, we cannot assume that an inverse *function* f^{-1} exists. In order to help alleviate this confusion, we place brackets around the subset B_0 to remind the reader that we are working with inverse images of sets rather than inverse functions. The same notational convention applies to foward images.

By our definition of function, if $f : A \to B$, then $f^{-1}[B] = A$. That is, every $a \in A$ must be mapped to *something* in B. On the other hand, $f[A]$ may not be all of B, unless f happens to be surjective. In fact, $f[A]$ is by definition the *range* of f.

Definition A.2.9. Suppose $f : A \to B$ is a function. The **range** of f is the forward image $f[A] \subseteq B$.

Example 154. Let $X = \{1, 2, 3, 4, 5, 6\}$ and $Y = \{1, 2, 3, 4, 5, 6, 7, 8\}$. Suppose $f : X \to Y$ is a function defined on $x \in X$ by the table below.

x	1	2	3	4	5	6
$f(x)$	4	5	1	7	5	4

Find the following values or subsets.

(a) $f[X]$ (b) $f[\{1, 2\}]$ (c) $f^{-1}[\{2, 3\}]$ (d) $f^{-1}[f[\{1, 2\}]]$

Solution:

(a) $f[X]$ is the range of the function, $\{1, 4, 5, 7\}$

(b) $f[\{1, 2\}] = \{4, 5\}$, since $f(1) = 4$ and $f(2) = 5$

(c) $f^{-1}[\{2, 3\}] = \emptyset$, since there are no x-values such that $f(x) = 2$ or $f(x) = 3$

(d) from part (b), we have $f[\{1,2\}] = \{4,5\}$; therefore

$$f^{-1}[f[\{1,2\}]] = f^{-1}[\{4,5\}] = \{1,2,5,6\}$$

Part (d) of the above example demonstrates that $f^{-1}[f[A_0]] \neq A_0$ in general. However, the following relationships regarding forward and inverse images are true.

> **Proposition A.2.10.** *Let* $f : A \to B$, $A_0 \subseteq A$, *and* $B_0 \subseteq B$. *The following relationships hold.*
>
> 1. $A_0 \subseteq f^{-1}[f[A_0]]$, *with equality if* f *is injective*
>
> 2. $f[f^{-1}[B_0]] \subseteq B_0$, *with equality if* f *is surjective*
>
> 3. *if* $A_0 \subseteq A_1 \subseteq A$, *then* $f[A_0] \subseteq f[A_1]$
>
> 4. *if* $B_0 \subseteq B_1 \subseteq B$, *then* $f^{-1}[B_0] \subseteq f^{-1}[B_1]$

We shall prove part 1 of Proposition A.2.10, leaving the other parts as exercises.

Proof. Let $x \in A_0$ be arbitrary. We must show that $x \in f^{-1}[f[A_0]]$. Since $x \in A_0$, $f(x) \in f[A_0]$, by definition of forward image. But by definition of inverse image,

$$f^{-1}[f[A_0]] = \{a \in A \mid f(a) \in f[A_0]\}.$$

Certainly the x we chose qualifies under this definition to be a member of $f^{-1}[f[A_0]]$. This proves the inclusion $A_0 \subseteq f^{-1}[f[A_0]]$. $\square$

Exercises

1. Let $d \geq 1$ be a whole number, and define a relation on the set of integers $\mathbb{Z}$ by $n \sim m \iff n - m$ is a multiple of d.

   (a) Prove that $\sim$ is an equivalence relation.

   (b) Write out the equivalence classes defined on $\mathbb{Z}$ by the relation $\sim$ when $d = 9$. How many classes are there?

2. Let $D = \{1,2,3\}$ and $E = \{a,b\}$. How many functions are there from D to E? List them all using bubble diagrams.

3. Decide whether each function below is injective, surjective, and/or bijective on the indicated domain and range. If bijective, find the inverse function.

(a) $f : \mathbb{R} \to \mathbb{R}, \ f(x) = 8x + 7$ (d) $f : (-\infty, 0] \to [0, \infty), \ f(x) = 3x^4$

(b) $f : \mathbb{R} \to \mathbb{R}, \ f(x) = 3x^4$ (e) $f : \mathbb{R} \to \mathbb{R}, \ f(x) = \sin x$

(c) $f : \mathbb{R} \to [0, \infty), \ f(x) = 3x^4$ (f) $f : [0, 2\pi] \to [-1, 1], \ f(x) = \sin x$

(g) $f : [-\pi/2, \pi/2] \to [-1, 1], \ f(x) = \sin x$

(h) $f : \mathbb{R} \setminus \{0\} \to \mathbb{R} \setminus \{0\}, \ f(x) = \dfrac{1}{x}$

(i) $f : \mathbb{R} \setminus \{0\} \to \mathbb{R} \setminus \{0\}, \ f(x) = \dfrac{1}{x^2}$

(j) $f : \mathbb{R} \setminus \{-2/5\} \to \mathbb{R} \setminus \{3/5\}, \ f(x) = \dfrac{3x - 7}{5x + 2}$

(k) $f : \mathbb{R} \to \mathbb{R}^+, \ f(x) = 7e^{2x}$

4. Suppose that $f : A \to B$ is a bijective function, and suppose that $(g \circ f)(x) = x$ for all $x \in A$. Let $a \in A$ be arbitrary. Prove that if $b = f(a)$, then $g(b) = a$ (completing the proof of Proposition A.2.7).

5. Let $f : X \to Y$ be defined as in Example 154. Find the following values or subsets, or if the expression is undefined, explain why.

   (a) $f(1)$ (b) $f[\{1\}]$ (c) $f^{-1}(1)$ (d) $f^{-1}[\{1\}]$

6. Let $f(x) = \dfrac{1}{x^2}$. Find the forward or inverse images.

   (a) $f[(1,3)]$ (b) $f[(-3,-1)]$ (c) $f^{-1}[(1,9)]$ (d) $f^{-1}[(-9,9)]$

7. Let $f : D \to C$. Assume that $A, B \subseteq D$ and $U, V \subseteq C$. Prove the following set identities involving forward or inverse images.

   (a) $f[A \cup B] = f[A] \cup f[B]$ (c) $f^{-1}[U \cap V] = f^{-1}[U] \cap f^{-1}[V]$

   (b) $f^{-1}[U \cup V] = f^{-1}[U] \cup f^{-1}[V]$ (d) $f^{-1}[U \setminus V] = f^{-1}[U] \setminus f^{-1}[V]$

8. Let $f : D \to C$. Assume that $A, B \subseteq D$. Prove that $f[A \cap B] \subseteq f[A] \cap f[B]$. Provide a specific counterexample showing that equality is not true in general. Then show that if f is injective, there is equality $f[A \cap B] = f[A] \cap f[B]$.

9. Prove parts 2, 3, and 4 of Proposition A.2.10.

10. Suppose A has two elements and B has four elements. How many injective functions are there from A to B? What if A has m elements and B has n elements? (*Hint:* consider $m \le n$ and $m > n$ separately.)

11. Assume $f : A \to B$ and $g : B \to C$ are functions. For each statement, if it is true, then carefully prove it, or if it is false, then find a specific counterexample.

(a) If both f and g are injective, then so is $g \circ f$.

(b) If both f and g are surjective, then so is $g \circ f$.

(c) If $g \circ f$ is injective, then so is f.

(d) If $g \circ f$ is injective, then so is g.

(e) If $g \circ f$ is surjective, then so is f.

(f) If $g \circ f$ is surjective, then so is g.

Supplemental Reading

- Epp [Epp10], Chapters 5 and 7.

- Halmos [Hal74]. Thorough treatment of set theory, a classic.

- Munkres [Mun00], Chapter 1. Foundational topics of set theory from the point of view of topology.

- Mendelson [Men90], Chapter 1.

- Smullyan [Smu94], Parts I–II. A rigorous source text for propositional and first-order logic.

Appendix B

Group Theory and Linear Algebra

In Chapter 7, we introduced the powerful topological invariant called the *fundamental group*. To properly understand how this tool works, we must be familiar with the basics of **group theory**. It is also important to have some knowledge of elementary **linear algebra** in order to understand and compute *homology* in Chapter 8. In this appendix, we review only the essential concepts of group theory and linear algebra that are useful for our purposes in this text. For more detailed treatments of group theory or linear algebra, consult the suggested reading list at the end of the appendix.

B.1   Groups

Definition of a Group

A *group* is a set together with a *product* operation defined on elements of that set satisfying certain axioms.

Definition B.1.1. A **group** is a set G together with a function $G \times G \to G$ called the **product**. The product of $g, h \in G$ is typically written $g \cdot h$, $g * h$, or gh. The product must satisfy three axioms:

   I. *(Associativity)* $\forall g, h, k \in G$, $(g \cdot h) \cdot k = g \cdot (h \cdot k)$.

  II. *(Existence of identity)* $\exists e \in G$ such that $\forall g \in G$, $g \cdot e = e \cdot g = g$. Such an element e is called an **identity** element.

 III. *(Existence of inverses)* $\forall g \in G$, $\exists h \in G$ such that $g \cdot h = h \cdot g = e$, where e is the identity element. The element h is called an **inverse** for g, and written $h = g^{-1}$.

The three group axioms imply that the identity element e of any given group G must be unique (hence it's justifiable to say "where e is *the* identity element" in the statement of Axiom III). Let's verify this important fact. Suppose e_1 and e_2 both meet the requirements of Axiom II. Then, since e_1 is an identity, we know that $e_1 \cdot e_2 = e_2$. But since e_2 is also an identity, we know that $e_1 \cdot e_2 = e_1$. Thus by transitivity, $e_1 = e_2$. Often we write the identity element as $e = 1$.

The product in a group is not exactly like the product of real numbers. We are used to the ability to rearrange the order of a product. For example, in algebra we know that $3 \cdot x \cdot 4 \cdot x^2 = 3 \cdot 4 \cdot x \cdot x^2 = 12 \cdot x^3$; and $xyx^{-1}y^{-1} = xx^{-1}yy^{-1} = 1 \cdot 1 = 1$. However, in group theory, we cannot assume that $g \cdot h \cdot g^{-1} \cdot h^{-1}$ simplifies in any straightforward way. Group products are typically not *commutative*. In other words, if $g, h \in G$, then generally $g \cdot h \neq h \cdot g$. Of course sometimes two elements may commute; for example, if $e \in G$ is the identity, then $e \cdot g = g \cdot e$ for all $g \in G$. If it turns out that *all* pairs of elements commute ($g \cdot h = h \cdot g$, $\forall g, h \in G$), then the group is called **commutative** or **abelian**, and in that case we often use the *plus* symbol $+$ for the group operation and $\mathbf{0}$ for the identity.

Example 155. The set of integers $\mathbb{Z}$ is a group under the addition operation. The identity element is $0 \in \mathbb{Z}$, and the inverse of $g \in \mathbb{Z}$ is its opposite, $-g$. Because addition is commutative ($g + h = h + g$), the group $\mathbb{Z}$ is abelian.

For an example of a nonabelian group, consider all of the possible ways to rearrange three letters. There are six 3-letter **permutations**:

$$S_3 = \{\mathtt{ABC}, \mathtt{ACB}, \mathtt{BAC}, \mathtt{BCA}, \mathtt{CAB}, \mathtt{CBA}\}.$$

Moreover, each permutation listed above may be thought of as an *action*.[1] For instance, "$\mathtt{ACB}$" may be interpreted as the action that leaves the first element of a three-term sequence the same while swapping the second and third. Thus we can *compose* permutations by applying the action of the first to the letters of the second, for example, $\mathtt{ACB} \cdot \mathtt{BAC} = \mathtt{BCA}$. It is straightforward to verify that S_3 is a group under the composition operation, with $e = \mathtt{ABC}$ as identity. S_3 is not abelian, though. Consider $\mathtt{BAC} \cdot \mathtt{ACB} = \mathtt{CAB} \neq \mathtt{BCA}$.

In general, for any $n \in \mathbb{N}$, the **symmetric group** S_n is defined as the group of all permutations on n letters under composition, which is nonabelian when $n \geq 3$.

The number of elements in a group G is called the **order** of G. A group of order 1 is called a **trivial** group, and is often denoted by 1 (or 0 in the context of abelian groups). According to the axioms, that single element must be the identity, that is, $1 = \{\mathbf{1}\}$.

Group Homomorphisms

Suppose G and H are two groups. Since groups are also sets, we may consider set functions $f : G \to H$; however, not all such functions may be very useful

---

[1]The term "group action" has a precise mathematical definition, which falls beyond the scope of this text.

in group theory (just as not all functions $f : \mathbb{R} \to \mathbb{R}$ may be useful in calculus; only the continuous, or differentiable, or some other subclass of functions may be considered). We will require that our functions between groups also respect the product structure of each group. To say that $(G, \cdot)$ is a group means that G is a group with operation $g \cdot h$.

Definition B.1.2. Suppose $(G, \cdot)$ and $(H, *)$ are two groups. A set function $\phi : G \to H$ is called a **group homomorphism** (or simply **homomorphism**) if for all $g_1, g_2 \in G$,

$$\phi(g_1 \cdot g_2) = \phi(g_1) * \phi(g_2).$$

In particular, a homomorphism must always send the identity element of one group to the identity of the other. To see why, let $e \in G$ and $u \in H$ be the identities of each group, and suppose $\phi : G \to H$ is a homomorphism. Try to track each group axiom or property of homomorphism as it is used in the derivation of $\phi(e) = u$ shown below.

$$
\begin{aligned}
e &= e \cdot e \\
\phi(e) &= \phi(e \cdot e) = \phi(e) \cdot \phi(e) \\
[\phi(e)]^{-1} \cdot \phi(e) &= [\phi(e)]^{-1} \cdot (\phi(e) \cdot \phi(e)) = \big([\phi(e)]^{-1} \cdot \phi(e)\big) \cdot \phi(e) \\
u &= u \cdot \phi(e) = \phi(e)
\end{aligned}
$$

For each group G there is a special homomorphism from G to itself called the **identity homomorphism**,

$$
\begin{aligned}
\mathrm{id}_G &: \quad G \to G \\
\mathrm{id}_G(g) &= \quad g, \quad \forall g \in G.
\end{aligned}
$$

A group homomorphism $\phi : G \to H$ is called **invertible** if there is another homomorphism $\psi : H \to G$ such that as set maps, $\psi \circ \phi = \mathrm{id}_G$ and $\phi \circ \psi = \mathrm{id}_H$; in this case, we call ψ the **inverse** of ϕ and write $\psi = \phi^{-1}$.

Group homomorphisms are called **injective**, **surjective**, or **bijective** if their underlying set functions are. It is useful to know that a group homomorphism is invertible *if and only if* it's bijective. Recall from §A.2 that a function ϕ is bijective if and only if both $\phi^{-1} \circ \phi$ and $\phi \circ \phi^{-1}$ are identity functions. Thus if f is an invertible group homomorphism, then f is clearly bijective. On the other hand, if ϕ is a homorphism that is bijective as a set function, then the only question is whether the inverse function ϕ^{-1} is a group homomorphism. The proof is not difficult (see Exercise 6). Two groups are *equivalent* in the realm of group theory whenever there is a bijective homomorphism from one to the other; this type of equivalence is called a **group isomorphism**.

> **Definition B.1.3.** Two groups G and H are **isomorphic** if there is an invertible group homomorphism $\phi : G \to H$. In this case, we write $G \cong H$, and ϕ is called an **isomorphism**.

Generators, Relations, and Presentations

Consider a set of elements $S = \{g_1, g_2, \ldots, g_n\}$ of a group G. By the closure property of groups, every possible finite product of elements of S and their inverses are also elements of the group. If every element $g \in G$ can be expressed as a product of elements in S, then S is a set of **generators** for G. However, some of these products may represent the same element $g \in G$. For example, it may be that $g_3 = g_2^{-1} g_1 g_2$ or that $g_1^5 = \mathbf{1}$. A **presentation** is a list of generators and expressions or equations that define the **relations** among generators that are sufficient to reconstruct the group G. The presentation is given using the following notation:

$$G = \langle g_1, g_2, \ldots, g_n \mid r_1, r_2, \ldots, r_m \rangle,$$

where $g_1, \ldots, g_n$ generate G, and each r_k is an equation involving the generators. It is a challenging problem to decide how many and which generators and relations might be sufficient to define any given group; vice versa, it is often difficult or impossible to decide whether a given presentation of a group actually represents a group given in a different form.[2]

Example 156. Let $n \in \mathbb{N}$. Define the **cyclic group** $\mathbb{Z}/n\mathbb{Z}$ by its presentation:

$$\mathbb{Z}/n\mathbb{Z} = \langle g \mid g^n = \mathbf{1} \rangle.$$

This group has exactly n elements, $\{\mathbf{1}, g, g^2, g^3, \ldots, g^{n-1}\}$, and G is abelian since $g^i \cdot g^j = g^{i+j} = g^{j+i} = g^j \cdot g^i$.

The cyclic group, presented below in an additive way, also occurs in the context of *modular arithmetic.*

$$\mathbb{Z}/n\mathbb{Z} = \langle 1 \mid n \cdot 1 = 0 \rangle$$

When written this way, the elements of $\mathbb{Z}/n\mathbb{Z}$ are: $\{0, 1, 2, 3, \ldots, n-1\}$, with $n = 0 \in \mathbb{Z}/n\mathbb{Z}$ (compare Example 149 in Appendix A).

Example 157. Let $G = \langle g, h \mid g^2 = \mathbf{1}, h^2 = \mathbf{1}, ghg = hgh \rangle$. Let's explore the elements of G. Certainly $\mathbf{1} \in G$ (because every group has an identity), $g \in G$, and $h \in G$, but what other distinct elements are there? Because $g^2 = h^2 = \mathbf{1}$, we do not list g^2 or h^2 as separate elements of the group. However, gh and hg have

---

[2]In fact, there is not even an algorithm that can decide whether any given presentation represents the trivial group. For more details on *combinatorial group theory*, see [CKC+13].

not been listed yet. Now, from gh, we could multiply either g or h to obtain ghg or gh^2, respectively. But $gh^2 = g\mathbf{1} = g$ is already listed. Doing the same starting with hg, we get $hg^2 = h$ and hgh. According to the presentation, $ghg = hgh$, so we have so far found six elements, $\{\mathbf{1}, g, h, gh, hg, ghg\}$, in the group G. All other products of g or h can be reduced to one of these forms. What about the inverses? From $g^2 = \mathbf{1}$, or $gg = \mathbf{1}$, we get $g^{-1} = g$. Similarly, $h^{-1} = h$. Thus $G = \{\mathbf{1}, g, h, gh, hg, ghg\}$, a group of order 6. It can be shown that $G \cong S_3$.

Exercises

1. Explain why each set listed below is *not* a group with respect to the operation.

   (a) $(\mathbb{N}, +)$ (b) $(\mathbb{Q}, \times)$ (c) $(\mathbb{Z}, -)$

2. Let G be a group and suppose $g \in G$ satisfies $g^n = \mathbf{1}$ for some $n \in \mathbb{N}$. Find g^{-1} in terms of g.

3. Let $g, h \in G$, a group. Show that $(gh)^{-1} = h^{-1}g^{-1}$.

4. Let $r \in \mathbb{N}$. Show that $\mathbb{Z}^r$ is an abelian group with respect to vector addition. What is the identity element?

5. Let S_3 be the symmetric group on three letters as defined in Example 155. Construct a multiplication table for S_3, paying special attention to the order of composition. Use your table to identify the inverse of each permutation. Then show that the operation of S_3 is associative by verifying all possible equations of the form $a \cdot (b \cdot c) = (a \cdot b) \cdot c$. (*Hint:* Look for time-saving arguments so that you don't have to write $6^3 - 216$ distinct equations.)

6. Let $(G, \cdot)$ and $(H, *)$ be groups. Suppose $\phi : G \to H$ is a bijective group homomorphism. Prove that the inverse function $\phi^{-1} : H \to G$ is a group homomorphism by verifying that $\phi^{-1}(h_1 * h_2) = \phi^{-1}(h_1) \cdot \phi^{-1}(h_2)$ for all $h_1, h_2 \in H$.

7. Consider the rotation matrix $R_\theta = \begin{pmatrix} \cos\theta & -\sin\theta \\ \sin\theta & \cos\theta \end{pmatrix}$. Let G_θ be the group generated by R_θ, that is, $G_\theta = \{I, R_\theta, R_\theta^2, R_\theta^3, \ldots\}$.

   (a) Write out all group elements and determine the order of $G_{2\pi/3}$, $G_{\pi/2}$, and $G_{2\pi/5}$.

   (b) Find a group isomorphism $G_{2\pi/3} \to \mathbb{Z}/3\mathbb{Z}$.

   (c) Let $n \in \mathbb{N}$ be arbitrary. Determine θ so that $G_\theta \cong \mathbb{Z}/n\mathbb{Z}$.

   (d) Under what condition on θ would G_θ be infinite?

8. The cyclic group $\mathbb{Z}/n\mathbb{Z}$ also admits a multiplication operation called *modular multiplication*, in which the product $ab \in \mathbb{Z}/n\mathbb{Z}$ is defined by first multiplying ab as integers and then reducing modulo n. Write the multiplication tables for $\mathbb{Z}/7\mathbb{Z}$ and $\mathbb{Z}/8\mathbb{Z}$. Show that $(\mathbb{Z}/7\mathbb{Z}) \setminus \{0\}$ is a group under multiplication, but $(\mathbb{Z}/8\mathbb{Z}) \setminus \{0\}$ is not.

9. Find an explicit isomorphism to prove that $S_3 \cong G$, where G is the group from Example 157.

10. Let G be a group, and suppose that $(gh)^2 = g^2h^2$ for all $g, h \in G$. Prove that G is abelian.

11. Let $G = \langle a, b \mid a^2 = \mathbf{1}, b^2 = \mathbf{1} \rangle$.

   (a) Show that $(ab)^{-1} = ba$.

   (b) Show that every $g \in G$ can be written as a product of only a and b alternating, so that $G = \{\ldots, bab, ba, b, \mathbf{1}, a, ab, aba, \ldots\}$.

B.2 Linear Algebra

The topic of linear algebra comprises two interrelated themes: computational techniques and a conceptual theory. The computations in linear algebra are typically *matrix operations*. Certain operations are used to solve systems of linear equations or to describe how linear transformations act. Conceptually, linear algebra can be used to show how diverse mathematical structures may be regarded as vectors in a suitable *vector space*. In this short section we only present the computational and conceptual ideas that are needed for this textbook. We assume a working knowledge of real numbers and integers.

Matrices

A matrix is nothing more than a table of numbers, but there are great advantages in arranging numbers in a rectangular array. For example, a system of equations is naturally represented by a matrix of coefficients and constants.

Definition B.2.1. Suppose $m, n \geq 1$ are integers. A (real, complex, integer, etc.) **matrix** A of dimensions $m \times n$ is a rectangular array of (real, complex, integer, etc.) numbers, a_{ij}, called **entries**, where

$$A = (a_{ij}) = \begin{pmatrix} a_{11} & a_{12} & \cdots & a_{1n} \\ a_{21} & a_{22} & \cdots & a_{2n} \\ \vdots & \vdots & \ddots & \vdots \\ a_{m1} & a_{m2} & \cdots & a_{mn} \end{pmatrix}.$$

The set of all $m \times n$ matrices with real-number entries is denoted $\mathcal{M}_{m \times n}$. The set of matrices having entries in another set R may be denoted by $\mathcal{M}_{m \times n}(R)$. If $A = (a_{ij})$ and $B = (b_{ij})$ are two matrices of the same dimensions (both in $\mathcal{M}_{m \times n}$), then their sum $A + B$ may be defined.

$$A + B = (a_{ij} + b_{ij}) = \begin{pmatrix} a_{11} + b_{11} & \cdots & a_{1n} + b_{1n} \\ \vdots & & \vdots \\ a_{m1} + b_{m1} & \cdots & a_{mn} + b_{mn} \end{pmatrix}$$

Moreover, if k is a number (real, complex, integer, etc., but matching the type of numbers in the matrix), then the **scalar product** kA is defined by:

$$kA = (ka_{ij}) = \begin{pmatrix} ka_{11} & \cdots & ka_{1n} \\ \vdots & & \vdots \\ ka_{m1} & \cdots & ka_{mn} \end{pmatrix}.$$

Then we may define matrix subtraction by $A - B = A + (-1)B$ whenever the dimensions of A and B are the same. Note that both $A + B$ and kA are also in $\mathcal{M}_{m \times n}$.

Matrix multiplication is not quite as straightforward as addition. Suppose $A \in \mathcal{M}_{m \times n}$ and $B \in \mathcal{M}_{p \times q}$. Then the product AB is only defined when $n = p$, in which case $AB \in \mathcal{M}_{m \times q}$ is defined by the formula

$$AB = (c_{ij}), \quad \text{where} \quad c_{ij} = \sum_{k=1}^{n} a_{ik} b_{kj}. \tag{B.1}$$

Example 158. If $A = \begin{pmatrix} 4 & -1 & 7 \\ 0 & 2 & -5 \end{pmatrix}$ and $B = \begin{pmatrix} 1 & -1 \\ 2 & 3 \end{pmatrix}$, then the product AB is undefined, while

$$\begin{aligned} BA &= \begin{pmatrix} 1 & -1 \\ 2 & 3 \end{pmatrix} \begin{pmatrix} 4 & -1 & 7 \\ 0 & 2 & -5 \end{pmatrix} \\ &= \begin{pmatrix} (1)(4) + (-1)(0) & (1)(-1) + (-1)(2) & (1)(7) + (-1)(-5) \\ (2)(4) + (3)(0) & (2)(-1) + (3)(2) & (2)(7) + (3)(-5) \end{pmatrix} \\ &= \begin{pmatrix} 4 & -3 & 12 \\ 8 & 4 & -1 \end{pmatrix}. \end{aligned}$$

For each $n \in \mathbb{N}$, define the $n \times n$ **identity matrix** by the rule $I_n = (\delta_{ij})$, where $\delta_{ij} = 1$ if $i = j$, and 0 if $i \neq j$. For instance,

$$I_4 = \begin{pmatrix} 1 & 0 & 0 & 0 \\ 0 & 1 & 0 & 0 \\ 0 & 0 & 1 & 0 \\ 0 & 0 & 0 & 1 \end{pmatrix}.$$

The identity matrices act as *multiplicative* identities, in the following sense. If $A \in \mathcal{M}_{m \times n}$, then $I_m A = A I_n = A$ (note that the proper size identity matrix must be used). Now if A is a square matrix $(n \times n)$, then we say that A is **invertible** if and only if there is a matrix B such that $AB = BA = I_n$. Such a matrix B must also be $n \times n$, and we use the notation $A^{-1} = B$ for the **inverse** of the matrix A when it exists. Not all matrices have an inverse, but the following tool is useful for determining when a matrix is invertible.

Definition B.2.2. The **determinant** of a matrix $A = (a_{ij}) \in \mathcal{M}_{n \times n}$ is a real number defined recursively by

$$\det(A) = \det([a_{11}]) = a_{11}, \quad \text{if } n = 1.$$
$$\text{If } n > 1, \quad \det(A) = a_{11} A_{11} - a_{12} A_{12} + \cdots + (-1)^{n+1} a_{1n} A_{1n},$$

where A_{1j} is obtained from A by deleting row 1 and column j.

Example 159.   Let $A = \begin{pmatrix} a & b \\ c & d \end{pmatrix}$. Then $A_{11} = [d]$ and $A_{12} = [c]$.

$$\det(A) = a \det([d]) - b \det([c]) = ad - bc$$

We often interpret a vector $\mathbf{x} \in \mathbb{R}^n$ as an $n \times 1$ (column) matrix. Then for an $m \times n$ matrix A, the product $A\mathbf{x}$ is an $m \times 1$ matrix, that is, a vector $\mathbf{y} = A\mathbf{x} \in \mathbb{R}^m$.

A system of m equations in n unknowns may be represented by an **augmented matrix** as follows:

$$\begin{cases} a_{11}x_1 + \cdots + a_{1n}x_n &= b_1 \\ \quad\vdots & \vdots \\ a_{m1}x_1 + \cdots + a_{mn}x_n &= b_m \end{cases} \quad \leftrightarrow \quad \left(\begin{array}{ccc|c} a_{11} & \cdots & a_{1n} & b_1 \\ \vdots & \ddots & \vdots & \vdots \\ a_{m1} & \cdots & a_{mn} & b_m \end{array} \right). \quad \text{(B.2)}$$

If $A = (a_{ij})$ is the matrix of coefficients and $\mathbf{b} \in \mathbb{R}^m$ is the column vector of constants for the system (B.2), then the solution to the system consists of all vectors $\mathbf{x} \in \mathbb{R}^n$ such that $A\mathbf{x} = \mathbf{b}$. To solve a system $A\mathbf{x} = \mathbf{b}$, it is usually most efficient to use **Gaussian elimination**. Gaussian elimination is an algorithm for reducing an augmented matrix into a form called **RREF (reduced row echelon form)** using a series of **row operations**. In the following, we use the notation R_i for the ith row $(a_{i1} \cdots a_{in})$ of a matrix $A = (a_{ij})$.

> **Proposition B.2.3.** *The following three* **row operations** *yield equivalent systems.*
>
> I. $(R_i \leftrightarrow R_j)$ *Swap the ith and jth rows.*
>
> II. $(kR_i \mapsto R_i)$ *If $k \neq 0$, replace row i with k times that row.*
>
> III. $(kR_i + R_j \mapsto R_j)$ *For any scalar k and if $i \neq j$, add k times row i to row j (which replaces R_j but leaves R_i unchanged in the matrix).*

Call the first nonzero entry in a given row the **leading entry** for that row. A row consisting entirely of zeros has no leading entry.

> **Definition B.2.4.** A matrix is said to be in **RREF (reduced row echelon form)** if all of the following conditions hold.
>
> 1. The leading entry of each row occurs strictly to the right of the leading entry of the previous row.
>
> 2. Each leading entry is equal to 1.
>
> 3. Every entry in the same column as a leading entry (except the leading entry itself) must be 0.

Note that condition 1 implies that any row above a row with a leading entry must itself have a leading entry; hence all rows having no leading entry (*zero* or *null* rows) must be at the bottom of the matrix.

Example 160. Interpret the system of equations as an augmented matrix and reduce that matrix to RREF. Then interpret the reduced matrix as a solution to the original system.

$$\begin{cases} x - 2y - z & = & 3 \\ 3x - 6y - 2z & = & 2 \end{cases}$$

Solution:

$$\begin{pmatrix} 1 & -2 & -1 & 3 \\ 3 & -6 & -2 & 2 \end{pmatrix} \quad -3R_1 + R_2 \mapsto R_2 \quad \begin{pmatrix} 1 & -2 & -1 & 3 \\ 0 & 0 & 1 & -7 \end{pmatrix}$$

$$(1)R_2 + R_1 \mapsto R_1 \quad \begin{pmatrix} 1 & -2 & 0 & -4 \\ 0 & 0 & 1 & -7 \end{pmatrix}$$

The final form of the matrix is interpreted as: $x - 2y = -4$, and $z = -7$. Solving the first equation for x, we find $x = 2y - 4$, so in terms of a parameter $y = t$, the solution to the system is the set of all points of the form $(x, y, z) = (2t - 4, t, -7)$ for arbitrary $t \in \mathbb{R}$.

The system from Example 160 does not have a *unique* solution; instead, there are infinitely many solutions, one for each choice of the variable that we did not isolate (the **free variable(s)**). The number of free variables is related to the *rank* of the coefficient matrix. Note that we may determine which variables are free without putting the matrix into RREF. All that is required is **REF**, or **row echelon form**. A matrix is said to be in REF if it satisfies only condition 1 of Definition B.2.4.

Definition B.2.5. The **rank** of a matrix $A \in \mathcal{M}_{m \times n}$ is the number of leading entries in the REF of A, and is denoted $\operatorname{rank}(A)$.

Equivalently, the rank of a matrix is the number of nonzero rows in the REF of the matrix.

Example 161.   Find the rank of $A = \begin{pmatrix} 1 & 2 & 3 \\ 4 & 5 & 6 \\ 7 & 8 & 9 \end{pmatrix}$.

Solution:

$$\begin{pmatrix} 1 & 2 & 3 \\ 4 & 5 & 6 \\ 7 & 8 & 9 \end{pmatrix} \quad \begin{matrix} -4R_1 + R_2 \mapsto R_2 \\ -7R_1 + R_3 \mapsto R_3 \end{matrix} \quad \begin{pmatrix} 1 & 2 & 3 \\ 0 & -3 & -6 \\ 0 & -6 & -12 \end{pmatrix}$$

$$-\frac{1}{3}R_2 \mapsto R_2 \quad \begin{pmatrix} 1 & 2 & 3 \\ 0 & 1 & 2 \\ 0 & -6 & -12 \end{pmatrix} \quad 6R_2 + R_3 \mapsto R_3 \quad \begin{pmatrix} 1 & 2 & 3 \\ 0 & 1 & 2 \\ 0 & 0 & 0 \end{pmatrix}$$

The REF form has two nonzero rows; hence $\operatorname{rank}(A) = 2$.

Vector Spaces and Subspaces

You may be familiar with **vectors** in $\mathbb{R}^2$ or $\mathbb{R}^3$ as arrows that represent both direction and magnitude. If $\mathbf{u}$ and $\mathbf{v}$ are vectors in the same space, then their sum $\mathbf{u} + \mathbf{v}$ is also a vector in that space. If c is any real number (or **scalar**), then $c\mathbf{u}$ is also a vector in that space. But there are other sets (besides $\mathbb{R}^n$) whose elements seem to behave like vectors in that there might be an addition operation and scalar multiplication. For example, two matrices A and B of the same dimension can be added to obtain a matrix of the same size, $A + B$, and the scalar multiple cA (for $c \in \mathbb{R}$) is also a matrix of the same size. So the set $\mathcal{M}_{m \times n}$, for example, *acts like* a set of vectors – what we call a *vector space*.

Before defining vector spaces, we should talk about the kinds of numbers that could serve as scalars. We are familiar with the set of real numbers $\mathbb{R}$, but

$\mathbb{R}$ is just one example of a *field*. The rational numbers ($\mathbb{Q}$) and complex numbers ($\mathbb{C}$) are other important examples of fields. Briefly, a **field** is any set of numbers R with addition and multiplication defined such that R is an abelian group with respect to addition, $R \setminus \{0\}$ is an abelian group with respect to multiplication, and multiplication distributes over addition: $a(b + c) = ab + ac, \forall a, b, c \in R$. Perhaps the most important feature in a field is the existence of multiplicative inverses for any nonzero number (which follows from the fact that $R \setminus \{0\}$ is a group).[3]

Definition B.2.6. A **vector space** over a field R is a set V satisfying the following 10 axioms.

1. There is an **addition** operation ($+$) such that V is closed under addition: $u, v \in V \implies u + v \in V$.

2. Addition is **commutative**: $u + v = v + u, \forall u, v \in V$.

3. Addition is **associative**: $(u + v) + w = u + (v + w), \forall u, v, w \in V$.

4. There is a **zero element**: $\exists 0 \in V$ such that $u + 0 = 0 + u = u$, $\forall u \in V$.

5. Each element of V has an **additive inverse**: $\forall u \in V, \exists\, -u \in V$ such that $u + (-u) = -u + u = 0$.

6. There is a **scalar multiplication** operation, and V is closed under this operation: $u \in V, c \in R \implies cu \in V$.

7. Scalar multiplication is **distributive over vector addition**: $c(u + v) = cu + cv, \forall u, v \in V, c \in R$.

8. Scalar multiplication is **distributive over scalar addition**: $(c + d)u = cu + du, \forall c, d \in R, u \in V$.

9. Scalar multiplication satisfies an **associative law**: $c(du) = (cd)u$, $\forall c, d \in R, u \in V$.

10. Scalar multiplication is **unital**: $1u = u, \forall u \in V$ (where $1 \in R$ is the unit).

Note that axioms 1–5 are equivalent to the statement that V is an abelian group with respect to addition.

---

[3]For more details, see, e.g., Bartle and Sherbert [BS11] or Dummit and Foote [DF04].

Definition B.2.7. A **subspace** of a vector space V is a subset $H \subseteq V$ satisfying the following three axioms.

1. The zero element $\mathbf{0} \in V$ is also in H.

2. H is closed under addition: $u, v \in H \implies u + v \in H$.

3. H is closed under scalar multiplication: $u \in H$, $c \in R \implies cu \in H$.

A subspace is just a vector space H within another vector space V. The reason that we only have to check 3 of the 10 axioms is that H inherits the addition and scalar multiplication operations of V, which already are supposed to satisfy all 10 axioms in V.

Example 162. Consider the set $H = \{(x, y, 0) \mid x, y \in \mathbb{R}\} \subseteq \mathbb{R}^3$. Geometrically, we see that H is the xy-plane within three-dimensional space. H is a subspace of $\mathbb{R}^3$, as we verify below:

- The zero element $\mathbf{0} = (0, 0, 0) \in V$ is also in H (just let $x = 0$ and $y = 0$).

- Suppose $u, v \in H$, say, $u = (x_1, y_1, 0)$ and $v = (x_2, y_2, 0)$. Then $u + v = (x_1 + x_2, y_1 + y_2, 0) \in H$.

- Suppose $u = (x, y, 0) \in H$ and $c \in \mathbb{R}$. Then $cu = (cx, cy, 0) \in H$.

Suppose $A = \{v_1, v_2, \ldots, v_r\}$ is a set of vectors in a vector space V. A **linear combination** of A is any expression of the form

$$\sum_{k=1}^{r} c_k v_k = c_1 v_1 + c_2 v_2 + \cdots + c_r v_r, \tag{B.3}$$

in which each coefficient $c_k \in R$ is a scalar. For example, let $u = \begin{pmatrix} 3 \\ -1 \end{pmatrix}$ and $v = \begin{pmatrix} -8 \\ 2 \end{pmatrix}$. Since $4u + \frac{1}{2}v = \begin{pmatrix} 8 \\ -3 \end{pmatrix}$, we can say that $\begin{pmatrix} 8 \\ -3 \end{pmatrix}$ is a linear combination of $B = \{u, v\}$. For any set of vectors $B = \{v_1, v_2, \ldots, v_r\}$, the **trivial combination** is defined as: $0v_1 + 0v_2 + \cdots + 0v_r$. Of course the trivial combination evaluates to the zero vector $\mathbf{0}$.

Definition B.2.8. A set of vectors B is called **independent** if the *only* linear combination of B that equals $\mathbf{0}$ is the trivial combination. In other words,

$$c_1 v_1 + c_2 v_2 + \cdots + c_r v_r = \mathbf{0} \implies c_1 = c_2 = \cdots = c_r = 0. \tag{B.4}$$

Definition B.2.8 may be tricky to use at first. There is an embedded "if-then" statement in (B.4). *If*, for some unspecified scalars c_k, we have $\sum c_k v_k = \mathbf{0}$, *then* it must have been the case that every $c_k = 0$. If a set of vectors B is not independent, then it is called **dependent**.

Example 163. Let $u = \begin{pmatrix} 3 \\ -4 \end{pmatrix}$, $v = \begin{pmatrix} -9 \\ 12 \end{pmatrix}$, and $w = \begin{pmatrix} 2 \\ -5 \end{pmatrix}$ be vectors in $\mathbb{R}^2$. Determine whether each set below is independent or dependent.

(a) $A = \{u, v\}$
(b) $B = \{u, w\}$
(c) $C = \{v, w, \mathbf{0}\}$

Solution: It should be easy to see that $v = -3u$; hence $3u + v = \mathbf{0}$. But this shows that A is a dependent set since there is a nontrivial linear combination evaluating to $\mathbf{0}$.

Next, consider B. Suppose $c_1 u + c_2 w = \mathbf{0}$. Then:

$$c_1 \begin{pmatrix} 3 \\ -4 \end{pmatrix} + c_2 \begin{pmatrix} 2 \\ -5 \end{pmatrix} = \begin{pmatrix} 0 \\ 0 \end{pmatrix} \quad \text{or} \quad \begin{pmatrix} 3 & 2 \\ -4 & -5 \end{pmatrix} \begin{pmatrix} c_1 \\ c_2 \end{pmatrix} = \begin{pmatrix} 0 \\ 0 \end{pmatrix}.$$

We may solve the system to find (c_1, c_2), using Gaussian elimination.

$$\left(\begin{array}{cc|c} 3 & 2 & 0 \\ -4 & -5 & 0 \end{array} \right) \quad R_2 + R_1 \mapsto R_1 \quad \left(\begin{array}{cc|c} -1 & -3 & 0 \\ -4 & -5 & 0 \end{array} \right) \quad (-1)R_1 \mapsto R_1$$

$$\left(\begin{array}{cc|c} 1 & 3 & 0 \\ -4 & -5 & 0 \end{array} \right) \quad 4R_1 + R_2 \mapsto R_2 \quad \left(\begin{array}{cc|c} 1 & 3 & 0 \\ 0 & 7 & 0 \end{array} \right)$$

The system need not be solved completely, as we can see there are leading entries in every column to the left of the bar. This implies that the *only* solution to the system is $(c_1, c_2) = (0, 0)$. Hence the set B is independent.

The set C is dependent, since $0v + 0w + (1)\mathbf{0} = \mathbf{0}$ is a nontrivial linear combination of $\{v, w, \mathbf{0}\}$. In fact, any set of vectors containing the zero vector is dependent for much the same reason.

We extend the definition of linear combination to include more abstract situations. Suppose now that $A = \{v_1, v_2, \ldots, v_r\}$ is simply a set (with no addition operation defined a priori on A). Then a **formal linear combination** of A with scalars in a field R is any expression of the form

$$\sum_{k=1}^{r} c_k v_k = c_1 v_1 + c_2 v_2 + \cdots + c_r v_r, \quad \text{where each } c_k \in R. \tag{B.5}$$

Definition B.2.9.

- Suppose A is a set of vectors in a vector space V (over a field R). The **span** of A is the set of all linear combinations of A.

- Given an arbitrary finite set A, the **span** of A over a field R is defined as the set of all formal linear combinations of A with scalars in R.

In either case, the span of $A = \{v_1, v_2, \ldots, v_r\}$ over R is written:

$$\mathrm{Span}_R A \quad \text{or} \quad \mathrm{Span}_R\{v_1, v_2, \ldots, v_n\}.$$

Regardless of whether A is a set of vectors or a general set, $\mathrm{Span}_R A$ is a vector space in its own right. The addition operation is given by

$$\sum_{k=1}^{r} c_k v_k + \sum_{k=1}^{r} d_k v_k = \sum_{k=1}^{r} (c_k + d_k) v_k, \tag{B.6}$$

and scalar multiplication by $d \in R$ by

$$d \sum_{k=1}^{r} c_k v_k = \sum_{k=1}^{r} (d c_k) v_k. \tag{B.7}$$

If a set of vectors $A = \{v_1, v_2, \ldots, v_r\}$ is independent, then we call the set A a **basis** for the span, and we say that $\mathrm{Span}_R A$ has **dimension** r. If A is simply a set, then it is by default independent when regarded as a set of vectors in $\mathrm{Span}_R A$. In other words, the only *formal* linear combination of a set A that equals the zero vector is the trivial combination.

Moreover, in any vector space (or subspace) V, if we can determine a set of independent vectors B whose span is equal to the entire space V, then we say that B is a **basis** for V. If the basis set consists of r vectors, then the vector space has dimension r, and we may identify $V \cong R^r$, where R is the field over which V is defined.[4]

Example 164. Let u, v, w be as in Example 163, and let $H = \mathrm{Span}\{u, v\}$ and $K = \mathrm{Span}\{u, w\}$. Since $\{u, v\}$ is dependent, the set is not a basis for H. Instead, since $v = -3u$, we have $v \in \mathrm{Span}\{u\}$, and since $\{u\}$ is independent, we find that $H = \mathrm{Span}\{u\}$ has dimension 1, and $H \cong \mathbb{R}^1$.

---

[4]If no finite set of vectors spans a given vector space, then we say the space is *infinite* dimensional. Example of infinite-dimensional spaces include $\mathbb{R}^\infty$ and the function spaces $L^p(\mathbb{R})$ mentioned in §2.5.

The Null Space of a Matrix

In a linear system (B.2), if all of the constants $b_1, b_2, \ldots, b_m$ are zero, we say the system is **homogeneous**. Homogeneous systems are especially important in applications. Note that a homogeneous system is equivalent to the matrix equation $Ax = 0$.

> **Definition B.2.10.** The set of all solutions to the homogeneous system $Au = 0$ is called the **null space** of A, which is denoted $\text{Nul}(A)$.

If the entries of an $m \times n$ matrix A are elements of a field R, then the $\text{Nul}(A)$ is a vector subspace of R^n. This fact is easily verified. Clearly $u = \mathbf{0} \in \text{Nul}(A)$, since $A\mathbf{0} = \mathbf{0}$. If $u, v \in \text{Nul}(A)$, then $Au = \mathbf{0}$ and $Av = \mathbf{0}$. Then

$$
\begin{aligned}
Au + Av &= \mathbf{0} + \mathbf{0} \\
A(u + v) &= \mathbf{0}.
\end{aligned}
$$

Therefore $u + v \in \text{Nul}(A)$, proving closure under addition. Closure under scalar multiplication is left as an exercise to the reader. To find a null space, we use the same methods as in solving a system of equations, typically Gaussian elimination.

Example 165. Find a basis for the null space of the matrix,

$$
A = \begin{pmatrix} 1 & 2 & -1 & -2 & -5 \\ 4 & 8 & -4 & -7 & -16 \\ -5 & -10 & 5 & 13 & 37 \end{pmatrix}.
$$

Solution: We must solve the homogeneous system,

$$
\left(\begin{array}{ccccc|c} 1 & 2 & -1 & -2 & -5 & 0 \\ 4 & 8 & -4 & -7 & -16 & 0 \\ -5 & -10 & 5 & 13 & 37 & 0 \end{array} \right).
$$

Use the following row operations, $-4R_1 + R_2 \mapsto R_2, \quad 5R_1 + R_3 \mapsto R_3, \quad 2R_2 + R_1 \mapsto R_1, \quad -3R_2 + R_3 \mapsto R_3$, and then interpret the solution.

$$
\begin{array}{ccccc} x_1 & x_2 & x_3 & x_4 & x_5 \end{array}
$$
$$
\begin{pmatrix} 1 & 2 & -1 & 0 & 3 \\ 0 & 0 & 0 & 1 & 4 \\ 0 & 0 & 0 & 0 & 0 \end{pmatrix} \implies \left\{ \begin{array}{r} x_1 + 2x_2 - x_3 + 3x_5 = 0 \\ x_4 + 4x_5 = 0 \end{array} \right.
$$

Therefore $x_1 = -2x_2 + x_3 - 3x_5, \ x_4 = -4x_5$, and x_2, x_3, x_5 are free variables. Thus every solution to the system has the form

$$
\mathbf{x} = \begin{pmatrix} x_1 \\ x_2 \\ x_3 \\ x_4 \\ x_5 \end{pmatrix} = \begin{pmatrix} -2x_2 + x_3 - 3x_5 \\ x_2 \\ x_3 \\ -4x_5 \\ x_5 \end{pmatrix} = x_2 \begin{pmatrix} -2 \\ 1 \\ 0 \\ 0 \\ 0 \end{pmatrix} + x_3 \begin{pmatrix} 1 \\ 0 \\ 1 \\ 0 \\ 0 \end{pmatrix} + x_5 \begin{pmatrix} -3 \\ 0 \\ 0 \\ -4 \\ 1 \end{pmatrix}.
$$

We find that $\mathrm{Nul}(A)$ is spanned by three vectors, which can easily be shown to be independent. Therefore $\{(-2, 1, 0, 0, 0), (1, 0, 1, 0, 0), (-3, 0, 0, -4, 1)\}$ is a basis for $\mathrm{Nul}(A) \cong \mathbb{R}^3$.

The dimension of the null space of a matrix is called the **nullity** of the matrix. The following important theorem can be used to determine the rank or nullity of a matrix if one or the other is already known.

Theorem B.2.11 (Rank Theorem). *For an $m \times n$ matrix A,*

$$\mathrm{rank}(A) + \mathrm{nullity}(A) = n.$$

Proof. Here $\mathrm{rank}(A)$ is equal to the number of leading entries, while $\mathrm{nullity}(A)$ is equal to the number of free columns, in other words, columns that do not have a leading entry in the REF of A. Therefore, since each column either has or does not have a leading entry, the sum must be equal to the number of columns, n. $\square$

Example 166.   The nullity of the matrix A from Example 165 is equal to 3. We have $n = 5$ (because A has five columns), and so the Rank Theorem implies that $\mathrm{rank}(A) = 2$. (Of course, once we have reduced A, the rank is clear without using the Rank Theorem.)

Exercises

1. Let $A = \begin{pmatrix} 1 & 2 & -3 \\ -1 & 0 & 7 \\ 0 & -2 & 5 \end{pmatrix}$, $B = \begin{pmatrix} 6 & -4 \\ 3 & 7 \\ -2 & 1 \end{pmatrix}$, $C = \begin{pmatrix} 1 & -4 & 0 \\ -2 & 0 & 1 \\ 2 & -7 & 3 \end{pmatrix}$,

   and $D = \begin{pmatrix} 3 \\ -2 \\ 5 \end{pmatrix}$. Evaluate the following expressions or explain why the expression is undefined.

   (a) $3A - 2C$ (c) BA (e) $\det(A)$

   (b) AB (d) AD (f) $\det(B)$

2. Find $\det(I_n)$ for arbitrary $n \in \mathbb{N}$.

3. Use Definition B.2.2 to show that

   $$\det \begin{pmatrix} a & b & c \\ d & e & f \\ g & h & i \end{pmatrix} = aei + bfg + cdh - bdi - afh - ceg.$$

4. Solve the system of equations by Gaussian elimination.

$$\left\{ \begin{array}{rcl} x + 2y + z & = & 0 \\ -3x - y + 2z & = & 1 \\ 5y + 3z & = & -1 \end{array} \right.$$

5. Let $A = \begin{pmatrix} 1 & 3 & 5 & 7 \\ 3 & 5 & 7 & 9 \\ 5 & 7 & 9 & 1 \end{pmatrix}$.

   (a) Put A into REF, using only type III operations.

   (b) Find the nullity of A and a basis for $\text{Nul}(A)$.

6. Let $m, n \in \mathbb{N}$ be fixed. Show that $\mathcal{M}_{m \times n}$ is a vector space.

7. Let V be a vector space over a field R. Prove each of the following using only Definition B.2.6.

   (a) $0u = \mathbf{0}, \forall u \in V$ (c) $-u = (-1)u, \forall u \in V$

   (b) $c\mathbf{0} = \mathbf{0}, \forall c \in R$

8. Equations (B.6) and (B.7) show that $\text{Span}_R A$ satisfies the closure axioms 1 and 6 of Definition B.2.6. Verify that $\text{Span}_R A$ satisfies the other axioms, thereby proving that a span is a vector space.

9. Let $A \in \mathcal{M}_{m \times n}$. Show that $\text{Nul}(A)$ is closed under scalar multiplication.

Supplemental Reading

- Armstrong [Arm88] for group theory.

- Dummit and Foote [DF04]; Part I for group theory, Part III for linear algebra and modules.

- Lay, Lay, and McDonald [LLM15] for linear algebra.

- Lang [Lan05] for a thorough treatment of group theory, linear algebra, and other aspects of abstract algebra.

- Poole and Lipsett [PL14] for linear algebra.

Appendix C

Selected Solutions

Problems of Chapter 1

- ### §1.1

  1. There are three sets, organized by the number of "holes" in the letter. No holes: $\{C, E, F, G, H, I, J, K, L, M, N, S, T, U, V, W, X, Y, Z\}$. One hole: $\{A, D, O, P, Q, R\}$. Two holes: $\{B\}$.

  2. Three pieces, no holes: $\{\Xi\}$. Two pieces, one with one hole and one with no holes) $\{\Theta\}$. One piece, no holes: $\{\Gamma, E, Z, H, I, K, \Lambda, M, N, \Pi, \Sigma, T, \Upsilon, X, \Psi, \Omega\}$. One piece, one hole: $\{A, \Delta, O, P\}$. One piece, two holes: $\{B, \Phi\}$.

  3. $A \approx C$ via a deformation that pushes the sides of the cylinder outward and flattens it. *Intermediate step shown.*

  5. Add edges as shown in order to ensure each face is *simply connected.* There are $v = 40$ vertices, $e = 96$ edges, and $f = 48$ faces, so $\chi = -8$.

- ### §1.2

  1. $\dfrac{9}{10} + \dfrac{9}{100} + \dfrac{9}{1000} + \cdots = \dfrac{9}{10} \sum_{n=0}^{\infty} \left(\dfrac{1}{10}\right)^n = \dfrac{9}{10} \cdot \dfrac{1}{1 - \frac{1}{10}} = 1$

6. *Left*, (b) Torus. *Right*, (c) Klein bottle.

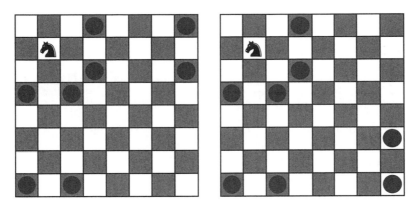

7. (a) C_n traces a circle $|n|$ times, counterclockwise if $n > 0$, clockwise if $n < 0$. (b) Here $\frac{1}{2\pi} \int_{C_n} \mathbf{V} \cdot d\mathbf{r} = n$. The integral computes the number of times the circle wraps around the origin.

Problems of Chapter 2

- ## §2.1

2. $\epsilon = \frac{1}{2} d(x, y)$

3. (a) Let U, V be open sets of $\mathbb{R}^n$. For any $x \in U \cap V$, we have $x \in U$ and $x \in V$. Since U is open, $\exists \epsilon_1 > 0$ such that $B_{\epsilon_1}(x) \subseteq U$. Similarly, $\exists \epsilon_2 > 0$ such that $B_{\epsilon_2}(x) \subseteq V$. Let $\epsilon = \min\{\epsilon_1, \epsilon_2\}$. Then $B_\epsilon(x) \subseteq U$ and $B_\epsilon(x) \subseteq V$; therefore $B_\epsilon(x) \subseteq U \cap V$. This proves $U \cap V$ open.

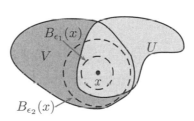

(b) Let $\{U_k\}_{k \in \mathcal{I}}$ be a set of open sets of $\mathbb{R}^n$. For any $x \in \bigcup_{k \in \mathcal{I}} U_k$, we know that $x \in U_k$ for some particular $k \in \mathcal{I}$. Thus there is an $\epsilon > 0$ such that $B_\epsilon(x) \subseteq U_k$. This implies that $B_\epsilon(x) \subseteq \bigcup_{k \in \mathcal{I}} U_k$; thus the arbitrary union of open sets is open.

6. (a) Open intervals: (a, ∞), $(-\infty, b)$, $(-\infty, \infty)$. (b) Closed intervals: $[a, \infty)$, $(-\infty, b]$, $(-\infty, \infty)$. (c) The set of limit points for both $(a, b]$ and $[a, b)$ is $[a, b]$; for both (a, ∞), $[a, \infty)$ is $[a, \infty)$; for both $(-\infty, b)$ and $(-\infty, b]$ is $(-\infty, b]$; and for $(-\infty, \infty)$ is $(-\infty, \infty)$.

9. $\text{int}((a, b]) = \text{int}([a, b)) = (a, b)$; $\text{ext}((a, b]) = \text{ext}([a, b)) = (-\infty, a) \cup (b, \infty)$; $\partial(a, b] = \partial[a, b) = \{a, b\}$; $\overline{(a, b]} = \overline{[a, b)} = [a, b]$. $\text{int}((a, \infty)) =$

$\text{int}([a, \infty)) = (a, \infty);$ $\text{ext}((a, \infty)) = \text{ext}([a, \infty)) = (-\infty, a);$ $\partial(a, \infty) = \partial[a, \infty) = \{a\};$ $\overline{(a, \infty)} = \overline{[a, \infty)} = [a, \infty).$ Similar for $(-\infty, b)$ and $(-\infty, b].$ $\text{int}((-\infty, \infty)) = \overline{(-\infty, \infty)} = (-\infty, \infty);$ $\text{ext}((-\infty, \infty)) = \partial(-\infty, \infty) = \emptyset.$

- ## §2.2

  5. Observe that $\left\| \frac{x}{\|x\|^2} \right\|^2 = \frac{\|x\|^2}{\|x\|^4} = \frac{1}{\|x\|^2}.$ Hence $(I \circ I)(x) = I\left(\frac{x}{\|x\|^2}\right) = \left(\frac{x}{\|x\|^2}\right) / \left(\frac{1}{\|x\|^2}\right) = x.$

  6. Translation T_r has inverse $T_r^{-1}(x) = x - r.$ Scaling S_k has inverse $S_k^{-1}(x) = \frac{1}{k}x$ (recall that $k \neq 0$). Rotation R_θ has inverse $R_{-\theta}.$ Reflection and inversion are their own inverses. Projection is not a homeomorphism in general. Stereographic projection has inverse defined by $F^{-1}(Q) = P$ if and only if $P \in \mathbb{S}^2 \setminus N$ such that P is on the segment $\overline{NQ}.$

  8. The map $f : (0, \infty) \to \mathbb{R}^1$ defined by $f(x) = \ln x$ is continuous, with continuous inverse $f^{-1}(x) = e^x.$

- ## §2.3

  1. If X is bounded in the sense of Definition 2.3.4, then clearly X satisfies $X \subseteq B_r(x)$ for $x = \mathbf{0}.$ Now suppose there is an $r \in \mathbb{R}^+$ and $x \in \mathbb{R}^n$ such that $X \subseteq B_r(x).$ Consider $p = \|x\| + r,$ and suppose $y \in B_r(x).$ By the triangle inequality, $d(y, \mathbf{0}) \leq d(y, x) + d(x, \mathbf{0}) < r + \|x\| = p.$ Thus $y \in B_p(\mathbf{0}),$ proving that $X \subseteq B_p(\mathbf{0})$ (where $p \in \mathbb{R}^+$) and so X is bounded.

  2. $(a, b),$ $(a, b],$ and $[a, b)$: noncompact and bounded. $[a, b]$: compact and bounded. $(a, \infty),$ $[a, \infty),$ $(-\infty, b),$ $(-\infty, b],$ and $(-\infty, \infty)$: noncompact and unbounded.

  3. (a) Not compact; bounded but not closed. (b) Compact. (c) Compact. (d) Not compact; bounded but not closed. (e) Not compact; closed but not bounded. (f) Not compact; bounded but not closed.

  6. Suppose $x \neq y$ and both x and y are limits of the sequence $(x_k).$ Let $\epsilon = \frac{1}{2}d(x, y),$ which is positive because $x \neq y.$ By §2.1, Exercise 2, $B_\epsilon(x) \cap B_\epsilon(y) = \emptyset.$ Because x is a limit of $(x_k),$ there exists $N_1 \in \mathbb{N}$ such that $x_k \in B_\epsilon(x)$ for all $k \geq N_1.$ Similarly, there exists $N_2 \in \mathbb{N}$ such that $x_k \in B_\epsilon(y)$ for all $k \geq N_2.$ This implies that $x_k \in B_\epsilon(x) \cap B_\epsilon(y)$ for any $k \geq \max\{N_1, N_2\},$ a contradiction.

- ## §2.4

  3. $\mathbb{S}^n$ is connected for every $n \geq 1.$ $\mathbb{S}^0 = \{-1, 1\}$ is not connected; there is a separation $U = (-2, 0),$ $V = (0, 2),$ for example.

  4. Let U, V be a separation of $X,$ and suppose $A \subseteq X$ is connected. $X \cap A$ is nonempty (A is in $X \cap A$), and $X \cap A = (U \cup V) \cap A = (U \cap A) \cup (V \cap A).$ Thus at least one of $U \cap A$ or $V \cap A$ must be nonempty. If both $U \cap A$ and $V \cap A$ are nonempty, then since U, V are open and disjoint, these two sets form a separation of $A,$ contradicting the connectedness of $A.$ So only

one of $U \cap A$ or $V \cap A$ can be nonempty. If $U \cap A = \emptyset$, then $A \subseteq V$. If $V \cap A = \emptyset$, then $A \subseteq U$.

6. (a) Connected. (b) Not connected; separation: $U = B_\epsilon(x)$ and $V = B_\epsilon(y)$, where $\epsilon = \frac{1}{2}d(x,y)$. (c) Connected. (d) Connected. (e) Connected when $n \geq 2$. When $n = 1$, we have $\text{ext}(B_\epsilon(x)) = \text{ext}((x - \epsilon, x + \epsilon))$, and there is a separation, $U = (-\infty, x - \epsilon/2)$ and $V = (x + \epsilon/2, \infty)$. (f) Connected when $n \geq 2$. When $n = 1$, we have $B_\epsilon(x) \setminus \{x\} \subseteq (-\infty, x) \cup (x, \infty)$, which defines the separation.

7. A and D are inside; B and C are outside.

- ## §2.5

3. (a) Let $x \in X$ and $\epsilon > 0$ be arbitrary. First show $f[B_{d_X,\epsilon}(x)] \subseteq B_{d_Y,\epsilon}(f(x))$. For any $u \in B_{d_X,\epsilon}(x)$, we have $d_X(x, u) < \epsilon$. Therefore, since f is an isometry, we have $d_Y(f(x), f(u)) < \epsilon$, which shows $f(u) \in B_{d_Y,\epsilon}(f(x))$. Hence $B_{d_X,\epsilon}(x) \subseteq f^{-1}[f[B_{d_X,\epsilon}(x)]] \subseteq f^{-1}[B_{d_Y,\epsilon}(f(x))]$. Thus f is continuous.

9. The image on the right shows metric circles for $p = 1, 2, 3, 10$, and 100 (from inner to outer). As p gets large, the metric circle approximates a square whose sides are parallel to the coordinate axes, which illustrates how the max norm d_∞ may be regarded as a limit of d_p as $p \to \infty$.

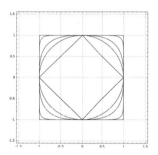

14.

	CAUCHY	COMPACT	CONNECTED	CONTINUITY	CONTINUOUS
(a) CAUCHY	0				
COMPACT	0.49219	0			
CONNECTED	0.49805	0.22461	0		
CONTINUITY	0.49902	0.24902	0.12402	0	
CONTINUOUS	0.49902	0.24902	0.12402	0.00684	0

(b) *Answers will vary.* WORDY ($d = 1/32$), WORK ($d = 1/16$), WORM ($d = 1/16$), WORTH ($d = 3/32$), WORRY ($d = 3/32$), etc.

(c) $d(\text{TEH}, \text{THE}) = 3/8$; $d(\text{TEH}, \text{TEN}) = 1/8$. TEH is most likely a misspelling of THE, but this metric would imply that TEN is a closer word to it. A better metric for autocorrect would give shorter distances between words that differ only by the transposing of two letters.

Problems of Chapter 3

- ## §3.1

  4. Each phase portrait may be found by sampling the vector field at sufficiently many (x, y) points. It may take many samples or the help of computer software[1] to accurately sketch each phase portrait.

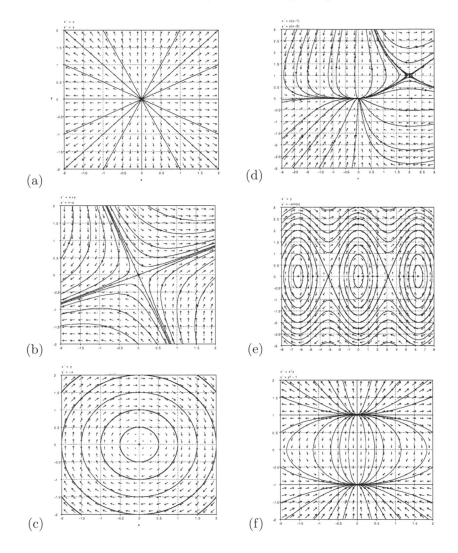

(a)

(d)

(b)

(e)

(c)

(f)

---

[1]Phase portraits created using pplane [Pol].

- **§3.2**

  1.

  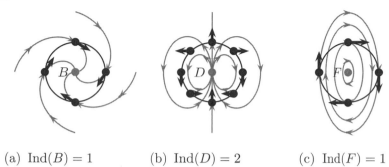

  | (a) Ind(B) = 1 | (b) Ind(D) = 2 | (c) Ind(F) = 1 |

  3. (a) $\tan\theta = \frac{v}{u}$, or $\theta = \arctan\frac{v}{u}$

  (b) So long as Δt is small enough so that the vectors do not make a complete turn in any subinterval $[t_{i-1}, t_i]$, then the total angle angle change around $\mathbf{C}$ is the sum of the individual angle changes, $\Delta\theta_i$:

  $$\sum_{i=1}^{n} \Delta\theta_i = \sum_{i=1}^{n} \frac{\Delta\theta_i}{\Delta t} \cdot \Delta t.$$

  (c) Since the sum from part (b) measures total change in angle, it must be divided by 2π to yield the number of complete revolutions. Then as $n \to \infty$, and $\Delta t \to 0$, the Riemann sum may be interpreted as an integral:

  $$\begin{aligned} I(C) &= \frac{1}{2\pi} \int_0^1 \theta' dt = \frac{1}{2\pi} \int_0^1 \frac{d}{dt}\left[\arctan\frac{v}{u}\right] dt \\ &= \frac{1}{2\pi} \int_0^1 \frac{1}{1 + \left(\frac{v}{u}\right)^2} \cdot \frac{u\frac{dv}{dt} - v\frac{du}{dt}}{u^2} dt = \frac{1}{2\pi} \int_0^1 \frac{u\frac{dv}{dt} - v\frac{du}{dt}}{u^2 + v^2} dt. \end{aligned}$$

  5. Index of a quadripole is 3. If $\mathbf{x}$ is a $2n$-pole, then $\text{Ind}(\mathbf{x}) = 1 + \frac{2n - 0}{2} = n + 1$.

  7. Set $5 = 1 + \frac{e(\mathbf{x}) - h(\mathbf{x})}{2} \implies e(\mathbf{x}) = h(\mathbf{x}) + 8$. Therefore, since $h(\mathbf{x}) \geq 0$, there must be at least eight elliptical sectors.

- **§3.3**

  2. (a) The line L has slope -1 and passes through $(2, 8.75)$, so the equation of the line is $y = -(x - 2) + 8.75 = -x + 10.75$. Substituting into (3.4),

  $$\begin{cases} dx/dt &= 1 - 3.5x + x^2(-x + 10.75) = 1 - 3.5x + 10.75x^2 - x^3, \\ dy/dt &= x(2.5 - x[-x + 10.75]) = 2.5x - 10.75x^2 + x^3. \end{cases}$$

  (C.1)

  (b) From (C.1), we get $\frac{dx}{dt} + \frac{dy}{dt} = 1 - x$, which is negative so long as $x > 1$. Thus, since every point on L has $x \geq 2$, we find that every vector on L has slope less than -1.

3.

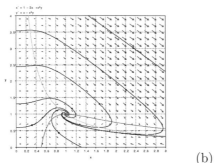

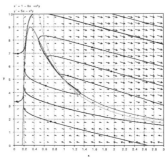

(a)   (b)

Problems of Chapter 4

- ## §4.1

  3. Suppose $C, D \subseteq X$ are closed in a topological space X. Then by definition $X \setminus C$ and $X \setminus D$ are open. Consider the union $E = C \cup D$. Now $X \setminus E = X \setminus (C \cup D) = (X \setminus C) \cap (X \setminus D)$. Since the intersection of two open sets is open, we find that E is closed.

  On the other hand, $\bigcup_{x \in (0,1)} \{x\} = (0, 1)$ is not closed in the Euclidean topology on $\mathbb{R}^1$, even though each point set $\{x\}$ is closed.

  7. Only a one-point space $X = \{x\}$.

  8. There are four topologies: $\mathscr{T}_1 = \{\emptyset, \{a, b\}\}$; $\mathscr{T}_2 = \{\emptyset, \{a\}, \{a, b\}\}$; $\mathscr{T}_3 = \{\emptyset, \{b\}, \{a, b\}\}$; and $\mathscr{T}_4 = \{\emptyset, \{a\}, \{b\}, \{a, b\}\}$.

- ## §4.2

  1. In what follows, assume f is a function $f : X \to Y$.

  (a) Suppose X is a discrete space. Then for every open set $V \subseteq Y$, we have $f^{-1}[V]$ open in X. Thus every function f is continuous.

  (b) Suppose Y is a discrete space. Then for every open set $U \subseteq Y$, we have $f[U]$ open in Y. Thus every function f is an open map.

  (c) Suppose Y is an indiscrete space. Then there are only two open sets, $\emptyset$ and Y. Now $f^{-1}[\emptyset] = \emptyset$ and $f^{-1}[Y] = X$, both of which are guaranteed to be open. Thus every surjective function f is continuous.

  (d) Suppose X is indiscrete and Y is discrete. If f is not constant, then there are points $x_1 \neq x_2$ in X such that $f(x_1) = y_1$ and $f(x_2) = y_2$ with $y_1 \neq y_2$. Now since Y is discrete, $U = \{y_1\}$ is open. Consider $Z = f^{-1}[U] \subseteq X$. Certainly $x_1 \in Z$, so $Z \neq \emptyset$, but also $x_2 \notin Z$, so $Z \neq X$. Since X is indiscrete, Z cannot be an open set. Thus any nonconstant function f cannot be continuous. On the other hand, every constant function is continuous with respect to any topology.

- ## §4.3

1. Typical basic open sets of H include: (A) open arcs of any circle C_k so long as the origin is not on the arc, and (B) the origin together with all circles and arcs of circles within a given distance from the orgin. No matter how small a neighborhood around $(0,0)$ is chosen, there will always be infinitely many distinct circles C_k contained in the neighborhood B.

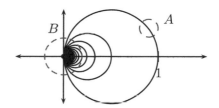

3. (a) Closed in both A and in $\mathbb{R}^1$. (b) Neither open nor closed in A or in $\mathbb{R}^1$. (c) Open in both A and in $\mathbb{R}^1$. (d) Closed in both A and in $\mathbb{R}^1$. (e) Closed in A but neither open nor closed in $\mathbb{R}^1$. (f) Closed in A, since $A \setminus \{1/k \mid k \in \mathbb{N}\} = (1,\infty) \cup (1/2,1) \cup (1/3,1/2) \cup (1/4,1/3) \cup \cdots$, but neither open nor closed in $\mathbb{R}^1$.

11. (a) For all $\mathbf{x} \in \mathbb{R}^{n+1}$, we have $\mathbf{x} = (1)\mathbf{x}$, so $\mathbf{x} \sim \mathbf{x}$ ($\sim$ is *reflexive*). For all $\mathbf{x}, \mathbf{y} \in \mathbb{R}^{n+1}$, if $\mathbf{x} \sim \mathbf{y}$, then there is $\lambda \neq 0$ such that $\mathbf{y} = \lambda \mathbf{x}$. It follows that $\mathbf{x} = \frac{1}{\lambda}\mathbf{y}$; hence $\mathbf{y} \sim \mathbf{x}$, since $\frac{1}{\lambda} \neq 0$ ($\sim$ is *symmetric*). For all $\mathbf{x}, \mathbf{y}, \mathbf{z} \in \mathbb{R}^{n+1}$, suppose $\mathbf{x} \sim \mathbf{y}$ and $\mathbf{y} \sim \mathbf{z}$. Then there are nonzero scalars λ_1, λ_2 such that $\mathbf{y} = \lambda_1 \mathbf{x}$ and $\mathbf{z} = \lambda_2 \mathbf{y}$. Then $\mathbf{z} = \lambda_2 \lambda_1 \mathbf{x}$, proving that $\mathbf{x} \sim \mathbf{z}$, since $\lambda_2 \lambda_1 \neq 0$ ($\sim$ is *transitive*).

- ## §4.4

1. Suppose X has the cofinite topology, and let $A \subseteq X$. Suppose $\mathscr{U}$ is an open cover of A, and choose a set $U_0 \in \mathscr{U}$. By definition, since U_0 is open, we have $U_0 = X \setminus F$ for some finite set of points $F = \{x_1, \ldots, x_n\}$. Now for every point $x_k \in F \cap A$, there must be at least one open set $U_k \in \mathscr{U}$ containing x_k. Let $\mathscr{V}$ be the collection of those finitely many sets U_k. Thus $\mathscr{U}' = \{U_0\} \cup \mathscr{V}$ is a finite subcover, proving that A is compact.

If X is infinite, then no separation can possibly exist for $A \subseteq X$, since every pair of open sets U, V must share all but finitely many points. This proves A connected.

- ## §4.5

3. It suffices to show that $p_1 : X \times Y \to X$ is continuous. Suppose that $U \subseteq X$ is open. Note that $p^{-1}[U] = \{(x,y) \in X \times Y \mid x \in U\} = U \times Y$. But since U is open in X, and Y is open in Y, $U \times Y$ is open with respect to the box topology, proving that p_1 is continuous.

6. $p_X \circ i_X = \mathrm{id}_X$. $p_Y \circ i_Y = \mathrm{id}_Y$. $p_X \circ i_Y$ is the constant map sending all of Y to the point $x_0 \in X$. $p_Y \circ i_X$ is the constant map sending all of X to the point $y_0 \in Y$.

- **§4.6**

  1. Note that $\mathbb{Z}$ is partitioned by $N_{0,b}$, $N_{1,b}$, ..., $N_{b-1,b}$, and so $N_{a,b} = \mathbb{Z}\backslash U$, where $U = N_{0,b} \cup \cdots \cup N_{a-1,b} \cup N_{a+1,b} \cup \cdots N_{b-1,b}$. U is open, being the union of open sets, and so $N_{a,b}$ is closed.

  3. Let $x \in \mathbb{Z}$, and consider the infinite intersection $C = \bigcap_{b\in\mathbb{N}} N_{x,b}$. We have $x \in C$ since $x \in N_{x,b}$ for all b. Now if $y \neq x$, then $y \notin N_{x,2|x-y|}$, so $y \notin C$; hence $C = \{x\}$. But since each $N_{x,b}$ is closed, so must C be.

Problems of Chapter 5

- **§5.1**

  1. (a) Compact surface-with-boundary. The boundary consists of the four edges of the square. (b) Compact surface-with-boundary. The boundary has two components, an inner circle of radius 1, and an outer circle of radius 2. (c) Noncompact surface. (d) Neither a surface nor surface-with-boundary; noncompact. (e) Noncompact surface.

  3. By "flattening" the wireframe, observe that n is always one less than the number of faces of the original polyhedron. Tetrahedron: 3$\mathbb{T}$. Cube: 5$\mathbb{T}$. Octahedron: 7$\mathbb{T}$. Dodecahedron: 11$\mathbb{T}$. Icosahedron: 19$\mathbb{T}$. Object in Figure 5.8: 25$\mathbb{T}$.

  5. A **manifold-with-boundary** of dimension n is a connected Hausdorff topological space M such that for each $x \in M$, there is an open set $U \subseteq M$ with $x \in M$, such that either $U \approx B_1(\mathbf{0}) \subseteq \mathbb{R}^n$ or $U \approx H = \{(x_1, x_2, \ldots, x_n) \in \mathbb{R}^n \mid x_n \geq 0\}$.

  8. (a) 3$\mathbb{T}$ (b) $(1 + 2 + 3 + \cdots + k)\mathbb{T} = \frac{k(k+1)}{2}\mathbb{T}$

- **§5.2**

  1. (a) Cone with no bottom surface, which is homeomorphic to a disc $\mathbb{D}^2$. Word: $aa^{-1}b \equiv b$. $P \sim Q$. (b) Sphere. Word: $abb^{-1}a^{-1} \equiv aa^{-1}$. $Q \sim S$. (c) After identifying edges a together, we obtain a plane model for the torus $\mathbb{T}$. Word: $a^{-1}abcb^{-1}c^{-1} \equiv bcb^{-1}c^{-1}$. $U \sim P \sim Q \sim R \sim S$. (d) After identifying edges a together, as well as edges w, the result is a plane model for a disk $\mathbb{D}^2$. Word: cb^{-1}. $U \sim S$ and $P \sim R$.

  4. $a_1 b_1 a_1^{-1} b_1^{-1} a_2 b_2 a_2^{-1} b_2^{-1} \cdots a_n b_n a_n^{-1} b_n^{-1}$

  5. To perform a connected sum $\mathbb{T}\#Q$, the letters in the word for Q must be distinct from those of $\mathbb{T}$. Therefore there is no $Q \in \mathcal{S}_c$ whose word can cancel the letters $aba^{-1}b^{-1}$ in the word for $\mathbb{T}$ and arrive at the (reduced) word ss^{-1} for $\mathbb{S}^2$.

  8. $abcd^{-1}c^{-1}d^{-1}e^{-1}bfa^{-1}ef$, or equivalent. Let y be any vertex not already labeled by x_k in Figure 5.16. Consider $b^+c^-d^+c^+d^-e^+f^-b^+$; since this curve bounds all of the rest of the vertices, we find there are only two equivalence classes, $[x]$ and $[y]$.

- **§5.3**

  1. $a_1^2 a_2^2 \cdots a_n^2$

  2. (a) Orientable. (b) Orientable. (c) Nonorientable. (d) Orientable. (e) Nonorientable. (f) Nonorientable. If a is an edge corresponding to a same-direction pair $\cdots a \cdots a \cdots$ in the word, any loop crossing a once will be orientation reversing (cases (c), (e), and (f)).

  4. If there are only factors $\mathbb{S}^2$, then the resulting sum is $\mathbb{S}^2$, so assume now that there are other factors and use $A \# \mathbb{S}^2 = A$ to get rid of all factors $\mathbb{S}^2$. Next, replace any factor $\mathbb{K}$ by the equivalent $2\mathbb{P}$. Since the connected sum is commutative, the result is homeomorphic to $n\mathbb{T} \# m\mathbb{P}$ for some $n, m \geq 0$ (but not both equal to 0). If $m = 0$, or if $n = 0$, then we are done (the surface is $n\mathbb{T}$ or $m\mathbb{P}$, respectively). If both $m, n > 0$, then replace pairs using $\mathbb{T} \# \mathbb{P} \approx 3\mathbb{P}$ as necessary.

  (a) $3\mathbb{T} \# 2\mathbb{P} \approx \mathbb{T} \# 6\mathbb{P} \approx 3\mathbb{P} \# 5\mathbb{P} \approx 8\mathbb{P}$. Genus 8, nonorientable.

  (b) $m\mathbb{K} \approx (2m)\mathbb{P}$. Genus $2m$, nonorientable.

  (c) $13\mathbb{S}^2 \# 5\mathbb{T} \# 6\mathbb{S}^2 \approx 5\mathbb{T}$. Genus 5, orientable.

  (d) $\mathbb{K} \# \mathbb{T} \# \mathbb{P} \approx \mathbb{T} \# 3\mathbb{P} \approx 3\mathbb{P} \# 2\mathbb{P} \approx 5\mathbb{P}$. Genus 5, nonorientable.

  (e) $a\mathbb{P} \# b\mathbb{T} \approx (a - b)\mathbb{P} \# (b\mathbb{P} \# b\mathbb{T}) \approx (a - b)\mathbb{P} \# 3b\mathbb{P} \approx (a + 2b)\mathbb{P}$. Genus $a + 2b$, nonorientable. (Note that this is also true for general $a, b \geq 0$.)

  (f) $k\mathbb{S}^2 \approx \mathbb{S}^2$. Genus 0, orientable.

- **§5.4**

  1. $v = 2 + 2n$, $e = 1 + 3n$, $f = n$; $\chi = 1$, for all $n \in \mathbb{N}$

  4. (a) With a single fixed 0-cell and a single additional n-cell (note that if $n = 0$, this implies a total of two disjoint 0-cells), we have $\chi(\mathbb{S}^n) = 1 + (-1)^n$.

  (b) With two cells in every dimension, $\chi(\mathbb{S}^n) = \sum_{k=0}^{n} (-1)^n \cdot 2$.

  In both (a) and (b), $\chi(\mathbb{S}^n) = 2$ if n is even and 0 if n is odd.

  5. (a) $2 - 2n$ (b) $2 - n$

  6. Let f be the number of faces for the triangulation of M. Then $\chi(M) = v - e + f \implies v - \chi(M) = e - f$. By exercise 9, $f = \frac{2e}{3}$. Thus $v - \chi(M) = e - \frac{2e}{3} = \frac{e}{3} \implies e = 3(v - \chi(M))$.

Problems of Chapter 6

- **§6.1**

  1. Every connected tree T has $V(T) = E(T) + 1$; hence $\chi(T) = 1$. If T is a forest of k components, then $\chi(T) = k$.

  2. $\{a, c, e, g, i\}$, $\{b, d, f, h\}$

3. $E(K_{k_1,k_2,k_3}) = k_1 k_2 + k_1 k_3 + k_2 k_3$. Since $V(K_{k_1,k_2,k_3}) = k_1 + k_2 + k_3$, $\chi(K_{k_1,k_2,k_3}) = k_1 + k_2 + k_3 - k_1 k_2 - k_1 k_3 - k_2 k_3$.

6. Suppose that $K_{3,3}$ is planar, and let f be the number of faces in an embedding $K_{3,3} \to \mathbb{S}^2$. Let $v = V(K_{3,3}) = 6$, and $e = E(K_{3,3}) = 9$. Then $\chi(\mathbb{S}^2) = \chi(K_{3,3}) + f$ implies $f = 5$. Now, since the smallest cycle in $K_{3,3}$ is length 4, we have $4f \le 2e$; however, $20 \not\le 18$. This contradiction implies that $K_{3,3}$ is nonplanar.

8. Answers will vary. Vertex labels are based on those in Figure 6.8. The cycle $jhfigj$ of P is present along the identified edge A of the plane diagram.

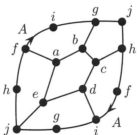

9. Of the two regions determined by the graph, one is not homeomorphic to a disc.

10. *From left to right*, $\mathbb{P}$ (six vertices); $\mathbb{T}$ (seven vertices); $\mathbb{K}$ (eight vertices).

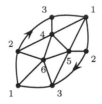 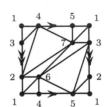 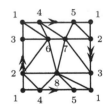

- ## §6.2

1. For each vertex, there are at least n edges adjacent, for a total of at least $n|V(G)|$. However, each edge is double counted, and so $2|E(G)| \ge n|V(G)|$, or $|V(G)| \le \frac{2}{n}|E(G)|$.

5. (a) Dual graph shown in black; original graph in light gray.

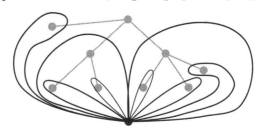

8. $\chi(\mathbb{P}) = 1$, so $\frac{2e}{f} \leq 6\left(1 - \frac{1}{f}\right) < 6$, which implies $\gamma(\mathbb{P}) \leq 6$. An embedding of the Petersen graph yields a map requiring six colors; therefore $\gamma(\mathbb{P}) = 6$.

9. (a) 2 (b) 2 (c) 2 (d) 2 if n is even; 3 if n is odd

- **§6.3**

7. Three half-twists: trefoil, or $\mathbb{T}(2,3)$. Five half-twists: $\mathbb{T}(2,5)$. Seven half-twists: $\mathbb{T}(2,7)$.

8. *Codes based on starting points and orientations as shown below.*

A B

– Gauss code:
   Knot A: $(1, -2, 3, -4, 2, -5, 6, -3, 4, -1, 5, -6)$
   Knot B: $(-1, 2, -3, 4, -5, 6, -7, 3, -2, 1, -4, 7, -6, 5)$
– Extended Gauss code:
   Knot A: $(1, -2, 3, -4, 2, -5, 6, 3, 4, -1, -5, -6)$
   Knot B: $(-1, 2, -3, 4, -5, 6, -7, 3, 2, 1, 4, 7, 6, 5)$

9.

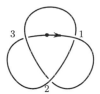

3   1
2

Gauss code:
$(1, -2, 3, -1, 2, -3)$
Extended Gauss code:
$(1, -2, 3, -1, -2, -3)$

- **§6.4**

1. (a) Hopf link: $\ell = 1$. Whitehead link: $\ell = 0$. Celtic design: $\ell = 3$.

(b) Reidemeister move R.I involves only a single component knot and so cannot contribute to the linking number. If move R.II involves strands from the same component, then there is nothing to check; if it involves strands from both components, then the linking number is still preserved because the signs of the two crossings are always opposite. After move R.III, the three crossings always keep their original sign (just their relative positions are changed), and so the linking number remains the same.

(c) Based only on linking number, the Hopf link and the Celtic design from Figure 6.31 are not equivalent to the unlink. (Note that the Whitehead link is also not an unlink, but we cannot prove that using the linking number.)

3. No, because that knot is equivalent to $\mathbb{U}$, which is not tricolorable.

7. 0; 3; 0; 7; 0

8. $\langle K \rangle = A^8 - A^4 + 1 - A^{-4} + A^{-8}$. The writhe is $w(K) = 0$, so $J(K) = (-A^3)^0 \langle K \rangle \big|_{A=t^{-1/4}} = t^2 - t + 1 - t^{-1} + t^{-2}$.

10. $\mathbb{R}^3 \setminus K$ is connected and Hausdorff. Since $\mathbb{S}^1$ is compact, so is $K = K[\mathbb{S}^1]$. Any compact subset of $\mathbb{R}^3$ is closed in the Euclidean topology, so $\mathbb{R}^3 \setminus K$ is open. Thus every point $x \in \mathbb{R}^3 \setminus K$ can be surrounded by an open ball $U = B_\epsilon(x) \approx \mathbb{R}^3$ (for some $\epsilon \in \mathbb{R}^+$).

Problems of Chapter 7

- ## §7.1

1. For any map $\gamma : Z \to \mathbb{R}^n$, there is a straight-line homotopy from γ to the constant map $\mathbf{0}$ defined by:

$$
\begin{aligned}
h &: \quad Z \times \mathbb{I} \to \mathbb{R}^n \\
h(s,t) &= \quad (1-t)\gamma(s).
\end{aligned}
$$

3. *Reflexivity:* Given any loop $\gamma : \mathbb{S}^1 \to X$, we have $\gamma \simeq \gamma$ via the homotopy $h : \mathbb{S}^1 \times \mathbb{I} \to X$ defined by $h(s,t) = \gamma(s)$ for all $t \in \mathbb{I}$. Observe that h is automatically basepoint preserving. *Symmetry:* If $\gamma \simeq \eta$ for loops γ, η, then by definition there is a homotopy h such that $h(s,0) = \gamma(s)$ and $h(s,1) = \eta(s)$ for all $s \in \mathbb{S}^1$. Define another function $h'(s,t) = h(s, 1-t)$. h' is continuous; $h'(s,0) = h(s,1) = \eta(s)$; and $h'(s,1) = h(s,0) = \gamma(s)$. Thus h' is a homotopy verifying that $\eta \simeq \gamma$. If h is basepoint preserving, then so will h' be, since $h'(0,t) = h(0, 1-t)$. *Transitivity:* Suppose $\gamma \simeq \eta$ and $\eta \simeq \theta$ for loops γ, η, θ. Then there are homotopies h_1 and h_2 such that $h_1(s,0) = \gamma(s)$, $h_1(s,1) = h_2(s,0) = \eta(s)$, and $h_2(s,1) = \theta(s)$ for all $s \in \mathbb{S}^1$. Define another function $h(s,t)$ via:

$$
h(s,t) = \begin{cases} h_1(s, 2t), & t \in [0, 1/2]; \\ h_2(s, 2t-1), & t \in (1/2, 1]. \end{cases}
$$

Note that $h(s,0) = h_1(s,0) = \gamma(s)$ and $h(s,1) = h_2(s,1) = \theta(s)$, and that h is continuous since $\lim_{t \to 1/2^-} h(s,t) = \lim_{t \to 1/2^+} h(s,t) = \eta(s)$. If both h_1 and h_2 are basepoint preserving, then so will h be, since $h(0,t)$ is either equal to $h_1(0, 2t)$ or $h_2(0, 2t-1)$.

- **§7.2**

  1. Let $[\gamma] \in \pi_1(X, x_0)$. Then

  $$
  \begin{aligned}
  (g \circ f)([\gamma]) = g\left(f([\gamma])\right) &= g\left([\theta^{-1} \cdot \gamma \cdot \theta]\right) \\
  &= [\theta \cdot (\theta^{-1} \cdot \gamma \cdot \theta) \cdot \theta^{-1}] \\
  &= [\theta] \cdot [\theta]^{-1} \cdot [\gamma] \cdot [\theta] \cdot [\theta]^{-1} \\
  &= [\mathbf{1}] \cdot [\gamma] \cdot [\mathbf{1}] = [\gamma].
  \end{aligned}
  $$

  Therefore $g \circ f = \mathrm{id}_{\pi_1(X, x_0)}$. The argument is analogous to show that $f \circ g = \mathrm{id}_{\pi_1(X, x_1)}$.

- **§7.3**

  1. *Proof of (a).* Take $Z = \{z_0\}$ to be any one-point space. There is an obvious homeomorphism $\mathbb{I} \approx \{z_0\} \times \mathbb{I}$, and so γ may be regarded as the map $Z \times \mathbb{I} \to X$ that sends $(z_0, t) \mapsto \gamma(t)$. Let $\gamma_0 : Z \times \{0\} \to X$ be the map sending $(z_0, 0) \mapsto x_0$, and consider the lift $\widetilde{\gamma_0} : Z \times \{0\} \to E$ sending $(z_0, 0) \mapsto e_0$. Then by Theorem 7.3.2, there is a unique lift $\widetilde{\gamma} : Z \times \mathbb{I} \approx \mathbb{I} \to E$ of γ satisfying $\widetilde{\gamma}(0) = \widetilde{\gamma_0}(0) = e_0$.

  Proof of (b). Take $Z = \mathbb{I}$. Then h may be regarded as a map $Z \times \mathbb{I} \to X$. Since h is a homotopy from γ to η, we have $h(s, 0) = h_0(s) = \gamma(s)$ ($\forall s \in \mathbb{I}$). Let $\widetilde{h_0} = \widetilde{\gamma}$ be the unique lift of γ to E. By Theorem 7.3.2, there is a unique lift $\widetilde{h} : Z \times \mathbb{I} \to E$ of h satisfying $\widetilde{h}(s, 0) = \widetilde{h_0}(s) = \widetilde{\gamma}(s)$ ($\forall s \in \mathbb{I}$). Note that $p(\widetilde{h}(s, 1)) = h(s, 1) = \eta(s)$ ($\forall s \in \mathbb{I}$), so $\widetilde{h}(-, 1)$ is a lift of η. By uniqueness of path lifting, then $\widetilde{h}(-, 1) = \widetilde{\eta}$, proving that $\widetilde{\gamma} \simeq \widetilde{\eta}$. Finally, since $h(0, t) = x_0$ for all $t \in \mathbb{I}$, and since $p^{-1}(\{x_0\})$ is a discrete set, it must be the case that $\widetilde{h}(0, t) = e_0$ ($\forall t \in \mathbb{I}$).

  3. $p^{-1}[U_1] = \bigcup_{k \in \mathbb{Z}} \left(n + \frac{1}{8}, n + \frac{7}{8}\right)$; $p^{-1}[U_2] = \bigcup_{k \in \mathbb{Z}} \left(n - \frac{3}{8}, n + \frac{3}{8}\right)$

- **§7.4**

  1. By deforming the loop within $\mathbb{T}$ so that the segments move toward either a or b, the (p, q)-torus knot is homotopic to $a^p b^q$ (note that the labeling and orientation of the edges affect the form of the answer).

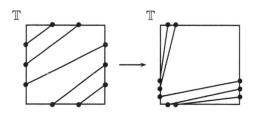

  2. (a) $\langle a_1, a_2, a_3, b_1, b_2, b_3 \mid a_1 b_1 a_1^{-1} b_1^{-1} a_2 b_2 a_2^{-1} b_2^{-1} a_3 b_3 a_3^{-1} b_3^{-1} = \mathbf{1} \rangle$

(b) $\langle a, b \mid abab^{-1} = \mathbf{1} \rangle$

(c) $\langle a_1, b_1, a_2, b_2 \mid a_1 b_1 a_1^{-1} b_1^{-1} a_2 b_2 a_2 b_2^{-1} = \mathbf{1} \rangle$

(d) $\langle a_1, \ldots a_n \mid a_1^2 \cdots a_n^2 = \mathbf{1} \rangle$

5. The figure-eight knot is shown with strands labeled a, b, c, d, and positive and negative crossings indicated. Before any simplifications, the knot group has the following presentation:

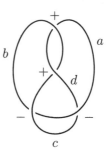

$$\langle a, b, c, d \mid a^{-1} b^{-1} db = c^{-1} d^{-1} bd = b^{-1} cac^{-1} = d^{-1} aca^{-1} = \mathbf{1} \rangle.$$

Problems of Chapter 8

- ## §8.1

  2. Betti numbers: (a) 1 (b) 10 (c) 3 (d) 0 (e) 6 (f) 1

  6. (a) $(\beta_k(\mathbb{T}))_{k \geq 0} = (1, 2, 1, 0, \ldots)$ (c) $(\beta_k(\mathbb{P}))_{k \geq 0} = (1, 0, 0, 0, \ldots)$

   (b) $(\beta_k(\mathbb{K}))_{k \geq 0} = (1, 1, 0, 0, \ldots)$ (d) $(\beta_k(\mathbb{S}^1 \times \mathbb{I}))_{k \geq 0} = (1, 1, 0, 0, \ldots)$

  8. For every $n \geq 0$, $\beta_0(\Delta^n) = 1$ and $\beta_k(\Delta^n) = 0$ when $k > 0$.

  9. For $n \in \mathbb{N}$, $\beta_k(\mathbb{S}^n) = 1$ if $k = 0$ or $k = n$, and $\beta_k(\mathbb{S}^n) = 0$ for $k \neq 0, n$.

- ## §8.2

  1. There are 33 distinct elements within the square, as shown below.

  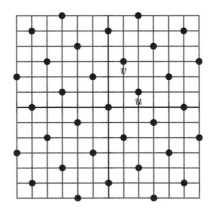

3. $5\mathbb{Z} = \{5n \mid n \in \mathbb{Z}\} = \{\ldots, -15, -10, -5, 0, 5, 10, 15, \ldots\}$. By definition of quotient module, $\mathbb{Z}/5\mathbb{Z}$ is equal to the set of equivalence classes $[m]$ for $m \in \mathbb{Z}$, with $[m_1] = [m_2] \iff m_1 - m_2 \in 5\mathbb{Z}$. But this is equivalent to $m_1 - m_2$ being a multiple of 5.

5.

$$H_k(\mathbb{K}; \mathbb{Q}) \approx \begin{cases} \mathbb{Q}, & k = 0 \\ \mathbb{Q}, & k = 1 \\ 0, & k \geq 2 \end{cases} \qquad H_k(\mathbb{K}; \mathbb{Z}) \approx \begin{cases} \mathbb{Z}, & k = 0 \\ \mathbb{Z} \times \mathbb{Z}/2\mathbb{Z}, & k = 1 \\ 0, & k \geq 2 \end{cases}$$

Problems of Appendix A

- ## §A.1

2. Suppose $x \in \mathbb{Q}$ and $y \in \mathbb{Q}$. Then there are numbers $a, c \in \mathbb{Z}$ and $b, d \in \mathbb{N}$ such that $x = \frac{a}{b}$ and $y = \frac{c}{d}$. Using algebra, $x + y = \frac{a}{b} + \frac{c}{d} = \frac{ad+bc}{bd}$. But since $a, b, c, d \in \mathbb{Z}$ (noting that $\mathbb{N} \subseteq \mathbb{Z}$), we know that $ad + bc \in \mathbb{Z}$ as well. Furthermore, since $b, d \in \mathbb{N}$, we have $bd \in \mathbb{N}$. Thus $\frac{ad+bc}{bd} = x + y \in \mathbb{Q}$.

Similarly, $x - y = \frac{a}{b} - \frac{c}{d} = \frac{ad-bc}{bd}$, proving that $x - y \in \mathbb{Q}$.

3. Suppose $x \in \mathbb{Q}$ and $y \in \mathbb{R} \setminus \mathbb{Q}$, and let $z = x + y$. Assume (to the contrary) that $z \notin \mathbb{R} \setminus \mathbb{Q}$. That is, $z \in \mathbb{Q}$. Now $z = x + y$ implies $z - x = y$, and since both z and x are rational numbers, so is their difference $z - x$ (by Exercise 2). Thus $y \in \mathbb{Q}$, which contradicts the hypothesis ($y \notin \mathbb{Q}$).

6. (a) *True*. (b) *False*. (c) *True*. (d) *False*. (e) *True*. (f) *True*. (g) *False*. (h) *True*. (i) *True*. (j) *False*. (k) *False*. (l) *True*.

7. Each subset B of A involves a choice for each element x, whether x should be included in B or excluded. Thus, if A has n elements, then $\mathcal{P}(A)$ has 2^n elements. Similarly, $\mathcal{P}(\mathcal{P}(A))$ has $2^{(2^n)}$ elements.

8.

(a) $\{n \in \mathbb{N} \mid n = 2m \text{ or } n = 3m \text{ for some } m \in \mathbb{N}, m > 1\}$

(b) $\{6, 12, 18, \cdots\} = \{6m \mid m \in \mathbb{N}\}$

(c) $\{30m \mid m \in \mathbb{N}\}$

(d) $\{(2m_1, 3m_2) \mid m_1, m_2 \in \mathbb{N}, m_1, m_2 > 1\}$

(e) $\{(2m_1, 3m_2, 4m_3) \mid m_1, m_2, m_3 \in \mathbb{N}, m_1, m_2, m_3 > 1\}$

(f) $\{(2m_1, 3m_2, 4m_3, \ldots) \mid m_k \subset \mathbb{N}, m_k > 1, \forall k \in \{2, 3, 4, \ldots\}\}$

13. Let $x \in \left(\bigcup_{k \in \mathcal{I}} A_k\right) \cap \left(\bigcup_{\ell \in \mathcal{J}} B_\ell\right)$. Then $x \in \bigcup_{k \in \mathcal{I}} A_k$ and $x \in \bigcup_{\ell \in \mathcal{J}} B_\ell$. Then there is some $k \in \mathcal{I}$ and $\ell \in \mathcal{J}$ such that $x \in A_k$ and $x \in B_\ell$. Thus $x \in A_k \cap B_\ell$, which implies $x \in \bigcup_{k \in \mathcal{I}, \ell \in \mathcal{J}} (A_k \cap B_\ell)$. This proves that $\left(\bigcup_{k \in \mathcal{I}} A_k\right) \cap \left(\bigcup_{\ell \in \mathcal{J}} B_\ell\right) \subseteq \bigcup_{k \in \mathcal{I}, \ell \in \mathcal{J}} (A_k \cap B_\ell)$. Reverse the steps to show the other inclusion.

- **§A.2**

  1. (a) Assume $d \geq 1$ is a fixed whole number. **Reflexivity.** For any $n \in \mathbb{Z}$, we have $n - n = 0 = 0 \cdot d$; thus $n \sim n$. **Symmetry.** Suppose $n \sim m$. Then $n - m = kd$ for some $k \in \mathbb{Z}$. But then $m - n = -kd$, which is also a multiple of d; thus $m \sim n$. **Transitivity.** Suppose $n \sim m$ and $m \sim p$. Then there are integers k and ℓ such that $n - m = kd$ and $m - p = \ell d$. Consider $n - p = (n - m) + (m - p) = kd + \ell d = (k + \ell)d$. Since $k + \ell \in \mathbb{Z}$, we have $n \sim p$.

  3.

     (a) Bijective: $f^{-1}(x) = \frac{x-7}{8}$.

     (b) Not injective since $3(-c)^4 = 3 = 3(c)^4$ for any $c \in \mathbb{R}$. Not surjective since no negative value is in the range of f.

     (c) Surjective, but not injective.

     (d) Bijective: $f^{-1}(x) = -\sqrt[4]{x/3}$.

     (e) Not injective since $\sin(c + 2\pi) = \sin c$ for any $c \in \mathbb{R}$. Not surjective; the range is $[-1, 1]$.

     (f) $f(x) = \sin x$ is surjective onto $[-1, 1]$, but not injective in the domain $[0, 2\pi]$, since, for example, $\sin(\pi/4) = \sin(3\pi/4)$.

     (g) Bijective: $f^{-1}(x) = \arcsin x$.

     (h) Bijective: $f^{-1}(x) = f(x) = \frac{1}{x}$.

     (i) Not injective since $\frac{1}{(-c)^2} = \frac{1}{c^2}$ for every $c \neq 0$. Not surjective; the range is $(0, \infty)$.

     (j) Bijective: $f^{-1}(x) = \dfrac{-2x - 7}{5x - 3}$.

     (k) Bijective: $f^{-1}(x) = \frac{1}{2} \ln \frac{x}{7}$.

  4. Given that $b = f(a)$, then $g(b) = g(f(a)) = (g \circ f)(a) = a$ by hypothesis.

  5. (a) $f(1) = 4$. (b) $f[\{1\}] = \{4\}$. (c) $f^{-1}(1)$ means "plug in 1 into the inverse function for f." However, since f is not bijective, f^{-1} does not exist. So $f^{-1}(1)$ is *undefined*. (d) $f^{-1}[\{1\}] = \{3\}$.

  6. (a) $f[(1,3)] = \left(\frac{1}{9}, 1\right)$ (b) $f[(-3,-1)] = \left(\frac{1}{9}, 1\right)$ (c) $f^{-1}[(1,9)] = \left(-1, -\frac{1}{3}\right) \cup \left(\frac{1}{3}, 1\right)$ (d) $f^{-1}[(-9,9)] = \left(-\infty, -\frac{1}{3}\right) \cup \left(\frac{1}{3}, \infty\right)$

Problems of Appendix B

- **§B.1**

  2. $g^n = \mathbf{1} \implies g^{-1} \cdot g^n = g^{-1} \cdot \mathbf{1} \implies (g^{-1} \cdot g) \cdot g^{n-1} = g^{-1} \implies \mathbf{1} \cdot g^{n-1} = g^{-1}$. Thus $g^{-1} = g^{n-1}$.

  5. The identity element is boxed within the table.

	ABC	ACB	BAC	BCA	CAB	CBA
ABC	ABC	ACB	BAC	BCA	CAB	CBA
ACB	ACB	ABC	BCA	BAC	CBA	CAB
BAC	BAC	CAB	ABC	CBA	ACB	BCA
BCA	BCA	CBA	ACB	CAB	ABC	BAC
CAB	CAB	BAC	CBA	ABC	BCA	ACB
CBA	CBA	BCA	CAB	ACB	BAC	ABC

According to the table, we have $\text{ABC}^{-1} = \text{ABC}$, $\text{ACB}^{-1} = \text{ACB}$, $\text{BAC}^{-1} = \text{BAC}$, $\text{BCA}^{-1} = \text{CAB}$, $\text{CAB}^{-1} = \text{BCA}$, and $\text{CBA}^{-1} = \text{CBA}$.

6. Suppose $h_1, h_2 \in H$ are arbitrary. Let $g = \phi^{-1}(h_1) \cdot \phi^{-1}(h_2)$. Since ϕ is a group homomorphism,

$$
\begin{aligned}
\phi(g) = \phi\left(\phi^{-1}(h_1) \cdot \phi^{-1}(h_2)\right) &= \phi\left(\phi^{-1}(h_1)\right) * \phi\left(\phi^{-1}(h_2)\right) \\
&= h_1 * h_2 \\
\Longrightarrow \quad g = \phi^{-1}(h_1) \cdot \phi^{-1}(h_2) &= \phi^{-1}(h_1 * h_2).
\end{aligned}
$$

The last line proves that ϕ^{-1} is a group homomorphism; hence ϕ is an invertible homomorphism.

- **§B.2**

1. (a) $\begin{pmatrix} 1 & 14 & -9 \\ 1 & 0 & 19 \\ -4 & 8 & 9 \end{pmatrix}$ (b) $\begin{pmatrix} 18 & 7 \\ -20 & 11 \\ -16 & -9 \end{pmatrix}$ (c) Does not exist because B has two columns while A has three rows. (d) $\begin{pmatrix} -16 \\ 32 \\ 29 \end{pmatrix}$ (e) 18 (f) Does not exist because B is not square.

4. $(3/5, -4/5, 1)$

5. (a) REF: $\begin{pmatrix} 1 & 3 & 5 & 7 \\ 0 & -4 & -8 & -12 \\ 0 & 0 & 0 & -10 \end{pmatrix}$. (Row operations used: $-3R_1 + R_2 \mapsto R_2$, $-5R_1 + R_3 \mapsto R_3$, $-2R_2 + R_3 \mapsto R_3$.) (b) Reduce to RREF, and interpret the solution. (Continued from REF, row operations used: $-\frac{1}{4}R_2 \mapsto R_2$, $-\frac{1}{10}R_3 \mapsto R_3$, $-3R_2 + R_1 \mapsto R_1$, $2R_3 + R_1 \mapsto R_1$, $-3R_3 + R_2 \mapsto R_2$.)

$$
\begin{pmatrix} 1 & 0 & -1 & 0 \\ 0 & 1 & 2 & 0 \\ 0 & 0 & 0 & 1 \end{pmatrix} \Longrightarrow
\begin{cases} x_1 - x_3 &= 0 \\ x_2 + 2x_3 &= 0 \\ x_4 &= 0 \end{cases} \Longrightarrow
\begin{pmatrix} x_1 \\ x_2 \\ x_3 \\ x_4 \end{pmatrix} = \begin{pmatrix} x_3 \\ -2x_3 \\ x_3 \\ 0 \end{pmatrix}
$$

Thus the nullity is 1, and a basis for $\text{Nul}(A)$ is $\{(1, -2, 1, 0)\}$.

6. We must verify that $\mathcal{M}_{m \times n}$ satisfies the 10 axioms (Definition B.2.6). $\mathcal{M}_{m \times n}$ is closed under addition and scalar multiplication (axioms 1, 6),

because these operations do not change the dimensions of the matrices. Matrix addition is commutative, associative, and the $m \times n$ zero matrix,

$$O_{m \times n} = \begin{pmatrix} 0 & \cdots & 0 \\ \vdots & \ddots & \vdots \\ 0 & \cdots & 0 \end{pmatrix},$$

acts as additive identity (axioms 2, 3, 4). The additive inverse of a matrix A is equal to $-A = (-1)A$ (axiom 5). Now suppose $A = (a_{ij})$, $B = (b_{ij})$, and let $c, d \in R$. Then $c(a_{ij} + b_{ij}) = ca_{ij} + cb_{ij}$ proves that $c(A + B) = cA + cB$ (axiom 7); $(c + d)a_{ij} = ca_{ij} + da_{ij}$ proves that $(c + d)A = cA + dA$ (axiom 8); $c(da_{ij}) = ((cd)a_{ij})$ proves that $c(dA) = (cd)A$ (axiom 9); finally, $1(a_{ij}) = (1a_{ij}) = (a_{ij})$ proves that $1A = A$ (axiom 10).

7. (a) $0u \in V$ because of axiom 6. Thus, by axiom 5, there is an additive inverse vector $-(0u) \in V$ (which will be used in the third line below).

$$
\begin{aligned}
0u &= (0 + 0)u & &\text{by real-number arithmetic} \\
0u &= 0u + 0u & &\text{by axiom 8} \\
-(0u) + 0u &= -(0u) + [0u + 0u] & &\text{adding } -(0u) \text{ to both sides} \\
-(0u) + 0u &= [-(0u) + 0u] + 0u & &\text{by axiom 3} \\
\mathbf{0} &= \mathbf{0} + 0u & &\text{by axiom 5} \\
\mathbf{0} &= 0u & &\text{by axiom 4}
\end{aligned}
$$

Thus we have proven $0u = \mathbf{0}$.

Appendix D

Notations

The following table identifies notation introduced in each chapter and appendix.

Chapter 1	
$\cong$	congruence (geometric equivalence)
$\approx$	homeomorphism (topological equivalence)
(a, b)	$\{x \mid a < x < b\}$
$[a, b)$	$\{x \mid a \leq x < b\}$
$(a, b]$	$\{x \mid a < x \leq b\}$
$[a, b]$	$\{x \mid a \leq x \leq b\}$
(a, ∞)	$\{x \mid x > a\}$
$[a, \infty)$	$\{x \mid x \geq a\}$
$(-\infty, b)$	$\{x \mid x < b\}$
$(-\infty, b]$	$\{x \mid x \leq b\}$
$(-\infty, \infty)$	all real numbers
$\mathbb{I}$	unit closed interval, $[0, 1]$
i, j, k	unit vectors in $\mathbb{R}^3$
Chapter 2	
$\mathbb{R}^n$	n-dimensional Euclidean space
$x \cdot y$	dot (inner) product of vectors x, y
$\|\mathbf{v}\|$	length of a vector $\mathbf{v}$
$d(x, y)$	distance between points x and y
$\mathbb{S}^n$	n-dimensional sphere
$B_\epsilon(x)$	open ϵ-ball around x
$D_\epsilon(x)$	closed ϵ-ball around x
$\text{int}(A)$	interior of a subset
$\text{ext}(A)$	exterior of a subset
∂A	boundary of a subset
$\overline{A}$	closure of A
$\lim_{k \to \infty} x_k$	limit of a sequence (x_k)
(X, d)	metric space on set X with metric d
$\|x\|_p$	p-norm of x

d_p	p-norm metric
ℓ^p	sequences converging under the p-norm
$L^p(\mathbb{R})$	real functions converging under the p-norm

Chapter 3

$\phi(\mathbf{x}, t)$	flow function
$\mathrm{Ind}(C)$	index or winding number of a curve C
$\mathrm{Ind}(\mathbf{x})$	index of a cricital point $\mathbf{x}$
$\mathcal{O}^+(\mathbf{x}),\ \mathcal{O}^-(\mathbf{x})$	positive, negative semiorbit of $\mathbf{x}$
$\omega(\mathbf{x})$	omega limit set of $\mathbf{x}$

Chapter 4

$\mathscr{T}_E$	Euclidean topology
$\mathscr{T}_D$	discrete topology
$\mathscr{T}_I$	indiscrete topology
$\mathscr{T}_C$	cofinite topology
$\mathscr{T}\vert_A$	subspace topology
$\sim$	identification of points
$[x]$	equivalence class of a point
$X/\sim$	quotient space
$\mathscr{T}/\sim$	quotient topology
$\mathbb{T}$	torus
$\mathbb{K}$	Klein bottle
$\mathbb{P}^n$	projective space
$\mathscr{T}_{\mathrm{box}}$	box product topology
$\mathscr{T}_{\mathrm{prod}}$	product topology

Chapter 5

$\#$	connected sum operation
$n\mathbb{T}$	n-holed torus
$\mathcal{S}_c$	compact surfaces, surfaces-with-boundary
$\equiv$	equivalence of plane model words
$\mathbb{P}$	projective plane
$\mathbb{D}^n$	closed disk in $\mathbb{R}^n$; n-cell

Chapter 6

$V(G)$	vertex set of a graph
$E(G)$	edge set of a graph
$F(G)$	regions determined by an embedding of a graph
$\vert V \vert, v$	number of vertices
$\vert E \vert, e$	number of edges
$\vert F \vert, f$	number of faces/regions
$\cong$	isomorphism of graphs
K_n	complete graph
$K_{m,n}$	complete bipartite graph
C_n	cycle graph
P_n	path graph
$X \hookrightarrow Y$	an embedding of X into Y
H_χ	Haewood's number

G^*	dual graph
$\mathbb{U}$	unknot
$\mathbb{T}(p, q)$	torus knot or link
$C(K)$	crossing number of a knot
K'	mirror image of knot K
$\langle L \rangle$	Kauffman bracket of L
χ^+, χ^-	positive, negative smoothing of χ
$L_1 \sqcup L_2$	separated union of knot diagrams
$J(L)$	Jones polynomial of L

Chapter 7

$\mathrm{Loop}(X)$	set of all loops in X
$\mathrm{Loop}(X, x_0)$	set of all loops in X with basepoint x_0
$\gamma \cdot \eta$	product loop; product of sphere maps
$\mathbf{1}_{x_0}$	constant loop at x_0
$\mathbf{1}$	constant loop (at x_0)
γ^{-1}	inverse loop
$\pi_1(X, x_0), \pi_1(X)$	fundamental group of X
$\pi_n(X)$	nth homotopy group of X
M_K	knot complement

Chapter 8

$C_n(X; \mathbb{Q}), C_n(X; \mathbb{Z})$	rational/integral n-chains of X
$Z_n(X; \mathbb{Q}), Z_n(X; \mathbb{Z})$	rational/integral n-cycles of X
$B_n(X; \mathbb{Q}), B_n(X; \mathbb{Z})$	rational/integral n-boundaries of X
$H_n(X; \mathbb{Q}), H_n(X; \mathbb{Z})$	rational/integral homology of X
∂_k	kth boundary map
d_n	nth boundary homomorphism
Δ^n	n-simplex
$[v_0, \ldots, v_n]$	n-simplex on a set of vertices
$\beta_n(X)$	nth Betti number of X
$\widehat{v_k}$	omit v_k from a bracket
$\mathrm{im}(f)$	image of f
$\ker(f)$	kernel of f

Appendix A

$\mathbb{N}$	set of natural numbers
$\mathbb{Z}$	set of integers
$\mathbb{Q}$	set of rational numbers
$\mathbb{R}$	set of real numbers
$\mathbb{C}$	set of complex numbers
$\mathbb{R}^+$	set of positive real numbers
$\mathbb{R}^-$	set of negative real numbers
$\emptyset, \{\}$	empty set
$\in$	element of a set
$\subseteq$	subset of a set
$\{\ldots \mid \ldots\}$	set-builder notation
$\mathcal{P}(A)$	power set of A

$\Longrightarrow$	implication
$\Longleftrightarrow$, iff	biconditional (logical equivalence)
$\forall$	universal quantifier
$\exists$	existential quantifier
$\cup$	union of sets
$\cap$	intersection of sets
$\setminus$	difference of sets
A^c	complement of A
$f : D \to C$	function from domain set D to codomain set C
$\mapsto$	maps to (by a function)
$f\|_A$	function restricted to domain A
$g \circ f$	composition of functions
f^{-1}, $f^{-1}(x)$	inverse function
$f[A_0]$	forward image (image) of a subset
$f^{-1}[B_0]$	inverse image (preimage) of a subset

Appendix B

e, $\mathbf{1}$, $\mathbf{0}$	identity element of a group
1	trivial group
0	trivial abelian group
id_G	identity group homomorphism
$\langle \ldots \mid \ldots \rangle$	group with specified generators and relations
$A = (a_{ij})$	matrix A with entries a_{ij}
$\mathcal{M}_{m \times n}$	set of all $m \times n$ matrices
I_n	identity matrix
$\det A$	determinant of a matrix A
$\mathrm{rank}(A)$	rank of a matrix A
$\mathrm{Span}_R A$	linear span of a set A over scalars R
$\mathrm{Nul}(A)$	null space of a matrix A

Bibliography

[Abb84]   E. A. Abbott, *Flatland: a romance of many dimensions*, 6th ed., Dover, New York, 1952 [1884].

[Ada01]   Colin C. Adams, *The knot book*, Henry Holt, New York, 2001.

[AH]   S. V. Ault and E. Holmgreen, *Dynamics of the brusselator*, unpublished.[1]

[Arm88]   M. A. Armstrong, *Groups and symmetry*, Undergraduate texts in mathematics, Springer-Verlag, New York, 1988.

[Arm10]   ———, *Basic topology*, Undergraduate texts in mathematics, Springer, New York, 2010.

[AS60]   L. V. Ahlfors and L. Sario, *Riemann surfaces*, Princeton mathematical series, Princeton University Press, Princeton, NJ, 1960.

[AZ10]   M. Aigner and G. M. Ziegler, *Proofs from the book*, Springer, New York, 2010.

[Bar64]   S. Barr, *Experiments in topology*, Dover, New York, 1964.

[BD08]   W. E. Boyce and R. C. DiPrima, *Elementary differential equations and boundary value problems*, Wiley, Hoboken, NJ, 2008.

[Bel85]   B. P. Belousov, *A periodic reaction and its mechanism*, Oscillations and traveling waves in chemical systems (R. J. Field and M. Burger, eds.), Wiley, New York, 1985.

[Blo97]   E. D. Bloch, *A first course in geometric topology and differential geometry*, Birkhäuser, Boston, 1997, Index.

[BM07]   J.-A. Bondy and U. S. R. Murty, *Graph theory*, Graduate texts in mathematics, Springer, New York, 2007, OHX.

[Bro06]   R. Brown, *Topology and groupoids*, BookSurge Publishing, Deganwy, UK, 2006.

[BS11]   R. G. Bartle and D. R. Sherbert, *Introduction to real analysis*, 4th ed., John Wiley & Sons, University of Illinois, Urbana-Champaign, 2011.

[Cai51]   S. S. Cairns, An elementary proof of the Jordan-Schoenflies theorem, *Proceedings of the American Mathematical Society* **2** (1951), no. 6, 860–867.

[Car09]   G. Carlsson, Topology and data, *Bulletin of the American Mathematical Society* **46** (2009), no. 2, 255–308.

[CF08]   R. H. Crowell and R.H. Fox, *Introduction to knot theory*, Dover books on mathematics, Dover, New York, 2008.

[CKC+13]   P. M. Cohn, A. I. Kostrikin, D. J. Collins, I. R. Shafarevich, R. I. Grigorchuk, P. F. Kurchanov, and H. Zieschang, *Algebra vii: Combinatorial group theory applications to geometry*, Encyclopaedia of mathematical sciences, Springer, Berlin, 2013.

[DD09]   M. M. Deza and E. Deza, *Encyclopedia of distances*, Springer Berlin, 2009.

---

[1]Available at: https://www.academia.edu/171695/Dynamics_of_the_Brusselator

[Dev03]   R. L. Devaney, *An introduction to chaotic dynamical systems*, 2nd ed., Addison-Wesley studies in nonlinearity, Westview Press, New York, 2003.

[DF04]   D. S. Dummit and R. M. Foote, *Abstract algebra*, 3rd ed., John Wiley & sons, Hoboken, NJ, 2004.

[DM68]   P. H. Doyle and D. A. Moran, A short proof that compact 2-manifolds can be triangulated, *Inventiones Mathematicae* **5** (1968), 160–162.

[EML45]   S. Eilenberg and S. Mac Lane, General theory of natural equivalences, *Transactions of the American Mathematical Society* **58** (1945), 231–294.

[Epp10]   S. S. Epp, *Discrete mathematics with applications*, Cengage Learning, New York, 2010.

[ES45]   S. Eilenberg and N. E. Steenrod, Axiomatic approach to homology theory, *Proceedings of the National Academy of Sciences of the United States of America* **31** (1945), no. 4, 117–120.

[Eul36]   L. Euler, Solutio problemetis ad geometriam situs pertinentis, *Commentarii Academiae Scientarium Petropolitanae* **8** (1736), 128–140. Reprinted in *Opera Omnia Series Prima* 7, 1–10, 1766.

[FYH$^+$85] P. Freyd, D. Yetter, J. Hoste, W. B. R. Lickorish, K. Millett, and A. Ocneanu, A new polynomial invariant of knots and links, *Bulletin of the American Mathematical Society* **12** (1985), no. 2, 239–246.

[Ges09]   M. Gessen, *Perfect rigor: a genius and the mathematical breakthrough of the century*, Houghton Mifflin Harcourt, Boston, 2009.

[Goo05]   S. E. Goodman, *Beginning topology*, American Mathematical Society, Providence, RI, 2005.

[Hal74]   P. R. Halmos, *Naive set theory*, Springer-Verlag, New York, 1974.

[Hat02]   A. Hatcher, *Algebraic topology*, Cambridge University Press, Cambridge, MA, 2002.

[Hau14]   F. Hausdorff, *Grundzüge der mengenlehre*, Das Hauptwerk von Felix Hausdorff, Veit, Leipzig, 1914.

[Hei41]   R. Heinlein, *And he built a crooked house*, Astounding Science Fiction, Street and Smith, New York, 1941.

[Hen79]   M. Henle, *A combinatorial introduction to topology*, W. H. Freeman, San Francisco, 1979.

[HHM08]   J. Harris, J. L. Hirst, and M. Mossinghoff, *Combinatorics and graph theory*, Undergraduate texts in mathematics, Springer, New York, 2008.

[Kau87]   L. H. Kauffman, State models and the Jones polynomial, *Topology* **26** (1987), 395–407.

[Kho00]   M. Khovanov, A categorification of the jones polynomial, *Duke Mathemathical Journal* **101** (2000), 359–426.

[Kur30]   K. Kuratowski, Sur le problème des courbes gauches en topologie, *Fundamenta Mathematica* **15** (1930), 271–283.

[Lan05]   S. Lang, *Algebra*, Graduate Texts in Mathematics, Springer, New York, 2005.

[L'E62]   M. L'Engle, *A wrinkle in time*, Farrar, Straus, and Giroux, New York, 1962.

[Lee00]   J. M. Lee, *Introduction to topological manifolds*, Graduate texts in mathematics, Springer, New York, 2000.

[LLM15]   D. C. Lay, S. R. Lay, and J. J. McDonald, *Linear algebra and its applications*, 5th ed., Pearson, London, 2015.

[Lor63]   E. N. Lorenz, The predictability of hydrodynamic flow, *Transactions of the New York Academy of Sciences* **25** (1963), no. 4, 409–432.

[Man04]   V. Manturov, *Knot theory*, CRC Press, Boca Raton, FL, 2004.

[Mas91]   W. S. Massey, *A basic course in algebraic topology*, Graduate texts in mathematics, Springer, New York, 1991.

[May99]   J. P. May, *A concise course in algebraic topology*, Chicago lectures in mathematics, University of Chicago Press, Chicago, 1999.

[Men90]   B. Mendelson, *Introduction to topology*, 3rd ed., Dover, New York, 1990.

[Mil15]   J. Milnor, Topology through the centuries: low dimensional manifolds, *Bulletin of the American Mathematical Society* **52** (2015), 545–584.

[ML71]   S. Mac Lane, *Categories for the working mathematician*, Springer-Verlag, New York, 1971.

[MS06]   R. Messer and P. Straffin, *Topology now!*, Classroom resource materials, Mathematical Association of America, Washington, DC, 2006.

[MT91]   W. Menasco and M. Thistlethwaite, The Tait flyping conjecture, *Bulletin of the American Mathematical Society* **25** (1991), 403–412.

[Mun00]   J. R. Munkres, *Topology*, 2nd ed., Prentice Hall, Upper Saddle River, NJ, 2000.

[Mur87a]   K. Murasugi, The Jones polynomial and classical conjectures in knot theory, *Topology* **26** (1987), 187–194.

[Mur87b]   _____, The Jones polynomial and classical conjectures in knot theory ii, *Mathematical Proceedings of the Cambridge Philosophical Society* **102** (1987), 317–318.

[PL14]   D. Poole and R. Lipsett, *Linear algebra: a modern introduction*, Cengage Learning, New York, 2014.

[Pol]   J. C. Polking, *dfield and pplane*, http://math.rice.edu/~dfield/dfpp.html, last accessed April, 2017.

[Pra95]   V. V. Prasolov, *Intuitive topology*, American Mathematical Society, Providence, RI, 1995.

[Rad25]   T. Radó, Uber den Begriff der Riemannshen Fläche, *Acta Scientiarum Mathematicarum (Szeged)* **2** (1925), 101–121.

[Rud76]   W. Rudin, *Principles of mathematical analysis*, International series in pure and applied mathematics, McGraw-Hill, New York, 1976.

[Rud87]   _____, *Real and complex analysis*, 3rd ed., McGraw-Hill, New York, 1987.

[Sch97]   E. Schechter, *Handbook of analysis and its foundations*, Academic Press, San Diego, 1997.

[Smu94]   R. M. Smullyan, *First-order logic*, Dover, New York, 1994.

[Spa94]   E. H. Spanier, *Algebraic topology*, Mathematics subject classifications, Springer, 1994.

[SS95]   L. A. Steen and J. A. Seebach, *Counterexamples in topology*, Dover books on mathematics, Dover, New York, 1995.

[Ste]   W. Stein, *Sagemath*, http://www.sagemath.org, last accessed April, 2017.

[Str94]   S. H. Strogatz, *Nonlinear dynamics and chaos: with applications to physics, biology, chemistry, and engineering*, Westview Press, 1994.

[Thi87]   M. B. Thistlethwaite, A spanning tree expansion of the Jones polynomial, *Topology* **26** (1987), 297–309.

[Thi88]   _____, Kauffman's polynomial and alternating links, *Topology* **27** (1988), 311–318.

[Tho92]   C. Thomassen, The Jordan-Schönflies theorem and the classification of surfaces, *American Mathematical Monthly* **99** (1992), 116–130.

[Thu94]   W. P. Thurston, On proof and progress in mathematics, *Bulletin of the American Mathematical Society* **30** (1994), no. 2, 161–177.

[TWH10]   G. B. Thomas, M. D. Weir, and J. R. Hass, *Thomas' calculus, early transcenden-tals*, 12th ed., Addison-Wesley, Boston, 2010.

[Wal72]   C. T. C. Wall, *A geometric introduction to topology*, Dover, New York, 1972.

[Wee02]   Jeffrey Weeks, *The shape of space*, CRC Press, New York, 2002.

[Zha64]   A. M. Zhabotinsky, Periodical oxidation of malonic acid in solution (a study of the belousov reaction kinetics), *Biofizika* **9** (1964), 306–11.

Index